Lecture Notes in Mathematics

Edited by J.-M. Morel, F. Takens and B. Teissier

Editorial Policy
for the publication of monographs

1. Lecture Notes aim to report new developments in all areas of mathematics – quickly, informally and at a high level. Monograph manuscripts should be reasonably self-contained and rounded off. Thus they may, and often will, present not only results of the author but also related work by other people. They may be based on specialized lecture courses. Furthermore, the manuscripts should provide sufficient motivation, examples and applications. This clearly distinguishes Lecture Notes from journal articles or technical reports which normally are very concise. Articles intended for a journal but too long to be accepted by most journals, usually do not have this "lecture notes" character. For similar reasons it is unusual for doctoral theses to be accepted for the Lecture Notes series.

2. Manuscripts should be submitted (preferably in duplicate) either to one of the series editors or to Springer-Verlag, Heidelberg. In general, manuscripts will be sent out to 2 external referees for evaluation. If a decision cannot yet be reached on the basis of the first 2 reports, further referees may be contacted: the author will be informed of this. A final decision to publish can be made only on the basis of the complete manuscript, however a refereeing process leading to a preliminary decision can be based on a pre-final or incomplete manuscript. The strict minimum amount of material that will be considered should include a detailed outline describing the planned contents of each chapter, a bibliography and several sample chapters.
Authors should be aware that incomplete or insufficiently close to final manuscripts almost always result in longer refereeing times and nevertheless unclear referees' recommendations, making further refereeing of a final draft necessary.
Authors should also be aware that parallel submission of their manuscript to another publisher while under consideration for LNM will in general lead to immediate rejection.

3. Manuscripts should in general be submitted in English.
Final manuscripts should contain at least 100 pages of mathematical text and should include
- a table of contents;
- an informative introduction, with adequate motivation and perhaps some historical remarks: it should be accessible to a reader not intimately familiar with the topic treated;
- a subject index: as a rule this is genuinely helpful for the reader.

Continued on inside back-cover

Lecture Notes in Mathematics 1758

Springer
Berlin
Heidelberg
New York
Barcelona
Hong Kong
London
Milan
Paris
Singapore
Tokyo

Nicolas Monod

Continuous Bounded Cohomology of Locally Compact Groups

Springer

Author

Nicolas Monod
ETH-Zürich
Department of Mathematics
Rämistrasse 101
8092 Zürich, Switzerland

E-mail: monod@math.ethz.ch

Cataloging-in-Publication Data applied for

Die Deutsche Bibliothek - CIP-Einheitsaufnahme

Monod, Nicolas:
Continuous bounded cohomology of locally compact groups / Nicolas Monod. - Berlin ; Heidelberg ; New York ; Barcelona ; Hong Kong ; London ; Milan ; Paris ; Singapore ; Tokyo : Springer, 2001
(Lecture notes in mathematics ; 1758)
ISBN 3-540-42054-1

Mathematics Subject Classification (2000): 22E41, 55N35, 20J05, 20J06, 22E40

ISSN 0075-8434
ISBN 3-540-42054-1 Springer-Verlag Berlin Heidelberg New York

Springer-Verlag Berlin Heidelberg New York
a member of BertelsmannSpringer Science+Business Media GmbH

http://www.springer.de

Typesetting: Camera-ready TeX output by the author
SPIN: 10836990 41/3142/du-543210 - Printed on acid-free paper

Preface

Recent research has repeatedly led to connections between important rigidity questions and bounded cohomology. However, the latter has remained by and large intractable. This monograph introduces the functorial study of the continuous bounded cohomology for topological groups, with coefficients in Banach modules. The powerful techniques of this more general theory have successfully solved a number of the original problems in bounded cohomology. As applications, one obtains, in particular, rigidity results for actions on the circle, for representations on complex hyperbolic spaces and on Teichmüller spaces.

This text was written at the end of the time I spent at the ETH under the supervision of Prof. Marc Burger ; I am indebted to him for his generous advice, but actually the debt I owe him extends further : several of the results presented here are in fact joint work with him (see the Introduction). I am also grateful to Étienne Ghys who had the kindness to accept spontaneously to be the external examiner of this manuscript.

I owe special thanks to Scot Adams and Freddy Delbaen for many useful and enjoyable discussions. The former had the patience to explain a couple of technical points about amenable actions to me. The latter generously answered innumerable questions on Banach spaces.

I also thank Gopal Prasad for kindly explaining a point in the cohomology of arithmetic lattices to me.

N.M., Montreux, February 2001.

Contents

Introduction

In his celebrated 1982 I.H.É.S. paper *Volume and bounded cohomology* [**69**], M. Gromov considers the singular bounded cohomology $\mathrm{H}_\mathrm{b}^\bullet(M)$ of a manifold M and proves a spray of deep and surprising theorems on the geometry of manifolds. This cohomology $\mathrm{H}_\mathrm{b}^\bullet(M)$ is defined exactly like usual singular cohomology, except that all cochains are required to be bounded. Similarly, one can define for a group Γ the bounded cohomology $\mathrm{H}_\mathrm{b}^\bullet(\Gamma)$ by the usual inhomogeneous complex with the only change that one restricts attention to bounded cochains :

$$0 \longrightarrow \mathbf{R} \longrightarrow \ell^\infty(\Gamma, \mathbf{R}) \longrightarrow \ell^\infty(\Gamma^2, \mathbf{R}) \longrightarrow \cdots$$

One of the remarkable theorems of Gromov in this connection is that the bounded cohomology of M is canonically isomorphic to the bounded cohomology of its fundamental group $\Gamma = \pi_1(M)$. Hence one can draw the somewhat misleading conclusion that this rich theory of singular bounded cohomology reduces completely to bounded cohomology of discrete groups.

Yet Gromov was by no means the first to consider bounded cohomology : from a completely different viewpoint, B.E. Johnson has studied in a couple of chapters of his 1972 memoir [**89**] the cohomology of the inhomogeneous complex

$$0 \longrightarrow A \longrightarrow L^\infty_{\mathrm{w*}}(G, A) \longrightarrow L^\infty_{\mathrm{w*}}(G^2, A) \longrightarrow \cdots$$

where G is a locally compact group and A a dual Banach G-module. We see that if G is discrete and A trivial, this coincides with the above bounded cohomology. Johnson derives a number of elegant general results about this cohomology, some of which will be rediscovered in partial forms and painfully reproved many years later by various authors.

The part of Johnson's memoir dealing with this cohomology however lacked applications, and it is really after Gromov's contribution that the interest in bounded cohomology increased suddenly ; later, more connections with various fields emerged : actions on the circle, commutator length, rigidity theory. However, almost no general results were known about bounded cohomology itself : there was a dire need of a systematic and functorial approach.

For discrete groups and trivial coefficients, this task has been undertaken by N.V. Ivanov [**86**], and some contributions in this directions for more general coefficients are due to G.A. Noskov [**117, 118**].

The purpose of our work is twofold :

(i) To develop such a theory in the general setting of *topological groups* and coefficients in Banach modules.

(ii) To provide sufficiently powerful techniques, so that we can derive from them substantial results in view of applications ; to establish applications.

The generality sought in (i) is not merely a Bourbakian compulsion, but indeed leads also to many new results in the case of discrete groups and trivial coefficients ; this is also true for the techniques alluded to in (ii). Let us illustrate this with an example :

It was conjectured for some time that a higher rank lattice Γ should not admit non trivial actions on the circle. É. Ghys discovered in [**63**] that it would basically be enough to understand $\mathrm{H}^2_{\mathrm{b}}(\Gamma)$ to solve this problem. This was done in [**35, 37**] by M. Burger and the author ; we show below :

THEOREM (Cor. 13.5.5).— *Let Γ be a lattice in $G = \mathbf{G}(k)$, where $\mathbf{G}$ is a connected, simply connected, almost simple k-isotropic group and k a local field.*

If $\mathrm{H}^2(\Gamma, \mathbf{R}) = 0$ and $\mathrm{rank}_k(\mathbf{G}) \geq 2$, then any Γ-action on the circle by orientation preserving C^1 diffeomorphisms factors through a finite group.

But this statement was obtained by relating $\mathrm{H}^2_{\mathrm{b}}(\Gamma)$ to the continuous bounded cohomology of $\mathbf{G}(k)$ with non-trivial coefficients and by studying the latter[1]. Thus the original problem led naturally to the setting (i) for locally compact groups ; if we consider yet more general groups it is because we are convinced that there should be an interest in studying the continuous bounded cohomology of, say, diffeomorphisms groups.

The cohomological results behind the above theorem can be generalized to irreducible lattices in products of algebraic groups over various local fields (see Section 14.3). Given the boundary theory, structure theory and representation theory involved in the arguments, it is much more surprising that these results can be extended to lattices in very general locally compact groups. This is achieved by means of more sophisticated techniques, as alluded to in (ii). In order to formulate a representative statement, we agree that a lattice

$$\Gamma < G = G_1 \times \cdots \times G_2$$

is *non-elementary* if its projection on each factor is dense (this generalizes irreducibility).

THEOREM (see Thm. 14.2.2).— *Let $G_1, \ldots, G_n$ be compactly generated locally compact second countable groups and let $\Gamma < G = G_1 \times \cdots \times G_n$ be a non elementary lattice. Let (π, E) be a separable coefficient Γ-module*[2]. *Then there are topological isomorphisms*

$$\mathrm{H}^2_{\mathrm{cb}}(\Gamma, E) \cong \bigoplus_{j=1}^{n} \mathrm{H}^2_{\mathrm{cb}}(G_j, E_j) \cong \mathrm{H}^2_{\mathrm{cb}}\big(G, \textstyle\sum_{j=1}^{n} E_j\big),$$

[1] Interestingly enough, É. Ghys simultaneously proved a similar result for real Lie groups by a completely different method [**64**].

[2] Technically, a *coefficient* module will be defined as admitting a separable continuous predual ; the continuity assumption is void here.

where E_j is the maximal submodule on which π extends continuously to G via the projection $G \to G_j$.

At first sight, this has a striking similarity with Y. Shalom's super-rigidity results [**130**]. However, it turns out that even though the results are very complementary for applications, both the content and the methods of proof are completely different.

Keeping the same illustration — actions on the circle — we obtain statements of the following kind (see also [**37**]).

THEOREM (Cor. 14.2.4).— *Let $G_1, \ldots, G_n$ be compactly generated locally compact second countable groups and let $\Gamma < G_1 \times \cdots \times G_n$ be an non elementary co-compact lattice. Assume $\mathrm{H}^2_{\mathrm{cb}}(G_j) = 0$ and $\mathrm{Hom_c}(G_j) = 0$ for $1 \leq j \leq n$.*

Then any Γ-action by orientation-preserving homeomorphisms of the circle has a finite orbit. Moreover, if the action is by C^1 diffeomorphisms, then it factors through a finite group.

The second bounded cohomology is relevant to many more issues than actions on the circle. To cite two striking connections with rigidity theory, M. Burger and A. Iozzi relate $\mathrm{H}^2_{\mathrm{b}}$ to representations on complex hyperbolic spaces [**34**], while M. Bestvina and K. Fujiwara [**12**] consider representations in the mapping class group $\mathrm{Mod}(\Sigma)$ of any compact orientable surface Σ. Using the above cohomological theorem, we prove :

THEOREM (see Cor. 14.2.8).— *Let $G_1, \ldots, G_n$ be compactly generated locally compact second countable groups and let $\Gamma < G_1 \times \cdots \times G_n$ be a non elementary finitely generated lattice. Assume $\mathrm{H}^2_{\mathrm{cb}}(G_j) = 0$ for $1 \leq j \leq n$.*

Then the image of any representation $\Gamma \to \mathrm{PSU}(n,1)$ either fixes a point in $\partial_\infty \mathbf{H}^n_{\mathbf{C}}$ or leaves a totally real subspace invariant.

THEOREM (see Cor. 14.2.7).— *Let $G_1, \ldots, G_n$ be connected locally compact groups and $\Gamma < G_1 \times \cdots \times G_n$ a non elementary lattice. Then the image of any representation $\Gamma \to \mathrm{Mod}(\Sigma)$ has virtually Abelian image.*

It is well known that if a lattice as above has finite Abelianization Γ_{Ab}, then a result such as the last theorem above implies that any isometric Γ-action on a Teichmüller space has a fixed point (see Cor. 14.2.6 below). Thus this theorem generalizes the main result of [**55**] about the arithmetic case.

We shall now say a few words about the techniques introduced. Unfortunately, bounded cohomology is extremely difficult to compute. It is not a cohomological functor stricto sensu : triangulations, Mayer-Vietoris sequences, in a word, everything that makes cohomology computable, breaks down completely for bounded cohomology.

Here is an example. The bounded cohomology of the group $\mathbf{Z}$ vanishes in every positive degree, whilst the bounded cohomology of the free group $\mathbf{Z} * \mathbf{Z}$ is not known as yet ! Moreover, the little that is known about it is rather disquieting : both spaces $\mathrm{H}^2_{\mathrm{b}}(\mathbf{Z} * \mathbf{Z})$ and $\mathrm{H}^3_{\mathrm{b}}(\mathbf{Z} * \mathbf{Z})$ have uncountable dimension.

One concept which will turn out to compensate a little bit these difficulties is R.J. Zimmer's notion of amenable actions. We establish a characterization of the latter in cohomological terms ; this gives then immediately a central place to amenable actions in continuous bounded cohomology (see also [**37**]) :

THEOREM (7.5.3).— *Let G be a locally compact second countable group, S an amenable regular G-space and E a coefficient G-module. Then*

$$0 \longrightarrow E \xrightarrow{\epsilon} L^\infty_{\mathrm{w*}}(S,E) \longrightarrow L^\infty_{\mathrm{w*}}(S^2,E) \longrightarrow L^\infty_{\mathrm{w*}}(S^3,E) \longrightarrow \cdots$$

is a strong augmented resolution of E by relatively injective G-modules.

Moreover, the cohomology of the complex

$$0 \longrightarrow L^\infty_{\mathrm{w*}}(S,E)^G \longrightarrow L^\infty_{\mathrm{w*}}(S^2,E)^G \longrightarrow L^\infty_{\mathrm{w*}}(S^3,E)^G \longrightarrow \cdots$$

is canonically isometrically isomorphic to $\mathrm{H}^\bullet_{\mathrm{cb}}(G,E)$.

The reason why such a result is useful is that we can produce interesting amenable spaces for most locally compact groups ; more precisely, we show (see also [**37**]) :

THEOREM (11.1.3).— *Let G be a compactly generated locally compact group. There exists a canonical topologically characteristic finite index open subgroup $G^* \lhd G$ and a regular G^*-space S such that*

(i) *The G^*-action on S is amenable.*

(ii) *The G^*-action on S is doubly $\mathfrak{X}^{\mathrm{sep}}$-ergodic.*

This result makes a heavy use of structure theory for locally compact groups, since it involves Hilbert's fifth problem, Poisson boundaries for random walks on Schreier graphs, the finiteness of the group of automorphisms of semi-simple groups, Furstenberg boundaries for simple Lie groups and Mautner's lemma.

Another powerful technique is a spectral sequence à la Hochschild-Serre, giving rise to an interesting exact sequence (see also [**37**]) :

THEOREM (12.0.3).— *Let G be a locally compact second countable group, $N \lhd G$ a compactly generated closed subgroup and $Q = G/N$ the quotient. Let (π, F) be a separable coefficient G-module.*

Then we have the exact sequence

$$0 \longrightarrow \mathrm{H}^2_{\mathrm{cb}}(Q,F^N) \xrightarrow{\mathrm{inf}} \mathrm{H}^2_{\mathrm{cb}}(G,F) \xrightarrow{\mathrm{res}} \mathrm{H}^2_{\mathrm{cb}}(N,F^{Z_G(N)})^Q \longrightarrow$$
$$\longrightarrow \mathrm{H}^3_{\mathrm{cb}}(Q,F^N) \xrightarrow{\mathrm{inf}} \mathrm{H}^3_{\mathrm{cb}}(G,F).$$

A first remark is that in usual continuous cohomology even the most basic analogues, such as a Künneth formula in degree one, may fail to be true for unitary representations of infinite dimension.

This sequence presents two important features absent from the Lyndon-Hochschild-Serre sequence in usual cohomology of abstract groups, and both are manifestations of double ergodicity :

The sequence goes up to degree three, even though the coefficients are quite general ; secondly, the term $\mathrm{H}^2_{\mathrm{b}}(N, F^{Z_G(N)})^Q$ is more precise than the expected $\mathrm{H}^2_{\mathrm{b}}(N, F)^Q$.

As the results mentioned above already suggest, the most useful techniques are not available in the complete generality of topological groups ; when amenable actions are concerned, we even restrict ourselves to locally compact second countable groups G, so as to have on G a standard measure space structure. We would however like to point out that this setting could be relaxed a little bit in asmuch as almost every statement we shall make about second countable groups can be extended to σ-compact groups because every locally compact σ-compact group admits a second countable quotient by a compact kernel, see Scholium 5.3.11 below.

Finally, we dedicate a few lines to the more elementary and straightforward techniques, the functorial rank-and-file, since quite some space is devoted to them in the pages below. Why do we carefully establish certain statements which seem to be, and sometimes are, direct adaptations of well known facts in classical cohomology ?

Well, first of all, such "obvious transliterations" often just fail to hold true. Furthermore, it happens also that the usual proofs do not yield the most accurate statement in our setting. This is illustrated e.g. by the fundamental lemma on comparison of resolutions. The corresponding result in continuous bounded cohomology (Lemma 7.2.4) leads indeed to topological isomorphisms in cohomology, and even to a control on the semi-norm ; but it does not yield *isometric* isomorphisms. The latter can however be obtained by a different method, as the reader can observe in the proofs of Theorem 7.4.5 and Lemma 7.5.6. Summarizing our practice in these matters, let us say that whilst we have tried to satisfy the less experienced, to which we belong, aswell as the trained Pedant, we contend to the intention of all other readers that Oscar Wilde's words

Talking nonsense is much cleverer than listening to it[3]

apply also to abstract nonsense.

N.M., Gordes, 20.10.2000.

A bird's eye view

Most chapters and some sections begin with a short summary of their content. Here is the overall structure :

In the first chapter, we get acquainted with the basic building blocks of the theory, namely Banach modules and coefficient modules, presented in Section 1. A remarkable family of such modules, spaces of L^∞ type, is studied in Section 2.

[3]O. Wilde, *The importance of being earnest.*

The third section presents integration techniques which, when available, will prove most useful.

The second chapter deals with the fundamental concept of relative injectivity. This notion is introduced in Section 4, where we establish it for spaces of continuous functions over proper G-spaces. Yet it is in the context of coefficient modules that relative injectivity turns out to be the most powerful ; indeed we establish in Section 5 the precise relation between this notion and R.J. Zimmer's amenable actions.

In the third chapter, we introduce continuous bounded cohomology. But before presenting it functorially with bell, book and candle, we define it very concretely in Section 6. The functorial characterization itself comes in Section 7, together with a number of examples. Section 8 deals with the elementary covariance and contravariance properties of the functors just introduced. The important relation between continuous bounded cohomology and the ordinary continuous cohomology, namely the comparison map, is introduced in Section 9 ; we seize the opportunity to fix exactly what we understand as continuous cohomology, as this is not so familiar an object for general (non locally compact) topological groups.

All three families of techniques studied in the fourth chapter are essential for applications. Section 10 is concerned with rather classical techniques, notably the straightforward induction, neat but not all so useful, and L^p induction, at first sight not quite adapted to continuous *bounded* cohomology, but very effective in degree two. In Section 11, we take advantage of the structure theory of locally compact groups and establish double ergodicity. This is an ingredient for the refined Lyndon-Hochschild-Serre exact sequences that we derive in Section 12 with the help of a Hochschild-Serre spectral sequence.

In the fifth chapter we cast a glance at applications. The connection of continuous bounded cohomology with rough actions, property (TT), quasi-morphisms, actions on the circle and other rigidity issues are presented in Section 13. Finally, in Section 14, we prove the theorems on irreducible lattices and deduce some consequences.

General conventions and notations

Most notations are introduced in the text ; a comprehensive index has been added for the convenience of the reader, locating both concepts and symbols.

Theorems, corollaries, definitions etc. are uniformly numbered and referred to with an upper case initial, even though the corresponding noun is written in lower case ; we would thus for instance write : "the first lemma is Lemma 1.1.1". Proofs begin with the indication PROOF and are concluded by the symbol □. If a proof does not follow immediately the corresponding statement, it will always be precised what is being proved.

The collection of natural, integer, rational, real and complex numbers are denoted as usual by $\mathbf{N}$, $\mathbf{Z}$, $\mathbf{Q}$, $\mathbf{R}$, $\mathbf{C}$. The French convention $0 \in \mathbf{N}$ has been adopted. The composition of two maps f, g is denoted both by $f \circ g$ and fg. The function with constant value one on some set, group, etc. X is $\mathbf{1}_X$.

For basic topological concepts, we follow mostly N. Bourbaki ; this applies in particular for the notions of locally compact and compact spaces, which are Hausdorff by definition. Also Bourbakian are uniform spaces and the left and right canonical uniform structures on topological groups. However, we follow the widespread consensus and write "σ-compact" rather than "countable at infinity". With a few exceptions, among which the introduction, we use $\mathbf{C}$ as basic field when no explicit mention is made. Thus for instance $\mathrm{C}(X)$ denotes the space of continuous functions $X \to \mathbf{C}$ unless otherwise stated. The use of nets and sub-nets as "generalized sequences" is standard, see [**91**, Chap. 2].

The notation $A \subset B$ for inclusions does not exclude the equality case. If moreover B is a group, vector space, etc. and that A is a subgroup, vector subspace, etc. then we write often $A < B$, again allowing the equality case.

We shall sometimes use loosely the word "identification" when referring to a canonical isomorphism given in a particularly obvious way.

CHAPTER I

Banach modules, L^∞ spaces

Let G be a topological group. We encompass "abstract" groups in this setting by endowing them with the discrete topology. Our characterization of continuous bounded cohomology will be formulated in a homological language for which the fundamental objects are various types of *Banach G-modules*. Whenever available, spaces of L^∞ type are distinguished representatives of the latter.

1. Banach modules

1.1. Notations and first definitions. Unless otherwise specified, all Banach spaces are over the field $\mathbf{C}$ of complex numbers, and by "dual" we understand the normed topological dual. The dual of a Banach space E is denoted by E^*, but an alternative notation $E^\sharp$ is to be introduced below, reflecting the G-structure. We use brackets

$$\langle\cdot|\cdot\rangle : \; E^* \times E \longrightarrow \mathbf{C}$$

as a standing notation for the canonical duality.

A *morphism* $\alpha : E \to F$ of Banach spaces E, F is a continuous linear map, and $\|\alpha\|$ denotes its operator norm. We write $\mathcal{L}(E,F)$ for the space of morphisms $E \to F$ and endow it with the Banach space structure determined by the operator norm.

A *Banach G-module* is a pair (π, E) where E is a Banach space and π is a group homomorphism from G to the group of isometric linear automorphisms of E. No a priori continuity assumption is made on π. We will often refer to the Banach G-module (π, E) merely as the underlying space E, and the action of $g \in G$ on $v \in E$ will occasionally be written gv instead of $\pi(g)v$. We call (π, E) a *trivial* Banach G-module if the map π is constant.

If E and F are Banach G-modules, a *G-morphism* $\alpha : E \to F$ is a G-equivariant morphism. We emphasize that our definition of morphism and G-morphism does not require them to have closed range, contrary to N. Bourbaki's definition of *homomorphism*.

The Banach G-module (π, E) is *continuous* if the structure map

$$G \times E \longrightarrow E$$
$$(g, v) \longmapsto \pi(g)v$$

is continuous, $G \times E$ being endowed with the product topology (whence the occasional use of the expression *jointly continuous*). Equivalent conditions are given by the following lemma.

LEMMA 1.1.1.— *Let (π, E) be a Banach G-module. The following assertions are equivalent.*

(i) *The Banach G-module (π, E) is continuous.*

(ii) *For all $v \in E$ the orbit map*

$$G \longrightarrow E$$
$$g \longmapsto \pi(g)v$$

is left uniformly continuous on G.

(iii) *For all $v \in E$ the orbit map*

$$G \longrightarrow E$$
$$g \longmapsto \pi(g)v$$

is continuous at $e \in G$.

PROOF. It follows from the very formulation that we have only to show (iii)$\Rightarrow$(i)&(ii). Let $(v_n)_{n\in\mathbf{N}}$ be a sequence converging to $v \in E$ and $(g_\alpha)_{\alpha\in\mathrm{A}}$ a net converging to $e \in G$; let $g \in G$. The estimate

$$\begin{aligned}\|\pi(gg_\alpha)v_n - \pi(g)v\|_E &\leq \|\pi(gg_\alpha)v_n - \pi(gg_\alpha)v\|_E \ + \ \|\pi(gg_\alpha)v - \pi(g)v\|_E \\ &= \|v_n - v\|_E \ + \ \|\pi(g_\alpha)v - v\|_E\end{aligned}$$

shows that $\pi(gg_\alpha)v_n$ converges to $\pi(g)v$ uniformly on $g \in G$, proving (i) and (ii) at once. □

The Banach G-module (π, E) is said separable, reflexive, etc. whenever the underlying Banach space E enjoys the corresponding property. In this context, we insist on the

NOTATION 1.1.2. Whenever a topological attribute is attached to a Banach space, it refers to the topology induced by the norm. Since weak and weak-* topologies will also frequently be considered, every epithet will in this case be decorated with the corresponding prefix, as for instance : *weak-* Borel, weakly continuous*, etc. Occasionally we shall even mention pedantically the continuity *in norm*, etc.

Using Baire's category theorem, one can give a further sufficient, and obviously necessary, continuity condition. We recall (N. Bourbaki [**28**, IX §5]) that a subset of a topological space is called *meager*, or *of the first category*, if it is a countable union of nowhere dense subsets ; a topological space is a *Baire space* if no non empty open set is meager.

All locally compact spaces aswell as all complete metric spaces are Baire spaces (Théorème 1 in [**28**, IX §5 N° 3]). Examples of non Baire spaces relevant to our matter are provided by non reflexive separable dual Banach spaces A^* endowed with the weak-* topology : indeed, the dual of such a topological vector space is isomorphic to the predual A of the Banach space by Proposition 1 in [**30**, IV §1 N° 2] ; however, if A^* were Baire in its weak-* topology, then an argument very similar to the proposition below would show that every norm continuous linear form on A^* is weak-* continuous, thus identifying A with A^{**}.

PROPOSITION 1.1.3.— *Let G be a Baire topological group (e.g. locally compact) and (π, E) a separable Banach G-module. If the orbital maps*

$$G \longrightarrow E, \quad g \longmapsto \pi(g)v$$

are Borel for all $v \in E$, then the Banach G-module (π, E) is continuous.

The proof uses the well known

LEMMA 1.1.4.— *Let A be a Borel subset of the Baire group G. If A is not meager in G, then AA^{-1} is a neighbourhood of e in G.*

ON THE PROOF OF THE LEMMA. This is the Lemme 9 in N. Bourbaki [**28**, IX §6 N° 8]. □

PROOF OF PROPOSITION 1.1.3. Fix $v \in E$. In view of Lemma 1.1.1, we shall prove that the orbital map

$$F: \ G \longrightarrow E, \quad F(g) = \pi(g)v$$

is continuous at $e \in G$. We first claim that for every open neighbourhood V of v in E, the Borel set $F^{-1}(V)$ is not meager in G.

Indeed, if $F^{-1}(V)$ were meager, then so would be $F^{-1}(\pi(g)V)$ for every $g \in G$ in view of the identity $F^{-1}(\pi(g)V) = gF^{-1}(V)$. Since E is separable metrizable and hence second countable, the set $\pi(G)V$ is also second countable. Thus Lindelöf's theorem (Proposition 1 (i) in [**28**, IX, Appendice 1]) provides us with a countable sub-cover of its open cover by all $\pi(g)V$; that is, there is a countable subset $D \subset G$ such that

$$\pi(G)V = \bigcup_{g \in D} \pi(g)V.$$

Since for all $g \in G$ the set $F^{-1}(\pi(g)V)$ contains g, we have

$$G = F^{-1}(\pi(G)V) = \bigcup_{g \in D} F^{-1}(\pi(g)V).$$

Thus G would be a countable union of meager sets, hence be meager itself, in contradiction with the assumption that G is a Baire space, whence the claim.

Given now any neighbourhood W of v in E, we have to prove that $F^{-1}(W)$ is a neighbourhood of e in G. Let thus $r > 0$ such that the ball $B(v, 2r)$ is in W. Applying the above claim to $V = B(v, r)$, we see that $F^{-1}(V)$ is not meager. Moreover, it is a Borel set by the assumption on (π, E). Hence, by Lemma 1.1.4, the set $F^{-1}(V)(F^{-1}(V))^{-1}$ is a neighbourhood of e in G. The choice of r implies that $F^{-1}(W)$ contains $F^{-1}(V)(F^{-1}(V))^{-1}$ because for every $x, y \in F^{-1}(V)$ one has

$$\begin{aligned}\|\pi(xy^{-1})v - v\|_E &\leq \|\pi(xy^{-1})v - \pi(x)v\|_E + \|\pi(x)v - v\| \\ &= \|v - \pi(y)v\|_E + \|\pi(x)v - v\| < r + r,\end{aligned}$$

so that $F(xy^{-1}) \in W$. Hence $F^{-1}(W)$ is a neighbourhood of e in G, and F is continuous. □

REMARK 1.1.5. This proof is a variation on the standard argument to the end that measurable homomorphisms should be continuous, see e.g. [**46**, Corollary 12.6]. In this reference, meager Borel sets are replaced with null sets and Baire's theorem with the Fubini-Lebesgue theorem, thus restricting the statement to locally compact groups. We prefer the more general setting of the category theorem.

A statement similar to Proposition 1.1.3 is made by B.E. Johnson in [**89**] on pages 25–26 but we could not follow the detail of his argument.

REMARKS 1.1.6.

(i) It is also possible to develop a theory of bounded cohomology within the setting of uniformly bounded Banach representations, that is, representations π of G in the group of continuous linear automorphisms of a Banach space E and satisfying

$$\sup_{g \in G} \|\pi(g)\| < \infty.$$

However, since the precise semi-norm on cohomology spaces is at the heart of certain applications, this approach does not suit properly our purposes. Moreover, a uniformly bounded representation space E can always be equipped with an equivalent norm

$$\|v\|_{\mathrm{is}} = \sup_{g \in G} \|\pi(g)v\|_E \qquad (v \in E)$$

for which the representation is isometric.

(ii) According to our definition, Banach modules are left modules. A *right Banach G-module* is by definition a Banach G^{op}-module. Here G^{op} is the opposite group, i.e. the topological space underlying G endowed with the multiplication law $(x, y) \mapsto yx$.

(iii) We shall not indulge much in the language of categories, because several of the cohomological definitions that we are to introduce later on reflect compound aspects of various origins. In this respect, the discussion preceding the definition of relative injectivity at the beginning of Section 4 is symptomatic. Occasional use of the language of categories will refer to Banach G-modules with G-morphisms, not rendering the topological and geometric issues involved in the very definition of continuous bounded cohomology.

EXAMPLES 1.1.7. A continuous unitary representation $(\pi, \mathfrak{H})$ of G in a separable Hilbert space $\mathfrak{H}$ is a continuous separable Banach G-module. Our favorite trivial module is $\mathbf{C}$.

Suppose G locally compact. Fixing a left Haar measure on G, each classical Lebesgue space $L^p(G)$, for $1 \le p \le \infty$, is a Banach G-module for the left regular representation. If $1 \le p < \infty$, it is a continuous Banach G-module, whilst $L^\infty(G)$ is not so unless G is discrete.

Much more examples will be introduced later on.

1.2. Elementary constructions. A Banach G-submodule F of the Banach G-module E is of course a subspace $F \subset E$ such that the induced structure turns it into a Banach G-module ; this is equivalent to saying that F is closed and G-invariant. The natural quotient structure turns E/F into a Banach G-module. If E is continuous, then so are both F and E/F.

If (π, E) and (ϱ, F) are Banach G-modules, we endow the space $\mathcal{L}(E, F)$ of morphisms with a natural Banach G-module structure by setting

$$g\alpha = \varrho(g)\alpha\pi(g^{-1}) \qquad (*)$$

for $\alpha \in \mathcal{L}(E, F)$ and $g \in G$. In the particular case $E = \mathbf{C}$, this structure is compatible with the canonical isometric identification $\mathcal{L}(\mathbf{C}, F) \cong F$.

The *contragredient* Banach G-module $(\pi^\sharp, E^\sharp)$ associated to (π, E) is by definition $\mathcal{L}(E, \mathbf{C})$ with the structure $(*)$ above. Thus, as a Banach space, $E^\sharp = E^*$; the notational distinction reflects the fact that the adjoint structure (π^*, E^*) to (π, E) is a right Banach G-module, since according to $(*)$ we have $\pi(g)^\sharp = \pi(g^{-1})^*$ for all $g \in G$.

In order to avoid heavy terminology, we introduce the following concept, which will be of very frequent use.

DEFINITION 1.2.1. A *coefficient G-module* is a Banach G-module (π, E) contragredient to some separable continuous Banach G-module denoted $(\pi^\flat, E^\flat)$. The choice of $E^\flat$ is part of the data.

A morphism or G-morphism of Banach modules $\alpha : E \to F$ between coefficient modules is called *adjoint* if it is the adjoint of a morphism $\alpha^\flat : F^\flat \to E^\flat$, or equivalently if it is weak-* continuous. We say synonymously that α is a *morphism (or G-morphism) of coefficient modules*.

REMARK 1.2.2. We insist that a coefficient module is not just a dual Banach space with some contragredient representation, but includes by definition the choice of a predual ; for it may happen that $(\pi^\flat, E^\flat)$ is not uniquely determined by its contragredient, even if the contragredient is separable. In this respect, the bare notation (π, E) is a slight *abus de langage*. All the same, the above definition entitles us to speak of *the* weak-* topology of a coefficient module, whilst there is in general no uniquely defined weak-* topology on a dual Banach space. In particular, the notion of adjoint morphism depends indeed on the preduals.

EXAMPLES 1.2.3. Returning to the Examples 1.1.7, we observe that a continuous unitary representation $(\pi, \mathfrak{H})$ of G in a separable Hilbert space $\mathfrak{H}$ is also a coefficient G-module. For G locally compact second countable, $L^p(G)$ is a coefficient G-module if $1 < p \leq \infty$, whilst $L^1(G)$ is not so unless G is discrete.

An important feature of coefficient modules is that the action is weak-* continuous, since the structure maps are adjoint and the predual continuous. Observe however that a coefficient module needs not be continuous nor separable. Actually, for general coefficient G-module E, we shall see (Remark 3.3.5) that the action map $G \times E \to E$ is not even weakly continuous, even though it is weak-* continuous. It will turn out later on (Corollary 3.3.1 and Proposition 3.3.2) that there is a large class of coefficient modules which are much better behaved.

Suppose F is a Banach G-submodule of a coefficient G-module E. If F is weak-* closed in the weak-* topology attached to $E^\flat$, then one has the canonical identification

$$F \cong \left(E^\flat / {}^\perp F\right)^\sharp,$$

where ${}^\perp F$ is the annihilator of F in $E^\flat$; this is the duality principle, see [**126**, I, Theorem 4.9]. Thus F is endowed with a *canonical* coefficient G-module structure. Accordingly, we shall define a coefficient G-submodule of E to be any weak-* closed G-invariant subspace F endowed with this structure. The duality principle provides also the quotient $E/F \cong ({}^\perp F)^\sharp$ with a canonical coefficient G-module structure.

If (π, E) and (ϱ, F) are Banach G-modules, we endow the completed projective tensor product $E\widehat{\otimes}F$ with the diagonal G-action. We refer to the book of H. Jarchow [**88**, III 15] or to A. Grothendieck [**72**, I §1.1]) for a description of the underlying Banach space $E\widehat{\otimes}F$. We remark in particular that the canonical form $E^\sharp\widehat{\otimes}E \longrightarrow \mathbf{C}$ is a norm one G-morphism.

We obtain a canonical isometric identification of Banach G-modules

$$(E\widehat{\otimes}F)^\sharp \cong \mathcal{L}(E, F^\sharp)$$

from the corresponding Banach space identification, which can be found for instance in [**44**, Corollary VIII.2.2]. Furthermore, the so-called associativity isomorphism

$$D\widehat{\otimes}(E\widehat{\otimes}F) \cong (D\widehat{\otimes}E)\widehat{\otimes}F$$

and commutativity isomorphism $E\widehat{\otimes}F \cong F\widehat{\otimes}E$ preserve the G-structures in presence, and so does the adjoint associativity

$$\mathcal{L}(D\widehat{\otimes}E, F) \cong \mathcal{L}\big(D, \mathcal{L}(E, F)\big).$$

For any Banach G-module E, we define its *maximal continuous submodule*

$$\begin{aligned} \mathcal{C}E &= \{v \in E : G \to E, g \mapsto gv \text{ is continuous }\} \\ &= \{v \in E : G \to E, g \mapsto gv \text{ is continuous at } e \in G\}. \end{aligned}$$

The terminology is justified by the following lemma.

LEMMA 1.2.4.— *The Banach G-module E induces on the set $\mathcal{C}E$ a structure of continuous Banach G-module. Moreover, $\mathcal{C}E$ contains all continuous Banach G-submodules of E.*

PROOF. We only show that $\mathcal{C}E$ is closed in E. Let $(v_n)_{n\in\mathbf{N}}$ be a sequence of $\mathcal{C}E$ converging to $v \in E$ in E and $(g_\alpha)_{\alpha\in\mathbf{A}}$ a net converging to g in G. The inequality

$$\begin{aligned}\|\pi(g_\alpha)v - \pi(g)v\|_E &\le \|\pi(g_\alpha)v - \pi(g_\alpha)v_n\|_E \;+\; \|\pi(g_\alpha)v_n - \pi(g)v_n\|_E \;+\\ &\qquad\qquad\qquad +\; \|\pi(g)v_n - \pi(g)v\|_E\\ &= 2\|v - v_n\|_E \;+\; \|\pi(g_\alpha)v_n - \pi(g)v_n\|_E\end{aligned}$$

shows that $\pi(g_\alpha)v$ converges to $\pi(g)v$. □

NOTATION 1.2.5. If there is a possible ambiguity as to the G-representation π, we write $\mathcal{C}_\pi E$ instead of $\mathcal{C}E$.

The following observation states that the subcategory of continuous Banach G-modules is a *full* subcategory onto which $\mathcal{C}$ is a retraction :

LEMMA 1.2.6.— *Any G-morphism $\alpha : E \to F$ of Banach G-modules restricts to $\mathcal{C}E \to \mathcal{C}F$.*

PROOF. Let $v \in E$. By G-equivariance of α, the orbit map $G \to F$ corresponding to $\alpha(v)$ factors as

$$G \to E \xrightarrow{\alpha} F,$$

wherein the first arrow represents the orbit map associated to v. The second arrow is continuous, and so is the first if $v \in \mathcal{C}E$. Therefore $\alpha(\mathcal{C}E) \subset \mathcal{C}F$. □

We record also the following observation :

LEMMA 1.2.7.— *Let G be a topological group and (π, E) a Banach G-module. If $H \lhd G$ is a normal subgroup of G, then $\mathcal{C}_{\pi|H}$ is a Banach G-submodule of E.*

PROOF. In view of Lemma 1.2.4, we have only to check that G preserves $\mathcal{C}_{\pi|H}$. Let $v \in \mathcal{C}_{\pi|H}$ and take a net $(h_\alpha)_{\alpha\in\mathbf{A}}$ converging to $e \in H$ in order to check the H-continuity of an element $\pi(g)v$, where $g \in G$. We have

$$\big\|\pi(h_\alpha)(\pi(g)v) - \pi(g)v\big\|_E = \big\|\pi(g^{-1}h_\alpha g)v - v\big\|_E$$

which converges to zero along α because $(g^{-1}h_\alpha g)_{\alpha\in\mathbf{A}}$ is a net of H converging to e. □

For any G-module (π, E), we denote by E^G the set of G-invariant elements ; observe that this is a closed subspace.

REMARK 1.2.8. In particular, in view of the action that we have defined on $\mathcal{L}(E, F)$, we see that $\mathcal{L}(E, F)^G$ is the space of G-morphisms. Therefore, whenever a Banach G-module A comes with a canonical isomorphism to some $\mathcal{L}(E, F)$, we shall freely term the elements of A^G either as *G-equivariant* or G-invariant, according to whether the stress is on their being a morphism or an element of the G-module A.

If now $H < G$ is a subgroup with $\pi(H)$ normal in $\pi(G)$, then E^H is G-invariant and therefore E^H is a Banach G-module in a natural way. We notice the following fact :

LEMMA 1.2.9.— *Let (π, E) be a coefficient G-module and $H < G$ a subgroup with $\pi(H) \lhd \pi(G)$ (e.g. $H \lhd G$). Then E^H is a coefficient G-submodule of E.*

PROOF. Since the G-action is contragredient, the kernel of $(\pi(g) - Id)$ is weak-* closed for all $g \in G$. Thus

$$E^H = \bigcap_{h \in H} \mathrm{Ker}(\pi(h) - Id)$$

is weak-* closed. □

Another immediate property of the invariants is the following

LEMMA 1.2.10.— *Let (π, E) be a Banach G-module. Then $(E/E^G)^G = 0$.*

PROOF. Suppose $v \in E$ is such that the element $v + E^G$ of E/E^G is G-invariant. Then there is for every $g \in G$ an element $u_g \in E^G$ such that $\pi(g)v = v + u_g$. Thus, for all $h, g \in G$,

$$\pi(hg)v = v + u_{hg}$$

and

$$\pi(hg)v = \pi(h)v + \pi(h)u_g = v + u_h + u_g.$$

Therefore we have $u_{hg} = u_h + u_g$ and hence $g \mapsto u_g$ is a group homomorphism from G, considered without its topology, to the additive group E^G. However, the latter has no non trivial bounded subgroups, whilst $\|u_g\| \leq 2\|v\|$ by definition. Thus $u_g = 0$ for all $g \in G$, so that v is in E^G and hence is trivial in E/E^G. □

REMARKS 1.2.11.

(i) The property stated in this lemma may fail for general linear representations. For instance, the ($\mathbf{R}$-linear) representation of $\mathbf{Z}$ on $\mathbf{R}^2$ defined by

$$n(x,y) = (x + ny, y) \qquad (n \in \mathbf{Z}, (x,y) \in \mathbf{R}^2)$$

has $\mathbf{R} \oplus 0$ as invariants, yet the quotient action is trivial.

(ii) Both the Lemma 1.2.10 and the example given in (i) can be considered as elementary applications of a cohomological long exact sequence. The corresponding sequence in bounded cohomology is to be established in Section 8.2 for suitable classes of Banach G-modules.

If (π, E) and (ϱ, F) are two Banach G-modules, then we denote by $E \oplus F$ the direct sum vector space endowed with its canonical topological vector space structure and with the G-action $\pi \oplus \varrho$. Whilst the topological vector space structure is well defined, there are many Banach space structures on $E \oplus F$ representing this topology and for which $\pi \oplus \varrho$ is isometric and hence yields a structure of Banach G-modules (e.g. all Minkowski ℓ^p combinations of the norms of E and F). We shall use the following

NOTATION 1.2.12. The notation $E \oplus F$ in a statement involving Banach structures indicates that the latter is valid for all norms on the topological vector space $E \oplus F$ which arise from a normic combination of the norms of E and F.

If E, F are submodules of some Banach G-module D, we use the standard notation $E + F \subset D$ for their sum.

1.3. Spaces of continuous functions. Let X be a topological space on which G acts by homeomorphisms and let (π, E) be a Banach G-module. We consider the Banach space

$$\mathrm{C_b}(X, E)$$

of bounded continuous maps $f : X \to E$ with the supremum norm

$$\|f\|_\infty = \sup_{x \in X} \|f(x)\|_E$$

and turn it into a Banach G-module by means of the *left regular representation* λ_π defined by

$$\Big(\lambda_\pi(g)f\Big)(x) = \pi(g)f(g^{-1}x)$$

for $x \in X$, $g \in G$ and $f \in \mathrm{C_b}(X, E)$. Notice that here again invariant elements are but equivariant maps (this is similar to Remark 1.2.8). If the Banach G-module (π, E) is trivial, then we write also λ for λ_π and call it the left *translation*.

The G-action on the topological space X is said *continuous* if the action map $G \times X \to X$ is continuous.

LEMMA 1.3.1.— *Let (π, E) be a Banach G-module and X a topological space with G-action by homeomorphisms. If the action is continuous, then any element of $\mathcal{C}\mathrm{C_b}(X, E)$ ranges in $\mathcal{C}E$.*

PROOF. We fix $f \in \mathcal{C}\mathrm{C_b}(X, E)$, $x \in X$ and check the criterion (iii) of Lemma 1.1.1 for $f(x)$. Let $(g_\alpha)_{\alpha \in \mathrm{A}}$ be a net converging to $e \in G$. We have

$$\begin{aligned} \|f(x) - \pi(g_\alpha)f(x)\|_E &\leq \|f(x) - \lambda_\pi(g_\alpha)f(x)\|_E + \\ &\qquad + \|\lambda_\pi(g_\alpha)f(x) - \pi(g_\alpha)f(x)\|_E \\ &\leq \|f - \lambda_\pi(g_\alpha)f\|_\infty + \|f(g_\alpha^{-1}x) - f(x)\|_E. \end{aligned}$$

The first term tends to zero because f is in the maximal continuous submodule, and the second tends to zero by the continuity of f and of the action. □

2. L^∞ SPACES

Let G be a topological group. We turn our attention toward an important source of Banach G-modules, namely L^∞ spaces attached to measure spaces on which G acts in an appropriate way.

2.1. Regular spaces. As a matter of fact, the natural objects for us to consider are spaces endowed merely with an invariant measure *class* rather than with a measure. At any rate, the measure class is the only information retained when passing to the space of L^∞ function classes, but we have also another reason for this choice :

Most of the time, our spaces will not admit a finite non-trivial invariant measure. Indeed, we shall be particularly interested in *amenable* G-spaces (for G locally compact second countable), in which case the invariance of a probability measure would force the group G itself to be amenable, as shown in [**141**, Proposition 4.3.3]. However, amenable groups turn out to be completely trivial from the point of view of bounded cohomology.

This being so, we do not find it natural to distinguish a particular measure in its G-orbit.

This discussion accounts for the following definition.

DEFINITION 2.1.1. Let G be a topological group. A *regular G-space* is a standard Borel space S on which G acts measurably, together with a G-invariant measure class with the following property :

The measure class contains a probability measure μ turning (S, μ) in a standard probability space such that the natural isometric G-action $\lambda^\flat$ on $L^1(\mu)$

$$(\lambda^\flat(g)\varphi)(s) = \varphi(g^{-1}s)\frac{dg^{-1}\mu}{d\mu}(s), \qquad (\varphi \in L^1(\mu), s \in S)$$

is continuous[1].

In the above definition, $dg^{-1}\mu/d\mu$ is the Radon-Nikodým derivative. Since our measure class is invariant, the Lebesgue-Radon-Nikodým theorem implies that $dg^{-1}\mu/d\mu$ is represented by a locally integrable function, hence is in $L^1(\mu)$ since μ is finite.

EXAMPLES 2.1.2.

(i) If G is locally compact and second countable, the first example is of course G itself endowed with the class of its Haar measures. More generally, for any closed subgroup $H < G$, the homogeneous space G/H with the class of the natural quasi-invariant measures is a regular G-space.

(ii) The product of finitely many or countably many regular G-spaces is still a regular G-space when endowed with the product structure.

(iii) The Poisson boundaries (see H. Furstenberg [**58**] and [**59**]) associated to a probability measure on a locally compact second countable group G, that is, the random walk boundaries, are regular G-spaces for the class of the corresponding stationary measure (we consider here the measure-theoretical Poisson boundaries, for the Poisson *spaces* see [**6**]).

(iv) If S is a regular G-space, H another topological group and $H \to G$ a continuous homomorphism, then by pull-back S is a regular H-space.

[1]The symbol $\lambda^\flat$ suggests already that we shall be more interested in the contragredient of this action.

2.2. Measurable maps. Given a regular G-space S, we endow the Banach space $L^\infty(S)$ of complex-valued essentially bounded measurable function classes with the G-module structure given by the *left translation* representation λ defined by

$$(\lambda(g)f)(s) = f(g^{-1}s) \qquad\qquad (f \in L^\infty(S), g \in G)$$

for almost every $s \in S$. The notation hints to the fact that for every choice of a probability μ as in Definition 2.1.1, integration against μ realizes the module $(\lambda, L^\infty(S))$ as contragredient to $(\lambda^\flat, L^1(\mu))$.

The terminology overlaps with the translation for spaces of continuous functions ; this is in part justified in Remark 2.2.2 below.

We proceed now to generalize this to Banach valued L^∞ function spaces ; that is, given a regular G-space S and a Banach G-module (π, E), we shall consider a Banach G-module defined by means of E-valued maps on S. In this setting, the notion of measurability has to be dealt with more carefully : indeed, if E is non-separable, it may happen that the collection of (norm-) measurable maps $S \to E$ is not even closed under pointwise addition [**92**, page 135] ; yet situations where E is non-separable will be essential to us. We are actually interested in the case where (π, E) is a coefficient G-module, so that in particular E is a dual Banach space and comes moreover with a distinguished weak-* topology and the associated weak-* Borel structure. The point is that in general this Borel structure will be different from the normic, and much better behaved.

DEFINITION 2.2.1. Let S be a regular G-space and (π, E) a coefficient G-module. We denote by $L^\infty_{\mathrm{w*}}(S, E)$ the space of classes of essentially bounded E-valued weak-* measurable maps on S endowed with the essential supremum norm. We endow this space with the Banach G-module structure determined by the *left regular representation* λ_π defined by

$$(\lambda_\pi(g)f)(s) = \pi(g)(f(g^{-1}s)) \qquad\qquad (f \in L^\infty_{\mathrm{w*}}(S, E), g \in G)$$

for almost every $s \in S$. If π is trivial, we write $\lambda_\pi = \lambda$, recovering the left translation representation.

Now $L^\infty_{\mathrm{w*}}(S, E)$ is a Banach space with the above norm ; this Banach space is identified more precisely in Proposition 2.3.1 below. Mind that the very definition of $L^\infty_{\mathrm{w*}}(S, E)$ depends upon the choice of the isomorphism class of $E^\flat$ contained in Definition 1.2.1.

In the Definition 2.2.1, the map classes are understood with respect to equality almost everywhere. As in the scalar case, this amounts to considering the separated quotient of the semi-normed space of essentially bounded E-valued weak-* measurable maps on S endowed with the essential supremum semi-norm.

REMARK 2.2.2. The left regular representation on $L^\infty_{\mathrm{w*}}(S, E)$ is defined in complete analogy with the homonymous structure for bounded continuous maps on topological spaces. In the case of trivial coefficients $E = \mathbf{C}$, this is more than an analogy since by the Gelfand spectral representation on can view $L^\infty(S)$ as the space $\mathrm{C}(\Omega)$ of continuous functions on its spectrum Ω.

In the general case, however, it seems that the best we can do is to identify $L^\infty_{w*}(S,E)$ with the space of morphisms from $E^\flat$ to $C(\Omega)$ — in the case $E=\mathbf{C}$ we recover of course $\mathcal{L}(\mathbf{C}, C(\Omega)) \cong C(\Omega)$.

The following basic observation is one of the motivations for the separability requirement in our definition of coefficient G-modules.

LEMMA 2.2.3.— *Let S be a regular G-space, (π, E) a coefficient G-module and $f \in L^\infty_{w*}(S,E)$. Then the function class defined almost everywhere by*

$$\begin{aligned} S &\longrightarrow \mathbf{R} \\ s &\longmapsto \|f(s)\|_E \end{aligned}$$

belongs to $L^\infty(S)$.

PROOF. Since $f(s)$ can be considered as a continuous linear form on $E^\flat$, we have

$$\|f(s)\|_E = \sup_{u\in D\setminus\{0\}} \frac{\langle f(s)|u\rangle}{\|u\|_{E^\flat}}$$

for every dense subset $D \subset E^\flat$. Each of the function classes

$$\begin{aligned} S &\longrightarrow \mathbf{R} \\ s &\longmapsto \langle f(s)|u\rangle \end{aligned}$$

is in $L^\infty(S)$ by weak-* measurability of f. Since the predual $E^\flat$ is separable, D can be chosen countable, so that the function of the lemma is the supremum of a uniformly bounded countable family in $L^\infty(S)$, whence the statement. □

2.3. Duality. Let (S,μ) be as in Definition 2.1.1. The representation $\lambda^\flat$ given there turns $L^1(\mu)$ into a separable continuous Banach G-module. If we keep track of the G-actions introduced so far, a well known result of functional analysis reads :

PROPOSITION 2.3.1.— *Let $S_1, \ldots, S_n$ be regular G-spaces and (π, E) a continuous separable Banach G-module. Then the Banach G-module*

$$L^\infty_{w*}(S_1 \times \cdots \times S_n, E^\sharp)$$

is canonically contragredient to

$$L^1(\mu_1)\widehat{\otimes}\cdots\widehat{\otimes}L^1(\mu_n)\widehat{\otimes}E$$

for any $\mu_1, \ldots, \mu_n$ as in Definition 2.1.1.

PROOF. This follows from the Dunford-Pettis theorem, see [**48**, VI.8], [**72**, I §2.2] and [**88**, III 17.6]. The duality is obtained by integrating the duality $E^\sharp \times E \to \mathbf{C}$ against the measures μ_i. □

In this context, we mention that according to a result of A. Grothendieck (Théorème 2 in [**72**, I §2.2], see also [**88**, III 15.7.4]), the space $L^1(\mu_i)\widehat{\otimes}E$ identifies isometrically with the Bochner-Lebesgue space $L^1(\mu_i, E)$, and more generally

$$L^1(\mu_1)\widehat{\otimes}\cdots\widehat{\otimes}L^1(\mu_n)\widehat{\otimes}E \cong L^1(\mu_1\otimes\cdots\otimes\mu_n, E),$$

where $\mu_1\otimes\cdots\otimes\mu_n$ is the product measure.

In view of the identification

$$L^\infty_{\mathrm{w*}}(S, E^\sharp) \cong \mathcal{L}\Big(L^1(\mu), E^\sharp\Big),$$

we recall that according to the terminological Remark 1.2.8, the G-invariance in $L^\infty_{\mathrm{w*}}(S, E^\sharp)$ for the regular representation amounts to G-equivariance of the maps $L^1(\mu) \to E^\sharp$.

Two direct consequences of the Proposition 2.3.1 and the Definition 2.1.1 are :

COROLLARY 2.3.2.— *Let S be a regular G-space and E a coefficient G-module. Then $L^\infty_{\mathrm{w*}}(S, E)$ is also a coefficient G-module in a canonical way.* □

COROLLARY 2.3.3.— *Let S_1, S_2 be regular G-spaces and E a coefficient G-module. Then there is a canonical isometric coefficient G-module identification*

$$L^\infty_{\mathrm{w*}}(S_1 \times S_2, E) \cong L^\infty_{\mathrm{w*}}\Big(S_1, L^\infty_{\mathrm{w*}}(S_2, E)\Big). \qquad \square$$

This latter corollary shows that weak-* measurability is particularly precious, since it allows such simple a form of the "exponential law". Needless to say, there is in general no corresponding statement for spaces of continuous bounded functions on topological spaces X_i, for in general the canonical inclusions

$$\mathrm{C_b}\big(X_1, \mathrm{C_b}(X_2, -)\big) \subset \mathrm{C_b}(X_1 \times X_2, -)$$

are strict.

2.4. Double structure on $L^\infty_{\mathrm{w*}}(G, E)$. Let (π, E) be a coefficient G-module and suppose G locally compact and second countable, so that we may consider the space $L^\infty_{\mathrm{w*}}(G, E)$ over the regular G-space G (first point in Examples 2.1.2). In view of the above discussion, this particular case $S = G$ is of especial interest, because we have the following further natural Banach G-module structure at hand.

DEFINITION 2.4.1. The *right translation* ϱ is the isometric representation of G in $L^\infty_{\mathrm{w*}}(G, E)$ defined by

$$(\varrho(g)f)(x) = f(xg)) \qquad \big(f \in L^\infty_{\mathrm{w*}}(G, E), g \in G\big)$$

for almost every $x \in G$.

Notice that π is not involved in this definition.

PROPOSITION 2.4.2.— *There is an isometric isomorphism of coefficient G-modules*

$$\Big(\lambda_\pi,\ L^\infty_{\mathrm{w*}}(G, E)\Big) \cong \Big(\varrho,\ L^\infty_{\mathrm{w*}}(G, E)\Big).$$

PROOF. Since E is a coefficient G-module, the action is weak-* continuous, hence for every weak-* measurable map $f : G \to E$ the map $\alpha(f) : G \to E$ defined by $\alpha(f)(x) = \pi(x)f(x^{-1})$ is still weak-* measurable. We obtain in this way a well defined linear operator

$$\alpha :\ L^\infty_{\mathrm{w*}}(G, E) \longrightarrow L^\infty_{\mathrm{w*}}(G, E)$$

which preserves the norm. Let now $g \in G$ and let f represent a class in $L^\infty_{\mathrm{w}*}(G,E)$. We have for almost every $x \in G$

$$\begin{aligned}\alpha(\lambda_\pi(g)f)(x) = \pi(x)\Big((\lambda_\pi(g)f)(x^{-1})\Big) &= \pi(x)\pi(g)f(g^{-1}x^{-1})\\ &= \pi(xg)f((xg)^{-1}) = \alpha(f)(xg) = \Big(\varrho(g)\alpha(f)\Big)(x),\end{aligned}$$

hence $\alpha\lambda_\pi(g) = \varrho(g)\alpha$. By its very definition, α is an involution, so that $\lambda_\pi(g)\alpha = \alpha\varrho(g)$ also holds. Thus α realizes the sought isometric G-isomorphism as well as its inverse □

SCHOLIUM 2.4.3. Since the two G-actions λ_π and ϱ commute, we can endow $L^\infty_{\mathrm{w}*}(G,E)$ with a Banach $G \times G$-module structure by means of the representation $\lambda_\pi \times \varrho$. We can then reformulate the observation of Proposition 2.4.2 as follows : if τ denotes the natural factor permutation automorphism of $G \times G$, then the above involutive isometric isomorphism α is a τ-equivariant isometric isomorphism of coefficient $G \times G$-modules.

COROLLARY 2.4.4.— *If E is given two coefficient G-module structures (π, E) and (π', E), then there is an isometric isomorphism of Banach G-modules*

$$\Big(\lambda_\pi,\ L^\infty_{\mathrm{w}*}(G,E)\Big) \cong \Big(\lambda_{\pi'},\ L^\infty_{\mathrm{w}*}(G,E)\Big).$$

If both structures admit the same predual space $E^\flat$, then the above isomorphism is adjoint. In particular, for any coefficient G-module (π, E) there is an isometric isomorphism of coefficient G-modules

$$\Big(\lambda_\pi,\ L^\infty_{\mathrm{w}*}(G,E)\Big) \cong \Big(\lambda,\ L^\infty_{\mathrm{w}*}(G,E)\Big).$$

PROOF. Apply Proposition 2.4.2 twice. □

Another useful application of the map α is :

COROLLARY 2.4.5.— *Let (π, E) be a coefficient G-module and $H < G$ a closed subgroup. Then there is a canonical isometric isomorphism*

$$\Big(\varrho,\ L^\infty_{\mathrm{w}*}(G,E)^H\Big) \cong \Big(\lambda_\pi,\ L^\infty_{\mathrm{w}*}(G/H,E)\Big),$$

where the H-invariants of the left hand side are taken with respect to λ_π.

PROOF. Viewing $L^\infty_{\mathrm{w}*}(G,E)$ both as a $G \times H$ and a $H \times G$-module in accordance with Scholium 2.4.3, the map α induces an isomorphism

$$\Big(\varrho,\ L^\infty_{\mathrm{w}*}(G,E)^{\lambda_\pi(H)}\Big) \cong \Big(\lambda_\pi,\ L^\infty_{\mathrm{w}*}(G,E)^{\varrho(H)}\Big).$$

Now the Fubini-Lebesgue theorem implies $L^\infty_{\mathrm{w}*}(G,E)^{\varrho(H)} \cong L^\infty_{\mathrm{w}*}(G/H,E)$. □

REMARK 2.4.6. If E is only a coefficient H-module, the above isomorphism of coefficient G-modules cannot be defined. However, the choice of a measurable section of the map $G \to G/H$ still yields an adjoint isometric isomorphism of Banach spaces.

The double structure $\lambda_\pi \times \varrho$ will be useful on certain occasions, notably when we come to induction matters (Section 10.1). Nevertheless, if not otherwise stated, we shall stay to the following convention :

NOTATION 2.4.7. If not otherwise specified, we shall always consider on $L^\infty_{\mathrm{w*}}(G, E)$ the left regular representation λ_π.

2.5. ℓ^∞ spaces. We end this Section 2 by mentionning briefly ℓ^∞ spaces, which are only of marginal interest to us.

Let Γ be a discrete group.

DEFINITION 2.5.1. Let X be a set and E a Banach space. Then $\ell^\infty(X, E)$ denotes the Banach space of all bounded maps $X \to E$ endowed with the sup-norm $\|\cdot\|_\infty$.

If (π, E) is a Banach Γ-module and we are given a Γ-action on X, then we can as before define the left regular representation λ_π on $\ell^\infty(X, E)$, thus turning it into a Banach Γ-module. Unless otherwise specified, this structure λ_π will always be understood.

In other words, if we endow X with the discrete topology, the Banach module $\ell^\infty(X, E)$ is exactly $\mathrm{C_b}(X, E)$. Yet the ℓ^∞ spaces owe their flexibility to features inherited both from the family of $\mathrm{C_b}$ spaces and from the family of L^∞ spaces (for the obvious measure class when X is countable) :

To the former, they owe the fact that $\ell^\infty(X, E)$ is a well defined Banach space irrespective of the nature of E, countrary to the case of L^∞ spaces, as we have mentioned in Section 2.2.

From the latter, they borrow the so useful "exponential law", which is a triviality in the present case as there are no measurability questions :

LEMMA 2.5.2.— *Let X_1, X_2 be Γ-sets and (π, E) a Banach Γ-module. Then there is a canonical isometric Γ-isomorphism*

$$\ell^\infty(X_1 \times X_2, E) \cong \ell^\infty\Big(X_1, \ell^\infty(X_2, E)\Big). \qquad \square$$

Just as straightforward are the following adaptations of Section 2.4. Let ϱ be the right translation action on $\ell^\infty(\Gamma, -)$.

LEMMA 2.5.3.— *There is an isometric isomorphism of Banach Γ-modules*

$$\Big(\lambda_\pi,\ \ell^\infty(\Gamma, E)\Big) \cong \Big(\varrho,\ \ell^\infty(\Gamma, E)\Big). \qquad \square$$

COROLLARY 2.5.4.— *If E is given two Banach Γ-module structures (π, E) and (π', E), then there is an isometric isomorphism of Banach Γ-modules*

$$\Big(\lambda_\pi,\ \ell^\infty(\Gamma, E)\Big) \cong \Big(\lambda_{\pi'},\ \ell^\infty(\Gamma, E)\Big).$$

In particular, for any Banach Γ-module (π, E) there is an isometric isomorphism of Banach Γ-modules

$$\Big(\lambda_\pi,\ \ell^\infty(\Gamma, E)\Big) \cong \Big(\lambda,\ \ell^\infty(\Gamma, E)\Big). \qquad \square$$

We have seen that ℓ^∞ spaces are much simpler to handle than spaces of type C_b and L^∞. We shall however not benefit from this simplicity, as they are not suited to the study of non discrete topological groups. It will only be on one or two occasions that we will make an observation about the special case of discrete groups.

3. Integration

Integration methods will be an important technical tool in subsequent chapters. We shall find occasions to integrate over abstract measure spaces aswell as over the topological groups themselves when they are locally compact and hence admit Haar measures. The power of the latter accounts for the further development and higher flexibility of the theory of continuous bounded cohoomlogy of locally compact groups as opposed to general topological groups.

It can however not be denied that the general theory of Banach valued integration presents a couple of pitfalls unknown to the classical Lebesgue integration.

For this reason, we collect below the basic facts that suffice to our needs. We shall always be in either one of two rather simple situations: the so-called *strong* or Bochner integration, and the weak-* integration in dual spaces, called Gelfand-Dunford integration.

One last word about measure spaces, where a schism opposes the followers of N. Bourbaki's integration over locally compact spaces to the supporters of "abstract" measure spaces :

We shall prefer N. Bourbaki's approach for Bochner's integral, as it gives a completely elementary treatment for locally compact groups, irrespective of the fact that they are not always standard measure spaces (e.g. when they are not σ-compact). As for weak integration, we will use it only over standard measure spaces, so that both approaches are equivalent.

At any rate, we use only elementary parts of the theory of Banach integration, so that this difference of viewpoints is of mere cosmetic nature.

3.1. Bochner's integral. Let μ be a Radon measure on a locally compact space X. Recall that a map $f : X \to Y$ to a Hausdorff topological space Y is called *μ-measurable* in Bourbaki's sense if for every compact set $K \subset X$ there is a countable family $(K_n)_{n\in\mathbf{N}}$ of disjoint compact subsets of K on each of which f is continuous and such that $\mu(K \setminus \bigcup_{n\in\mathbf{N}} K_n) = 0$. The connection with abstract measurability with respect to the σ-algebra of Borel sets is the following :

PROPOSITION 3.1.1.— *Let Y be a metrizable space. Then $f : X \to Y$ is μ-measurable in Bourbaki's sense if and only if simultaneously :*

(i) *f is measurable in the abstract sense.*

(ii) *f is almost separably valued, i.e. there is $X' \subset X$ with $\mu(X \setminus X') = 0$ and $f(X')$ is contained in a separable subset of Y.*

ON THE PROOF. Théorème 4 in [**24**, IV §5 N° 5]. □

Let now $f : X \to E$ be a μ-measurable map to a Banach space E. Then $\|f\|_E$ is measurable and moreover f is integrable if and only if $\int_X \|f(x)\|_E \, \mu(x) \leq \infty$. This is Théorème 5 in [**24**, IV §5 N° 6]. We denote by

$$\int_X^{(\mathrm{B})} f(x) \, d\mu(x)$$

its integral in E and call it the *Borchner integral* of f.

Bochner's integral has two major avails.

The first is that it commutes with all continuous linear maps : if $\alpha : E \to F$ is a morphism of Banach spaces, then αf is also Bochner integrable and one has (Théorème 1 in [**24**, IV §4 N° 2]) :

$$\int_X^{(\mathrm{B})} \alpha f(x) \, d\mu(x) = \alpha \int_X^{(\mathrm{B})} f(x) \, d\mu(x).$$

Secondly, it is subadditive in norm, namely (Proposition 2 in [**24**, IV §4 N° 2]) :

$$\left\| \int_X^{(\mathrm{B})} f(x) \, d\mu(x) \right\|_E \leq \int_X \|f(x)\|_E \, d\mu(x).$$

Both properties fail for weak-* integration.

Take now a locally compact group G, fix a left Haar measure m and denote by Δ_G the modular function.

Let (π, E) be a continuous Banach G-module and pick $\psi \in L^1(G)$. The map

$$\begin{aligned} G &\longrightarrow E \\ g &\longmapsto \psi(g)\pi(g)v \end{aligned}$$

is m-measurable because ψ is an almost everywhere limit of continuous functions and the Banach module (π, E) is continuous. We observe incidentally that this implies that the above map is almost separably valued even though both G and E may be non separable. Since ψ is in $L^1(G)$, the above map is moreover integrable in Bochner's sense and we denote by

$$\pi(\psi)v = \int_G^{(\mathrm{B})} \psi(g)\pi(g)v \, dm(g) \tag{$*$}$$

its Bochner integral. We have thus defined on E a structure of $L^1(G)$-module that satisfies for all $g \in G$

$$\pi(g)\pi(\psi) = \pi(\lambda^1(g)\psi), \qquad \pi(\psi)\pi(g) = \pi(\varrho^1(g^{-1})\psi),$$

where $(\lambda^1(g)\psi)(h) = \psi(g^{-1}h)$ and $(\varrho^1(g)\psi)(h) = \Delta_G(g)\psi(hg)$ are the two isometric translation actions.

For the reader willing to restrict his attention to finite, thus also σ-finite measure spaces, all the above references to N. Bourbaki can be replaced, with a saving of time, by a few pages in the second chapter of J. Diestel and J. Uhl's book [**44**].

3.2. Gelfand-Dunford integral. We have introduced the $L^1(G)$-action on a continuous Banach G-module. There is now a first obvious way to let $L^1(G)$ act on the contragredient $E^\sharp$:

The canonical inversion isomorphism $G \to G^{\mathrm{op}}$ induces an isomorphism $L^1(G) \to L^1(G^{\mathrm{op}}) \cong (L^1(G))^\sim$, where $\psi^\sim(g) = \Delta_G(g^{-1})\psi(g^{-1})$. Therefore, $E^\sharp$ has a natural $L^1(G)$-module structure defined by $\pi^\sharp(\psi) = (\pi(\psi^\sim))^*$. In particular the action map $L^1(G) \times E^\sharp \to E^\sharp$ is continuous — even though in general $G \times E^\sharp \to E^\sharp$ is not continuous, nor measurable, nor weakly continuous (Remark 3.3.5 below).

We need a more general strategy to integrate maps ranging in dual Banach spaces, including maps that are not measurable for the norm Borel structure and maps that are not almost separably valued. This is the purpose of the so-called Gelfand-Dunford integral.

Let (S, μ) be a standard measure space and E a dual Banach space with chosen predual $E^\flat$. A map $f : S \to E$ is called *weak-* integrable* if for every $u \in E^\flat$ the function $\langle f(\cdot)|u\rangle$

$$\begin{aligned} S &\longrightarrow E \\ s &\longmapsto \langle f(s)|u\rangle \end{aligned}$$

is integrable. If f is weak-* integrable, one can define an element $\int_S^{(\mathrm{GD})} f(s)\,d\mu(s)$ of the *algebraic* dual $(E^\flat)' \supset E$ of $E^\flat$ by the formula

$$\langle \int_S^{(\mathrm{GD})} f(s)\,d\mu(s)|u\rangle = \int_S \langle f(s)|u\rangle\,d\mu(s). \qquad (u \in E^\flat)$$

The Gelfand-Dunford theorem (Théorème 1 in [**25**, VI §1 N° 4]) precisely states that $\int_S^{(\mathrm{GD})} f(s)\,d\mu(s)$ belongs actually to the topological dual E.

The following two observations follow immediately from the above definition.

LEMMA 3.2.1.—

(i) *If T is a weak-* continuous linear operator, then*

$$\int_S^{(\mathrm{GD})} Tf(s)\,d\mu(s) = T\int_S^{(\mathrm{GD})} f(s)\,d\mu(s).$$

(ii) *If f is bounded and $\psi \in L^1(\mu)$, then*

$$\left\|\int_S^{(\mathrm{GD})} \psi(s)f(s)\,d\mu(s)\right\| \leq \|\psi\|_1 \cdot \|f\|_\infty. \qquad \square$$

REMARKS 3.2.2. Concerning (i), we emphasize that contrary to Bochner's integral, the Gelfand-Dunford integral does not always commute with morphisms. The weak-* continuity assumption is really an essential restriction, as we shall see later : indeed, it implies that the morphism is adjoint, which is only the case in trivial situations as far as amenability issues are concerned.

As for (ii), we mention that another drawback of the Gelfand-Dunford integral as compared to Bochner's is that there is no general norm sub-additivity statement. There is an obstruction at the outset for such statements as the norm of a Gelfand-Dunford integrable function need not be measurable. The only and not so useful statement that can be made is that the norm of a Gelfand-Dunford integral is bounded by the *exterior* integral of the norm (Proposition 6 in [**25**, VI §1 N° 2]).

Suppose now (π, E) is a coefficient G-module, with G second countable. Let $v \in E$ and $\psi \in L^1(G)$. Since π is weak-* continuous, the function

$$\begin{aligned} G &\longrightarrow E \\ g &\longmapsto \psi(g)\pi(g)v \end{aligned}$$

is weak-* integrable, so that we may again define

$$\pi(\psi)v = \int_G^{(\mathrm{GD})} \psi(g)\pi(g)v \, dm(g),$$

thus giving E a $L^1(G)$-module structure. It is apparent on this definition that this structure coincides with the contragredient structure defined before by means of Bochner integration in the predual. Moreover, we can use this characterization in order to describe the maximal continuous submodule $\mathcal{C}E$:

PROPOSITION 3.2.3.— *Let (π, E) be a coefficient G-module.*

(i) *$\mathcal{C}E$ coincides with the image $L^1(G).E$ of $L^1(G) \times E$ under π.*

(ii) *$\mathcal{C}E$ is weak-* dense in E.*

As the proof uses an approximate identity, we fix our notation now once and for all :

DEFINITION 3.2.4. An ***strict approximate identity*** on G is a two-sided approximate identity consisting of a net of non-negative continuous functions of L^1 norm one.

Any locally compact group admits such a strict approximate identity [**47**, Theorem 13.4].

PROOF OF PROPOSITION 3.2.3. Let $(\pi^\flat, E^\flat)$ be the separable continuous Banach G-module $(\pi^\flat, E^\flat)$ to which (π, E) is contragredient given by convention in Definition 1.2.1.

Point (i) : fix $w \in E$, $\varphi \in L^1(G)$ and a net $(g_\alpha)_{\alpha\in A}$ converging to $e \in G$. Let us check that $\pi(g_\alpha)\pi(\varphi)w$ converges to $\pi(\varphi)w$ in norm. We have

$$\begin{aligned}\Big\|\pi(g_\alpha)\pi(\varphi)w - \pi(\varphi)w\Big\|_E &= \sup_{\|u\|_{E^\flat}=1} \Big|\langle(\pi(\lambda(g_\alpha)\varphi) - \pi(\varphi))w|u\rangle\Big| \\ &= \sup_{\|u\|_{E^\flat}=1} \Big|\langle w|\pi^\flat((\lambda(g_\alpha)\varphi - \varphi)^\sim)u\rangle\Big| \\ &\leq \sup_{\|u\|_{E^\flat}=1} \Big(\|w\|_E \cdot \|\pi^\flat((\lambda(g_\alpha)\varphi - \varphi)^\sim)u\|_{E^\flat}\Big).\end{aligned}$$

Since we have now a Bochner integral, we may use norm sub-additivity and bound this expression by

$$\|w\|_E \cdot \|(\lambda(g_\alpha)\varphi - \varphi)^\sim\|_1 = \|w\|_E \cdot \|\lambda(g_\alpha)\varphi - \varphi\|_1,$$

which converges to zero since $L^1(G)$ is a continuous Banach G-module.

Thus we have already $L^1(G).E \subset CE$. Fix a strict approximate identity $(\psi_\alpha)_{\alpha\in A}$. Since CE is continuous, $\pi(\psi_\alpha)w$ converges to w for all $w \in CE$, hence $L^1(G).CE$ is dense in CE. But P.J. Cohen's factorization theorem, as stated in [**47**, Theorem 16.1], implies that $L^1(G).CE$ is norm closed in CE. Therefore $L^1(G).CE = CE$, which completes the proof of (i).

Point (ii) : let $(\psi_\alpha)_{\alpha\in A}$ be a strict approximate identity for $L^1(G)$. Since $\pi^\flat(\psi_\alpha)u$ converges to u in norm for all $u \in E^\flat$, we see that $\pi(\psi_\alpha^\sim)w$, which is in CE by (i), weak-* converges to w for all $w \in E$, whence (ii). □

REMARKS 3.2.5.

(i) In the language of modules over Banach algebras, the first point of Proposition 3.2.3 states that the *essential submodule* (as defined e.g. in [**47**]) of the coefficient G-module E is CE. In particular, CE is *neo-unital* over $L^1(G)$ in the sense of [**89**, 1.c].

(ii) The weak-* density of CE in Proposition 3.2.3 stands in contrast to the fact that it is closed in the norm topology.

EXAMPLE 3.2.6. In order to illustrate the Proposition 3.2.3 and the second comment above, the simplest non-trivial example is perhaps the following : if G is second countable and E is the coefficient G-module $L^\infty(G)$, then CE is the space of bounded right uniformly continuous functions on G. The latter differs from $L^\infty(G)$ unless G is discrete.

REMARK 3.2.7. There is a more general theory of weak integration in complete locally convex topological vector spaces, presented by N. Bourbaki in [**25**, VI]. We do not use it here because anyway the Gelfand-Dunford integral is not a purely formal instance of it, since the weak-* topology is in general not complete, so that one still need the Gelfand-Dunford theorem to show the existence of this integral.

3.3. More on continuity. The second point of Proposition 3.2.3 has the following immediate consequence :

COROLLARY 3.3.1.— *Suppose E is a reflexive coefficient module. Then E is a continuous separable Banach G-module.*

PROOF. Since E is reflexive with separable predual, the dual E^* is separable hence E itself is so. As for continuity, we have seen in Proposition 3.2.3 point (ii) that CE is weak-* dense in E. Since the weak and weak-* topologies coincide in a reflexive space, CE is weakly dense. But CE being convex, weak density implies norm density [**126**, I, Theorem 3.12]. Since CE is a norm-closed subspace of E, we conclude $CE = E$. □

This corollary can also be considered as a special case of the following general fact, based on Proposition 1.1.3 :

PROPOSITION 3.3.2.— *Suppose (π, E) is a separable coefficient G-module, G being a Baire topological group (e.g. locally compact). Then E is continuous.*

We need the following elementary lemma.

LEMMA 3.3.3.— *Let E be a separable dual Banach space. Then the Borel structures induced by the norm topology, the weak topology and the weak-* topology all coincide.*

PROOF OF THE LEMMA. Since the topologies listed above are in decreasing containment order, it is enough to show that a Borel set for the norm topology is weak-* Borel, or equivalently that any norm open set $A \subset E$ is weak-* Borel. Since E is separable metrizable, A is second countable. Therefore Lindelöf's theorem (Proposition 1 (i) in [**28**, IX, Appendice 1]) ensures that one can write the open set A as a countable family of open balls, say $B(d, R_d)$ with d ranging over a countable set $D \subset A$ and $R_d > 0$. Choose a sequence $(R_{d,n})_{n\in\mathbf{N}}$ of radii $0 < R_{d,n} < R_d$ converging to R_d. Now A is a countable union of (norm-) closed balls

$$A = \bigcup_{d\in D} \bigcup_{n\in\mathbf{N}} \overline{B}(d, R_{d,n}).$$

The Banach-Alaoğlu theorem [**126**, I.3.15] states that each $\overline{B}(d, R_{d,n})$ is weak-* compact, hence weak-* Borel. Thus A is also weak-* Borel. □

PROOF OF PROPOSITION 3.3.2. Since (π, E) is a coefficient module, the orbital maps are weak-* continuous, hence in particular weak-* Borel. The Lemma 3.3.3 thus implies that the orbital maps are Borel in the sense of the norm topology. We may therefore apply Proposition 1.1.3 and deduce that (π, E) is continuous. □

Next we quote a different continuity statement, also generalizing Corollary 3.3.1. We indicate and it for the completeness of the picture but shall not make use of it in the sequel.

PROPOSITION 3.3.4.— *Let G be a locally compact group and E a Banach G-module such that all orbital maps are weakly continuous. Then E is a continuous Banach G-module.*

ON THE PROOF. One can trace back the argument in the literature at least to I. Glicksberg and K. de Leeuw, Theorem 2.8 in [**65**]. It consists in using an approximate identity as for Proposition 3.2.3 (ii) but integrating in the weak sense alluded to in Remark 3.2.7, thus obtaining convergence in the norm topology. The proof is also given in [**100**, Chap. 3.1] and [**89**, pages 26–27]. □

REMARK 3.3.5. The above proposition shows incidentally that $L^\infty(G)$ provides an example of a coefficient G-module in which orbital maps are not weakly continuous, unless G is discrete.

CHAPTER II

Relative injectivity and amenable actions

In this chapter, we introduce and study a notion of relative injectivity for Banach G-modules, a topic which has already an unmistakable cohomological flavour.

As before, modules of L^∞ functions attached to regular G-spaces have an important place in the discussion. It will turn out that these modules link relative injectivity to a dynamical concept introduced by R. Zimmer : amenability of group actions.

As long as we consider merely relative injectivity, there will be no assumption on the topological group G. Then, for L^∞ spaces and amenable actions, we shall have to assume local compactness and second countability.

4. Relative injectivity

Let G be a topological group. As is familiar in homological algebra, relative injectivity is a property related to an extension problem

$$\begin{array}{ccc} A & \hookrightarrow & B \\ & \searrow & \swarrow \\ & E & \end{array}$$

for G-morphisms.

However, the precise definition of relative injectivity given below may seem cumbersome at first sight. This is due to the fact that our definition has to take simultaneously into account three aspects of the G-morphism extension problem, and to balance them against each other :

a) The purely cohomological aspect : barely whether a given G-morphism can be extended or not.

b) The topological aspect : the definition should not retain the topological obstructions that are unrelated to the G-structure. The most obvious source of such obstructions is the lack of cokernels in the category of Banach spaces. This aspect accounts for the analogies with the so-called relative homological algebra (as defined by G. Hochschild in [**82**] ; see also [**83**]).

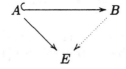

c) The geometric aspect : in order to suit the needs of geometric applications (Gromov norm, simplicial volume, etc.), relative injectivity should keep track of the precise norms involved, not only of the topology.

4.1. Definitions. In order to cope with points b) and c) above, we introduce the following notion. Recall that a *morphism* is just a continuous linear map, in general neither open nor equivariant.

DEFINITION 4.1.1. A morphism $\eta : A \to B$ of Banach spaces is *admissible* if there is a morphism $\sigma : B \to A$ with $\|\sigma\| \leq 1$ and $\eta\sigma\eta = \eta$.

A G-morphism of Banach G-modules is said admissible if the underlying morphism is so.

In other words, we require a strong sort of a splitting for η in the category of Banach spaces. We mention two particular cases : if η is an injective G-morphism, it is admissible if and only if it admits a left inverse morphism of norm one, hence in particular is a co-retraction of Banach spaces. If η is an surjective G-morphism, it is admissible if and only if it admits a right inverse morphism of norm one, hence in particular is a retraction of Banach spaces.

DEFINITION 4.1.2. A Banach G-module E is *relatively injective* if for every injective admissible G-morphism $\iota : A \to B$ of continuous Banach G-modules A, B and every G-morphism $\alpha : A \to E$ there is a G-morphism $\beta : B \to E$ satisfying $\beta\iota = \alpha$ and $\|\beta\| \leq \|\alpha\|$.

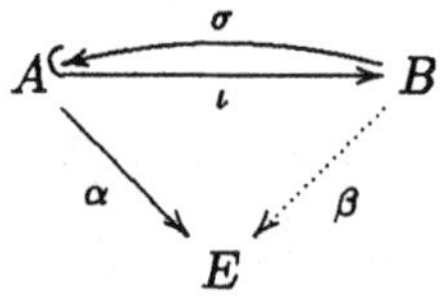

If there is any ambiguity as to the group, we say that E is G-relatively injective.

REMARK 4.1.3. In the case of *discrete* groups, a similar definition has been given by N.V. Ivanov [**86**, 3.2], following R. Brooks' idea [**31**, § 2].

DEFINITION 4.1.4. The data of A, B, ι, σ and α as in Definition 4.1.2 will be called an *extension problem for E over G*. A G-morphism β as in Definition 4.1.2 will be called a *solution* to the extension problem.

Because of the observation of Lemma 1.2.6, the above Definition 4.1.2 can be restated within one and the same category :

LEMMA 4.1.5.— *A Banach G-module E is relatively injective if and only if CE is so.*

PROOF. Suppose CE relatively injective and let

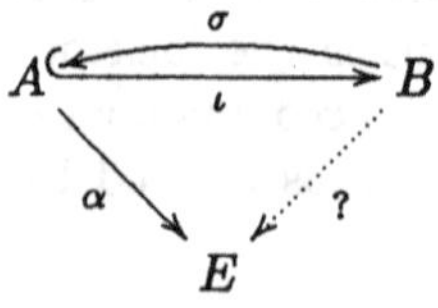

be an extension problem for E. By Lemma 1.2.6, α ranges in CE and hence its co-restriction gives an extension problem for CE. A solution $\beta : B \to CE$ also solves the original problem. Conversely, if E is relatively injective, any extension problem for CE extends to

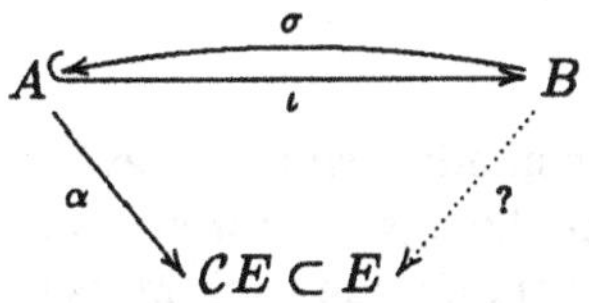

Now any solution $\beta : B \to E$ co-restricts to $\beta : B \to CE$ by Lemma 1.2.6. □

4.2. Comments on admissible G-morphisms. Our definition of an admissible G-morphism $\eta : A \to B$ implies in particular that its image is complemented. More generally, forgetting the exact norm requirement, we record the following elementary observation :

PROPOSITION 4.2.1.— *Let $\eta : A \to B$ be a morphism of Banach spaces. The following assertions are equivalent :*

(i) *There is a morphism $\sigma : B \to A$ with $\eta\sigma\eta = \eta$.*

(ii) *The linear subspaces* $\mathrm{Ker}(\eta) \subset A$ *and* $\mathrm{Im}(\eta) \subset B$ *are complemented Banach subspaces.*

Recall that a linear subspace E of a Banach space F is said *complemented* if there is a linear subspace $L \subset F$ so that the canonical sum map $E \oplus L \to F$ is an isomorphism of topological vector spaces. In particular, E and L must be closed (i.e. they are Banach subspaces). An equivalent condition for E to be complemented in F is the existence of an idempotent morphism $p : F \to F$ with range $\mathrm{Im(p)} = E$.

PROOF OF PROPOSITION 4.2.1. (i)⇒(ii) : observe that $Id - \sigma\eta : A \to A$ is an idempotent morphism, since

$$(Id - \sigma\eta)(Id - \sigma\eta) = Id - 2\sigma\eta + \sigma\eta\sigma\eta = Id - \sigma\eta.$$

Moreover, $\eta(Id - \sigma\eta) = \eta - \eta\sigma\eta = 0$, so that the image of $Id - \sigma\eta$ is in the kernel of η. Conversely, if $a \in \mathrm{Ker}(\eta)$, then $(Id - \sigma\eta)a = a$. Thus $\mathrm{Ker}(\eta)$ is complemented in A. Likewise, $\eta\sigma : B \to B$ is an idempotent morphism ranging exactly on $\mathrm{Im}(\eta)$.

(ii)⇒(i) : let $p : A \to \mathrm{Ker}(\eta)$ and $q : B \to \mathrm{Im}(\eta)$ be surjective idempotent morphisms. Let $\pi : A \to A/\mathrm{Ker}(\eta)$ be the canonical projection and $\overline{\eta} : A/\mathrm{Ker}(\eta) \to \mathrm{Im}(\eta)$ the induced map. Since $\mathrm{Im}(\eta)$ is closed, the open mapping theorem [**126**, I, Corollary 2.12] implies that $\overline{\eta}$ is an isomorphism of topological vector spaces, hence admits an inverse morphism $\overline{\eta}^{-1}$. The map $Id - p$ vanishes on $\mathrm{Ker}(\eta)$ and hence induces a morphism $\overline{Id - p} : A/\mathrm{Ker}(\eta) \to A$. We define $\sigma : B \to A$ by

$$\sigma = (\overline{Id - p})\overline{\eta}^{-1} q$$

and use the relations $\overline{\eta}^{-1}\eta = \pi$ and $(\overline{Id - p})\pi = Id - p$ in order to check

$$\eta\sigma\eta = \eta(\overline{Id - p})\overline{\eta}^{-1}q\eta = \eta(\overline{Id - p})\overline{\eta}^{-1}\eta =$$
$$= \eta(\overline{Id - p})\pi = \eta(Id - p) = \eta.$$

□

However, the presence of non-admissible and even non-open G-morphisms is not just another unfortunate pathology from which we should avert our eyes. On the contrary, according to a criterion due to S. Matsumoto and S. Morita [**105**, Theorem 2.3], such morphisms can provide an obstruction to the vanishing of bounded cohomology.

Even though our characterization of bounded cohomology will rely on the more constraining notion of admisible morphisms, complementation of the kernel and image will sometimes be relevant issues (e.g. in Section 8.2). Therefore, we propose the

DEFINITION 4.2.2. A morphism $\eta : A \to B$ of Banach spaces is said ***weakly admissible*** if it satisfies the equivalent conditions of Proposition 4.2.1.

EXAMPLE 4.2.3. Observe that if $A < B$ is a Banach subspace of the Banach space B, then the inclusion map $A \to B$ is weakly admissible if and only if A is complemented in B.

In the setup used in the books of A. Guichardet [**74**] and Borel-Wallach [**19**] in order to introduce the usual continuous cohomology of locally compact groups, a similar notion appears. In these references, a ***strong*** morphism (s-morphism) is a linear continuous map of topological vector spaces satisfying the conditions of Definition 4.2.2 translated to the category of Hausdorff locally convex topological vector spaces (Définition D.1 in [**74**, Appendice D] and [**19**, IX 1.5]).

Here is an immediate consequence of the above Proposition 4.2.1.

COROLLARY 4.2.4.— *Let*

$$0 \to A \xrightarrow{\alpha} B \xrightarrow{\beta} C \to 0$$

be a short exact sequence of morphisms of Banach spaces. Then α is weakly admissible if and only if β is weakly admissible.

PROOF. By Proposition 4.2.1, both conditions are equivalent to $\mathrm{Im}(\alpha) = \mathrm{Ker}(\beta)$ being complemented in B. □

DEFINITION 4.2.5. A short exact sequence is ***weakly admissible*** if it satisfies the equivalent conditions of Corollary 4.2.4.

Such short exact sequences thus split in the category of Banach spaces. For sequences that split equivariantly, one can repeat every single word of the preceding argumentation and deduce :

PROPOSITION 4.2.6.— *Let*

$$0 \to A \xrightarrow{\alpha} B \xrightarrow{\beta} C \to 0$$

be a short exact sequence of G-morphisms of Banach G-modules. The following assertions are equivalent :

(i) *There is a left inverse G-morphism for* α.

(ii) *There is a right inverse G-morphism for* β.

(iii) $\mathrm{Im}(\alpha) = \mathrm{Ker}(\beta)$ *admits a G-invariant complement in B.* □

DEFINITION 4.2.7. A short exact sequence satisfying the equivalent conditions of Proposition 4.2.6 will be said *G-split.*

The characterizations (i) and (ii) of Proposition 4.2.6, together with Lemma 1.2.6, imply immediately the following :

LEMMA 4.2.8.— *Let* $0 \to A \xrightarrow{\alpha} B \xrightarrow{\beta} C \to 0$ *be a G-split short exact sequence of G-morphisms of Banach G-modules. Then the sequence*

$$0 \longrightarrow CA \xrightarrow{\alpha} CB \xrightarrow{\beta} CC \longrightarrow 0$$

is also exact. □

4.3. First properties. The following elementary observation is of frequent use.

PROPOSITION 4.3.1.— *Let* $\vartheta : E \to F$ *be a norm one G-morphism of Banach G-modules admitting a left inverse G-morphism of norm one.*

If F is relatively injective, then so is E.

PROOF. Let τ be a norm one left inverse G-morphism for ϑ. We consider an extension problem for E and complete it as follows :

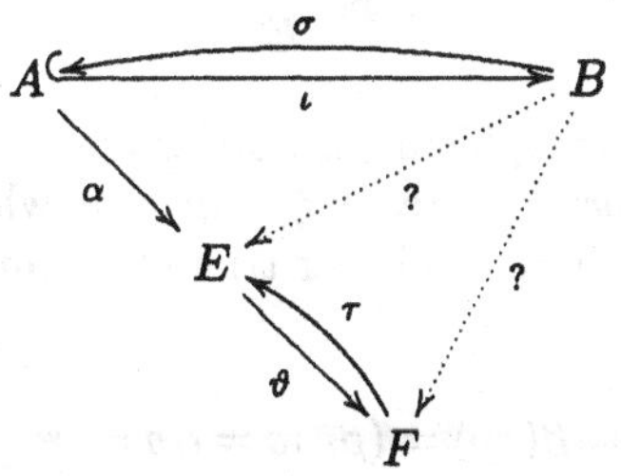

The big triangle is an extension problem for F for which we fix a solution $\beta : B \to F$. We claim that $\tau\beta$ is a solution to the extension problem for E. Indeed, we have

$$\|\tau\beta\| \le \|\beta\| \le \|\vartheta\alpha\| \le \|\alpha\|$$

and $\tau(\beta\iota) = \tau(\vartheta\alpha) = (\tau\vartheta)\alpha = \alpha$. □

Relative injectivity will most of the time be used in the following guise :

PROPOSITION 4.3.2.— *Let $\eta : A \to B$ be an admissible G-morphism of continuous Banach G-modules and let E be a relatively injective Banach G-module.*

For any G-morphism $\alpha : A \to E$ with $\mathrm{Ker}(\alpha) \supset \mathrm{Ker}(\eta)$ there is a G-morphism $\beta : B \to E$ with $\beta\eta = \alpha$ and $\|\beta\| \leq \|\alpha\|$.

The commutative diagram depicting this more general extension problem is

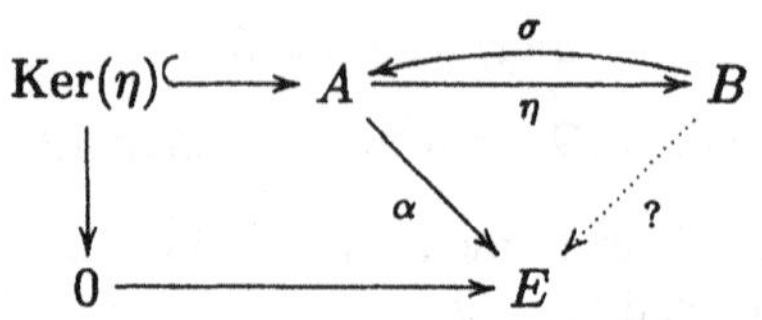

DEFINITION 4.3.3. The data of the triangular diagram above with the hypothesis $\mathrm{Ker}(\alpha) \supset \mathrm{Ker}(\eta)$ is called a ***generalized extension problem.*** A G-morphism β as in Proposition 4.3.2 is again called a ***solution*** to it.

PROOF OF THE PROPOSITION. We endow $\overline{A} = A/\mathrm{Ker}(\eta)$ with the quotient norm and Banach G-module structure and denote by p the quotient map. Let $\overline{\eta} : \overline{A} \to B$ be the unique G-morphism such that $\overline{\eta}p = \eta$. Since $\mathrm{Ker}(\alpha) \supset \mathrm{Ker}(\eta)$, there is a G-morphism $\overline{\alpha} : \overline{A} \to E$ such that $\overline{\alpha}p = \alpha$. Since $\overline{A}$ is endowed with the quotient norm, we have $\|\overline{\alpha}\| = \|\alpha\|$. The G-morphism η being admissible, there is a morphism $\sigma : B \to A$ of norm at most one with $\eta\sigma\eta = \eta$.

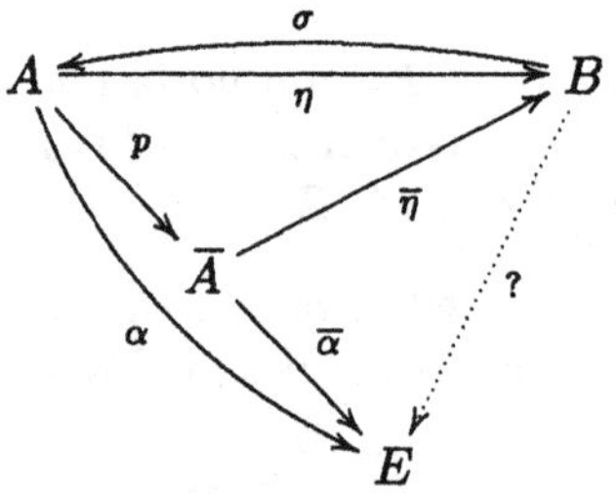

Observe that the injective G-morphism $\overline{\eta}$ is admissible, since $p\sigma$ is a morphism of norm at most one and since $\eta\sigma\eta = \eta$ reads $\overline{\eta}p\sigma\overline{\eta}p = \overline{\eta}p$ which implies $\overline{\eta}(p\sigma)\overline{\eta} = \overline{\eta}$ because p is surjective. Therefore the lower triangle is an extension problem for E. Any solution β satisfies

$$\beta\eta = \beta(\overline{\eta}p) = (\beta\overline{\eta})p = \overline{\alpha}p = \alpha$$

and $\|\beta\| \leq \|\overline{\alpha}\| = \|\alpha\|$, as required by the proposition. □

Another simple observation is :

PROPOSITION 4.3.4.— *Let $N \lhd G$ be a closed normal subgroup and $Q = G/N$ the quotient. For any G-relatively injective Banach G-module E, the Banach Q-module E^N is Q-relatively injective.*

PROOF. Let

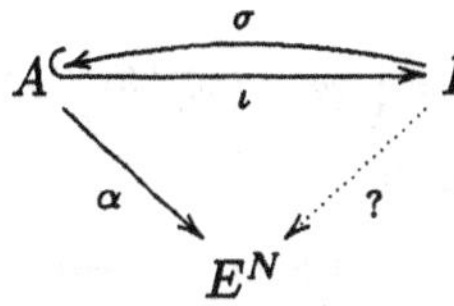

be an extension problem for E^N over Q and denote by i the inclusion map $E^N \to E$. Pulling back the Banach module structures through the quotient map, we get an extension problem

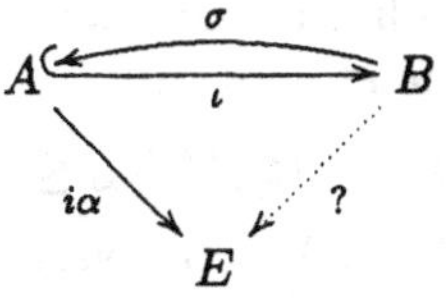

for E over G for which we fix a solution β. Since β is equivariant and B is trivial as Banach N-module, β ranges in E^N and hence its co-restriction is a solution for the original extension problem. □

A category-theoretical gloss on Proposition 4.3.4 would be to say that the functor $(-)^N$ and the pull-back through $G \to Q$ are adjoint, i.e. for every Banach G-module A and Banach Q-module B there is a canonical identification

$$\mathcal{L}(B, A)^G \cong \mathcal{L}(B, A^N)^Q.$$

The dual issue, namely the relation between G-relative injectivity and H-relative injectivity for subgroups $H < G$, will be addressed in Propositions 5.2.1 and 5.8.1 below.

4.4. Fundamental examples. Our first examples of relatively injective Banach G-modules will be spaces of bounded continuous maps (compare also Section 4.5). The most flexible examples, however, will turn out to be L^∞ spaces over G, when G is locally compact second countable. The latter are reminiscent of the classical notion of co-free modules in the usual cohomology theory of discrete groups.

Recall that for now G is still any topological group.

PROPOSITION 4.4.1.— *For any Banach G-module (π, E) and all $n \geq 0$, the Banach G-module $\mathrm{C_b}(G^{n+1}, E)$ is relatively injective.*

PROOF. Consider the extension problem

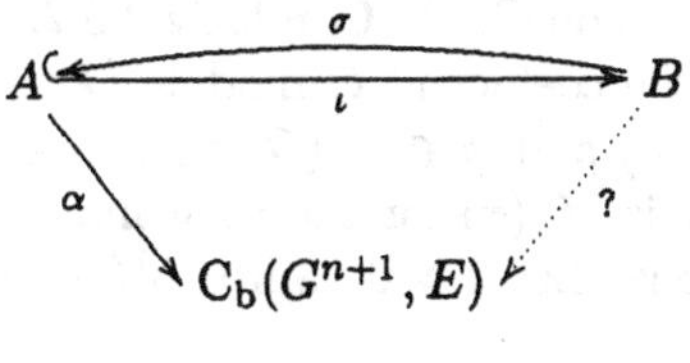

For all $x = (x_0, \ldots, x_n)$ in G^{n+1} and $b \in B$, we define an element $\beta(b)(x)$ of E by

$$\beta(b)(x) = \Big(\alpha x_0 \sigma(x_0^{-1} b)\Big)(x).$$

We observe that the evaluation map

$$\mathrm{C_b}(G^{n+1}, E) \times G^{n+1} \longrightarrow E$$

is continuous. Therefore, as A and B are continuous Banach G-modules, we can for each fixed b view the map $\beta(b)$ as a composition

$$G^{n+1} \longrightarrow B \times G^{n+1} \longrightarrow A \times G^{n+1} \longrightarrow \mathrm{C_b}(G^{n+1}, E) \times G^{n+1} \longrightarrow E$$

of continuous maps, so that β is a linear map ranging in the space of continuous maps $G^{n+1} \to E$. On the other hand,

$$\|\beta(b)(x)\|_E \leq \|\alpha x_0 \sigma(x_0^{-1} b)\|_\infty \leq \|\alpha\| \cdot \|b\|_E \tag{$*$}$$

implies that β ranges in $\mathrm{C_b}(G^{n+1}, E)$. For every $x = (x_0, \ldots, x_n)$ in G^{n+1}, $g \in G$ and $b \in B$ we have

$$\begin{aligned}
\beta(gb)(x) &= \Big(\alpha x_0 \sigma(x_0^{-1} g b)\Big)(x) = \Big(\alpha g (g^{-1} x_0) \sigma\big((g^{-1} x_0)^{-1} b\big)\Big)(x) \\
&= \Big(\lambda_\pi(g) \alpha (g^{-1} x_0) \sigma\big((g^{-1} x_0)^{-1} b\big)\Big)(x) \\
&= \pi(g) \Big(\alpha (g^{-1} x_0) \sigma\big((g^{-1} x_0)^{-1} b\big)\Big)(g^{-1} x) \\
&= \pi(g) \beta(b)(g^{-1} x) = \Big(\lambda_\pi(g) \beta(b)\Big)(x),
\end{aligned}$$

so that β is a G-morphism. The condition $\|\beta\| \leq \|\alpha\|$ follows from ($*$). In order to check the remaining condition $\beta\iota = \alpha$, we pick $a \in A$ and $x \in G^{n+1}$ and compute

$$\beta\iota(a)(x) = \Big(\alpha x_0 \sigma(x_0^{-1} \iota a)\Big)(x) = \Big(\alpha x_0 \sigma\iota(x_0^{-1} a)\Big)(x) = \alpha(a)(x),$$

completing the proof that β is a solution to the extension problem. □

PROPOSITION 4.4.2.— *Let E be a coefficient G-module, G locally compact second countable. Then $L^\infty_{w*}(G, E)$ is a relatively injective Banach G-module.*

The proof relies on the following lemma about left *translation* :

LEMMA 4.4.3.— *Every function class in the maximal continuous submodule $\mathcal{C}_\lambda L^\infty_{w*}(G, E)$ can be represented by a right uniformly continuous function $G \to E$.*

PROOF OF THE LEMMA. Let $f : G \to E$ represent a function class

$$f \in \mathcal{C}_\lambda L^\infty_{w*}(G, E)$$

and fix a left Haar measure m on G. By Corollary 2.3.2, we may apply the Proposition 3.2.3 point (i) to the coefficient G-module $\big(\lambda, L^\infty_{w*}(G, E)\big)$, and deduce that there is f' in $L^\infty_{w*}(G, E)$ and $\psi \in L^1(G)$ such that $f(x) = \lambda(\psi) f'(x)$ for almost every $x \in G$. We write $f''(x)$ for the right hand side and claim that f'' is right uniformly continuous. Let $\varepsilon > 0$. Since $L^1(G)$ is a continuous Banach

G-module, there is a neighbourhood U of $e \in G$ such that for any $x, y \in G$ with $xy^{-1} \in U$ we have $\|f'\|_\infty \cdot \|\psi - \lambda(xy^{-1})\psi\|_1 \leq \varepsilon$. By left invariance of m we have

$$f''(x) - f''(y) = \int_G^{(\mathrm{GD})} \Big(\psi(g) - \psi(yx^{-1}g)\Big) f'(g^{-1}x)\, dm(g).$$

Now the sub-additivity statement (ii) of Lemma 3.2.1 implies

$$\|f''(x) - f''(y)\|_E \leq \|\psi - \lambda(xy^{-1})\psi\|_1 \cdot \|f'\|_\infty \leq \varepsilon,$$

as was to be shown. □

PROOF OF PROPOSITION 4.4.2. By Corollary 2.4.4 we may suppose π trivial. Now the Lemma 4.4.3 implies that the canonical embedding of $\mathrm{C_b}(G, E)$ into $L^\infty_{\mathrm{w*}}(G, E)$ restricts to an identification

$$\mathcal{C}\mathrm{C_b}(G, E) \cong \mathcal{C}L^\infty_{\mathrm{w*}}(G, E),$$

so that by Lemma 4.1.5 the Proposition 4.4.2 reduces to the case $n = 0$ and $\pi = Id$ of Proposition 4.4.1. □

The proof of Proposition 4.4.2 benefits from the fact that the elements of the maximal continuous submodule of $L^\infty_{\mathrm{w*}}(G, E)$ with respect to the left translation are continuous maps. Although this fails completely already in the case of $L^\infty_{\mathrm{w*}}(G^2, E)$, the conclusion of Proposition 4.4.2 can immediately be extended to the following more general setting :

COROLLARY 4.4.4.— *Let S be a regular G-space and E a coefficient G-module, G locally compact second countable. Then $L^\infty_{\mathrm{w*}}(G \times S, E)$ is a relatively injective Banach G-module.*

PROOF. We invoque Corollary 2.3.3 for the canonical isometric Banach G-module identification

$$L^\infty_{\mathrm{w*}}(G \times S) \cong L^\infty_{\mathrm{w*}}\big(G, L^\infty_{\mathrm{w*}}(S, E)\big).$$

We may now apply the Proposition 4.4.2 with $L^\infty_{\mathrm{w*}}(S, E)$ instead of E. □

Because of its importance, we state explicitly the following particular case :

COROLLARY 4.4.5.— *Let E be a coefficient G-module, G locally compact second countable. Then $L^\infty_{\mathrm{w*}}(G^{n+1}, E)$ is a relatively injective Banach G-module for each $n \geq 0$.* □

4.5. Proper G-spaces. For locally compact groups, we can considerably generalize the Proposition 4.4.1, owing to the existence of generalized Bruhat functions recalled in Lemma 4.5.4 below. Let us first recall some terminology along the lines of N. Bourbaki [**27**, III §4] in the general setting.

A continuous action (Section 1.3) of a topological group G on a topological space X is said *proper* if the map

$$\begin{aligned} G \times X &\longrightarrow X \times X \\ (g, x) &\longmapsto (gx, x) \end{aligned}$$

is proper (for properness of maps, see Remark 4.5.1 below). Then X is also called a ***proper*** G***-space***. In presence of such a continuous proper action, both X and $G\backslash X$ are Hausdorff as soon as G is Hausdorff (Proposition 3 in [**27**, III §4 N° 2]). If moreover X is locally compact, then so is $G\backslash X$ (Proposition 9 in [**27**, III §4 N° 5]).

If G is locally compact and X a Haudorff topological space, then properness of the continuous G-action is equivalent to the following condition (Proposition 7 in [**27**, III §4 N° 4]) :

for all $x, y \in X$ there are neighbourhoods V_x, V_y of x respectively y such that the set of those $g \in G$ with $gV_x \cap V_y \neq \varnothing$ is relatively compact in G.

In this setting, another reference is the first chapter of J.-L. Koszul's book [**94**].

REMARK 4.5.1. Among the various characterizations of proper maps, we recall the following simple criterion, whose restricted generality is completely sufficient for our needs :

A continuous map from a Hausdorff space to a localy compact space is proper if and only if the pre-image of any compact set is compact.

This is Proposition 7 in [**27**, I §10 N° 3].

Here is a fundamental result on proper actions :

THEOREM 4.5.2.— *Let G be a locally compact group with a continuous proper action on a locally compact space X such that $G\backslash X$ is paracompact.*

Then $\mathrm{C_b}(X, E)$ is relatively injective for any Banach G-module (π, E).

REMARK 4.5.3 (On the paracompactness assumption). Recall that by definition a Hausdorff topological space is said *paracompact* if every open cover admits a locally finite open refinement. The paracompactnes assumption is needed for the Lemma 4.5.4 below, but is rather mild, as the following facts should recall :

Every metrizable space is paracompact (Théorème 4 in [**28**, IX §4 N° 5]). Every locally compact group is paracompact [**84**, II.8.13], and more generally every Hausdorff homogeneous space of a locally compact group is paracompact (Proposition 13 in [**27**, III §4 N° 6]). Every regular Lindelöf space is paracompact (Proposition 2 in [**28**, IX, Appendice 1]). A locally compact space, as is $G\backslash X$ in the above theorem by the remarks preceding it, is paracompact if and only if it is a topological sum of (arbitrarily many) σ-compact components (Théorème 5 in [**27**, I §9 N° 10]). The latter together with a result of E. Michael [**108**] implies that the product of two paracompact spaces is paracompact if one of the factors is locally compact.

The proof of Theorem 4.5.2 is based on the existence on X of a *generalized Bruhat function*, which is so to say a G-transverse partition of the unit. In the case where X is a supergroup of G, the the existence of a Bruhat function is proved e.g. in H. Reiter's book [**123**, VIII §1.9], generalizing the original statement of F. Bruhat [**33**]. We take the general case from N. Bourbaki :

LEMMA 4.5.4.— *Let G be a locally compact group with a continuous proper action on a locally compact space X such that $G\backslash X$ is paracompact. For every left Haar measure m on G there is a non-negative real-valued continuous function h on X such that*

(i) *For all $x \in X$ we have $\int_G h(g^{-1}x)\,dm(g) = 1$.*

(ii) *For every compact subset $K \subset X$ the intersection of the support of h with the saturation GK is compact.*

ON THE PROOF. This is Proposition 8 in [**26**, VII §2 N° 4], with the difference that in this reference G acts on X from the right. It is however important for the proof of Theorem 4.5.2 to observe that nonobstant the translation into the language of left actions, the *left* Haar measure of N. Bourbaki's statement remains left invariant in the above formulation. □

The properties (i) and (ii) of the above lemma have the following consequence, which we state and prove in a general form for later reference (for this section, the case where Y is trivial will suffice).

LEMMA 4.5.5.— *Keep the notations of Lemma 4.5.4 and let Y be a topological space. Let $F : G \times X \times Y \to V$ be a bounded continuous function with values in a Banach space V. Then the Bochner integral*

$$\int_G^{(\mathrm{B})} h(g^{-1}x)F(g,x,y)\,dm(g)$$

defines a continuous function of $(x, y) \in X \times Y$.

PROOF. Since Bochner's integral is sub-additive with respect to the norm, we are immediately reduced to the case $V = \mathbf{C}$ of scalar integration. Let $(x, y) \in X \times Y$ be fixed throughout the proof, and let $\varepsilon > 0$. Since X is Hausdorff, the set $\{x\}$ is compact. Hence by Lemma 4.5.4 (ii) the intersection of $\mathrm{Supp}(h)$ with the orbit Gx is compact. Since $X \times X$ is locally compact, the criterion of Remark 4.5.1 implies readily that there is a compact subset $K \subset G$ with

$$\mathrm{Supp}(h) \cap Gx \ \subset\ K^{-1}x.$$

For all $k \in K$, the continuity at (k, x) of $(g, x') \mapsto h(g^{-1}x')$ and of F at (k, x, y) entails in particular that there are open neighbourhoods U_k of (x, y) in $X \times Y$ and W_k of k in G such that for all $(x', y') \in U_k$ and $g \in W_k$ both

$$m(K)\Big|h(g^{-1}x)F(g,x,y) - h(g^{-1}x')F(g,x',y')\Big| \leq \varepsilon/2$$

and

$$\|F\|_\infty m(K)\Big|h(g^{-1}x) - h(g^{-1}x')\Big| \leq \varepsilon/2$$

hold. Since $\{W_k : k \in K\}$ is an open cover of K, there is a finite subset $F \subset K$ with $K \subset \bigcup_{k\in F} W_k$. Now $U = \bigcap_{k\in F} U_k$ is a neighbourhood of (x, y). Let

$(x', y') \in U$. We have

$$\left|\int_G h(g^{-1}x)F(g,x,y)\,dm(g) - \int_G h(g^{-1}x')F(g,x',y')\,dm(g)\right| \leq$$

$$\leq \int_K \left|h(g^{-1}x)F(g,x,y) - h(g^{-1}x')F(g,x',y')\right| dm(g) +$$

$$+ \int_{G\setminus K} \left|h(g^{-1}x)F(g,x,y) - h(g^{-1}x')F(g,x',y')\right| dm(g)$$

$$\leq \varepsilon/2 + \int_{G\setminus K} h(g^{-1}x')\left|F(g,x',y')\right| dm(g).$$

The second term is bounded by

$$\|F\|_\infty \int_{G\setminus K} h(g^{-1}x')\,dm(g),$$

and the condition (i) of Lemma 4.5.4 applied successively to x' and x yields

$$\int_{G\setminus K} h(g^{-1}x')\,dm(g) = 1 - \int_K h(g^{-1}x')\,dm(g) =$$

$$= \int_K \left(h(g^{-1}x) - h(g^{-1}x')\right) dm(g),$$

which is again bounded by $\varepsilon/2$. □

With these two lemmata at hand, we can give the proof of Theorem 4.5.2 within a certain analogy to the proof of Proposition 4.4.1 :

PROOF OF THEOREM 4.5.2. Fix h as in Lemma 4.5.4 and consider an extension problem

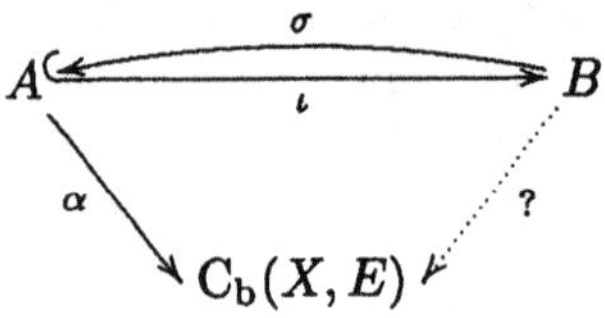

As in the proof of Proposition 4.4.1, the evaluation map $\mathrm{C_b}(X,E) \times X \to E$ is continuous, so that the map $G \times X \to E$ defined by

$$(g,x) \longmapsto \Big(\alpha g\sigma(g^{-1}b)\Big)(x)$$

is continuous. The continuity in the first variable implies that for all $x \in X$ the map

$$g \longmapsto h(g^{-1}x)\cdot\Big(\alpha g\sigma(g^{-1}b)\Big)(x)$$

is m-measurable. Since moreover

$$\left\|\Big(\alpha g\sigma(g^{-1}b)\Big)(x)\right\|_E \leq \|\alpha\|\cdot\|b\|_B,$$

the condition (i) of Lemma 4.5.4 ensures that the Bochner integral

$$\beta(b)(x) = \int_G^{(B)} h(g^{-1}x) \cdot \Big(\alpha g \sigma(g^{-1}b)\Big)(x)\, dm(g)$$

exists and satisfies

$$\|\beta(b)(x)\|_E \leq \|\alpha\| \cdot \|b\|_B.$$

The Lemma 4.5.5 with Y trivial implies that this defines for each $b \in B$ a continuous map $\beta(b) : X \to E$. Therefore we have a linear map

$$\beta : B \longrightarrow \mathrm{C_b}(X, E)$$

of norm at most $\|\alpha\|$. For all $a \in A$ and $x \in X$ we check

$$\beta\iota(a)(x) = \int_G^{(B)} h(g^{-1}x) \cdot \Big(\alpha g \sigma g^{-1}\iota(a)\Big)(x)\, dm(g)$$
$$= \int_G^{(B)} h(g^{-1}x) \cdot \Big(\alpha g \sigma\iota(g^{-1}a)\Big)(x)\, dm(g)$$
$$= \int_G^{(B)} h(g^{-1}x)\alpha(a)(x)\, dm(g),$$

which by condition (i) in Lemma 4.5.4 is $\alpha(a)(x)$. It remains only to check the equivariance of β, so let $k \in G$. We have

$$\beta(kb)(x) = \int_G^{(B)} h(g^{-1}x) \cdot \Big(\alpha g \sigma(g^{-1}kb)\Big)(x)\, dm(g)$$
$$= \int_G^{(B)} h(g^{-1}k^{-1}x) \cdot \Big(\alpha k g \sigma(g^{-1}b)\Big)(x)\, dm(g)$$

by left invariance of m. Further, this is

$$\int_G^{(B)} h(g^{-1}k^{-1}x) \cdot \Big(\lambda_\pi(k)\alpha g \sigma(g^{-1}b)\Big)(x)\, dm(g)$$
$$= \pi(k) \int_G^{(B)} h(g^{-1}k^{-1}x) \cdot \Big(\alpha g \sigma(g^{-1}b)\Big)(k^{-1}x)\, dm(g),$$

wherein we used Bochner integrability to commute $\pi(k)$ with the integral. What we have obtained is by definition

$$\pi(k)\Big(\beta(b)(k^{-1}x)\Big) = \Big(\lambda_\pi(k)\beta(b)\Big)(x),$$

as was to be shown. □

An important particular case is the following :

COROLLARY 4.5.6.— *Let G be a locally compact group, $H < G$ a closed subgroup and (π, E) a Banach H-module.*

Then $\mathrm{C_b}(G^{n+1}, E)$ is H-relatively injective for all $n \geq 0$.

(It is understood that H acts on G^{n+1} by left multiplication via the diagonal embedding $\Delta : H \to G^{n+1}$.)

PROOF. We want to apply Theorem 4.5.2 with G^{n+1} instead of X and H instead of G. The H-action is indeed continuous, and its properness is actually equivalent to $\Delta(H)$ being closed in G^{n+1} (*Exercice 5 du §4* in [**27**, III]). The quotient $\Delta(H)\backslash G^{n+1}$ is paracompact by the Proposition 13 in [**27**, III §4 N° 6] mentioned in Remark 4.5.3. Thus all hypotheses of Theorem 4.5.2 are satisfied. □

A slightly more general application of Theorem 4.5.2 is the following :

COROLLARY 4.5.7.— *Let G be a locally compact group, $H < G$ a closed subgroup, $K < G$ a compact subgroup and (π, E) a Banach H-module.*
Then $C_b\big((G/K)^{n+1}, E\big)$ is H-relatively injective for all $n \geq 0$.

EXAMPLE 4.5.8. Let M be a locally symmetric space, Γ its fundamental group and E any Banach Γ-module. Denote by $\widetilde{M}$ the universal cover of M. Then $C_b\big(\widetilde{M}^{n+1}, E\big)$ is Γ-relatively injective for all $n \geq 0$.

Indeed, we can apply Corollary 4.5.7 with the group of isometries of $\widetilde{M}$ for G, $H = \Gamma$ and K the stabilizer of a point in $\widetilde{M}$.

For the proof of Corollary 4.5.7 we need an elementary lemma in general topology.

LEMMA 4.5.9.— *Let Y be a locally compact space with an open equivalence relation R such that the canonical quotient map $Y \to R\backslash Y$ is proper.*
If Y is paracompact, then so is $R\backslash Y$.

Actually, E. Michael [**110**] has showed that the quotient of a paracompact space by a closed equivalence relation is again paracompact, and indeed the properness of $Y \to R\backslash Y$ implies that R is closed (Proposition 1 in [**27**, I §10 N° 1]). However the above lemma is much more elementary, so we prove it by lack of appropriate reference.

PROOF OF LEMMA 4.5.9. Let $\mathcal{U}$ be an open cover of $R\backslash Y$ and denote by p the canonical map $Y \to R\backslash Y$. Then

$$\mathcal{V} = \{p^{-1}(U) : U \in \mathcal{U}\}$$

is an open cover of Y. Let $\mathcal{W}$ be a locally finite open refinement of $\mathcal{V}$. Since $\mathcal{W}$ is a refinement, we may for all $W \in \mathcal{W}$ fix $U_W \in \mathcal{U}$ with $W \subset p^{-1}(U_W)$. Now

$$\mathcal{U}' = \{p(W) : W \in \mathcal{W}\}$$

is a cover of $R\backslash Y$ which is open because p is an open map. Moreover, $\mathcal{U}'$ is a refinement of $\mathcal{U}$ since for all $W \in \mathcal{W}$ we have $p(W) \subset U_W$.

In order to check that $\mathcal{U}'$ is locally finite, let $p(y) \in R\backslash Y$ and let A be a compact neighbourhood of y in Y. For all $z \in p^{-1}(p(A))$ there is a neighbourhood A_z of z in Y meeting only finitely many $W \in \mathcal{W}$. Since p is proper, $p^{-1}(p(A))$ is compact by the criterion of Remark 4.5.1, hence can be covered by a finite number of the sets A_z. Therefore, $p^{-1}(p(A))$ meets only finitely many $W \in \mathcal{W}$. Since p is open, $p(A)$ is a neighbourhood of $p(y)$. Moreover, the relation $p(W) \cap p(A) \neq \varnothing$ is equivalent to $W \cap p^{-1}(p(A)) \neq \varnothing$, so that $p(A)$ meets only finitely many $p(W) \in \mathcal{U}'$. □

PROOF OF COROLLARY 4.5.7. We want to apply Theorem 4.5.2 with

$$X = (G/K)^{n+1}$$

and with H instead of G. The compactness of K^{n+1} implies that G^{n+1} acts properly on $G^{n+1}/K^{n+1} \cong (G/K)^{n+1}$ [**27**, III §4 N° 2, Corollaire]. Hence, as in the proof of Corollary 4.5.6, the diagonal H-action is also proper since $\Delta(H)$ is closed in G^{n+1}. It remains only to show that $\Delta(H)\backslash G^{n+1}/K^{n+1}$ is paracompact. We know already that $Y = \Delta(H)\backslash G^{n+1}$ is paracompact (Remark 4.5.3). The canonical quotient map associated to the right action of K^{n+1} on Y is proper by Proposition 2 c) in [**27**, III §4 N° 1]. Therefore we may apply Lemma 4.5.9 and conclude that Y/K^{n+1} is paracompact. □

5. AMENABILITY AND AMENABLE ACTIONS

In order to put into context R. Zimmer's concept of amenable actions, we recall two definitions of group amenability which are relevant to the discussion. A locally compact group G is said *amenable* if the following equivalent conditions are satisfied :

(A_1) *(fixed point property)* For every (jointly) continuous linear G-action on a Hausdorff locally convex topological vector space F and every non-empty compact convex G-invariant subset $K \subset F$, there is a G-fixed point in K.

(A_2) *(invariant mean property)* There is an *invariant mean* on $L^\infty(G)$, that is, a norm one G-morphism $\mathfrak{m} : L^\infty(G) \to \mathbf{C}$ such that $\mathfrak{m}(\mathbf{1}_G) = 1$.

(See e.g. J.-P. Pier's book [**122**], in particular Section 1.4 and Theorem 5.4.)

REMARK 5.0.1. Since we implicitly always consider the left translation representation, our notion of invariant mean represents what many authors call a *left* invariant mean.

REMARK 5.0.2. The theory of amenable actions has been investigated almost exclusively for *locally compact second countable groups*, including of course countable abstract groups. Therefore, from Section 5.3 through the end of Section 5, we shall explicitly adopt this framework. The preliminary discussions of the next two sections are partly more general.

5.1. Prelude on relative injectivity. Before we consider amenable actions and their connection to relative injectivity, we present two simple observations.

Let G be a topological group.

PROPOSITION 5.1.1.— *Let E be a Banach G-module. If the canonical embedding $E \to \mathrm{C_b}(G, E)$ has a norm one left inverse G-morphism, then E is relatively injective.*

PROOF. Since by Proposition 4.4.1 the Banach G-module $\mathrm{C_b}(G, E)$ is relatively injective, this follows from Proposition 4.3.1. □

A partial converse is :

PROPOSITION 5.1.2.— *Let E be a Banach G-module. If E is relatively injective, then the canonical embedding $\mathcal{C}E \to \mathcal{C}\mathrm{C_b}(G, E)$ has a norm one left inverse G-morphism.*

PROOF. For any fixed $x \in G$, we obtain a map

$$\begin{aligned} \sigma : \mathcal{C}\mathrm{C_b}(G, E) &\longrightarrow \mathcal{C}E \\ f &\longmapsto f(x), \end{aligned}$$

ranging in $\mathcal{C}E$ by Lemma 1.3.1. Therefore, the canonical embedding $\mathcal{C}E \to \mathcal{C}\mathrm{C_b}(G, E)$ is admissible. We get the sought left inverse by virtue of the following extension problem :

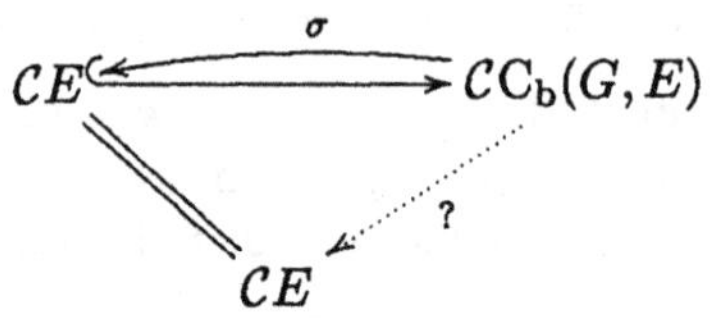

because $\mathcal{C}E$ is also relatively injective by Lemma 4.1.5. □

In the case of continuous Banach modules, we can put both propositions together :

COROLLARY 5.1.3.— *Let G be a topological group and E be a continuous Banach G-module. The following assertions are equivalent :*

(i) *The canonical embedding $E \to \mathcal{C}\mathrm{C_b}(G, E)$ has a norm one left inverse G-morphism.*

(ii) *E is relatively injective.* □

We recall now the following equivalent variation on the criterion ($\mathrm{A_2}$) for amenability of a *locally compact group* G, see [**122**, Theorem 4.19] :

($\mathrm{A_{2'}}$) There is an invariant mean on the space $\mathrm{C_{ub}}(G)$ of right uniformly continuous bounded functions on G.

Amenability has been less studied for general topological groups G. Recall however that a not necessarily locally compact topological group is called amenable if there is an invariant mean on $\mathrm{C_{ub}}(G)$ as in ($\mathrm{A_{2'}}$), see the remark preceding Theorem 5.4 in [**122**]. We deduce

COROLLARY 5.1.4.— *For any topological group G, the following assertions are equivalent :*

(i) *The group G is amenable.*

(ii) *The trivial Banach G-module $\mathbf{C}$ is relatively injective.*

□

PROOF. Since the space of bounded right uniformly continuous functions is exactly the maximal continuous submodule of $\mathrm{C_b}(G)$ for the left translation representation, this is a particular case of Corollary 5.1.3. □

Later in this section, we shall generalize this statement to amenable actions (Theorem 5.7.1). Moreover, the module $\mathbf{C}$ can be replaced by any coefficient module ; the only crucial point being that the module is contragredient, because in general an amenable group admits non relatively injective Banach modules. This does however not happen in the following case :

PROPOSITION 5.1.5.— *Let G be a compact group. Then any Banach G-module is relatively injective.*

PROOF. Let (π, E) be a Banach G-module and let m be the normalized Haar measure on G. The the Bochner integration against m gives a norm one left inverse to the canonical inclusion $E \to \mathrm{C_b}(G, E)$. This left inverse is a G-morphism because m is invariant and π can be commuted with the integral. Therefore, the condition of Proposition 5.1.1 is satisfied. □

5.2. Relative injectivity for subgroups, I. We are incidentally in position to derive a first statement as to the issue mentioned after the Proposition 4.3.4, namely how relative injectivity behaves with respect to subgroups. Actually, the main ingredient of the following proposition was already available with Corollary 4.5.6.

PROPOSITION 5.2.1.— *Let H be an open subgroup of a locally compact group G. Then every G-relatively injective Banach G-module is H-relatively injective.*

PROOF. Since H is open, we see from criterion (iii) in Lemma 1.1.1 that any Banach G-module is continuous if and only if the induced H-module structure is continuous. Thus we may write unambiguously C for the maximal G- or H-continuous submodule functor. Recall on the other hand that an open subgroup is also closed [**27**, III §2 N° 1].

Let now

$$\mathfrak{m} : C\mathrm{C_b}(G, CE) \longrightarrow CE$$

be a norm one G-morphism as granted by Proposition 5.1.2, that is, $\mathfrak{m}$ is left inverse to the canonical inclusion. The Corollary 4.5.6 tells us that $\mathrm{C_b}(G, CE)$ is H-relatively injective. Hence, by Lemma 4.1.5, $C\mathrm{C_b}(G, CE)$ is also H-relatively injective. Since $\mathfrak{m}$ is in particular a H-morphism, the Proposition 4.3.2 implies that CE is H-relatively injective, hence so is E by Lemma 4.1.5. □

SCHOLIUM 5.2.2. If we analyze the argument given above, we see that the relevance of H being open is the following. Let ι^* be the pull-back functor turning Banach G-modules into Banach H-modules. Then the diagram

$$\begin{array}{ccc} \{\text{Ban. } G\text{-modules}\} & \xrightarrow{\iota^*} & \{\text{Ban. } H\text{-modules}\} \\ \big\downarrow C & & \big\downarrow C \\ \{\text{cont. Ban. } G\text{-modules}\} & \xrightarrow{\iota^*} & \{\text{cont. Ban. } H\text{-modules}\} \end{array}$$

does not commute unless H is open in G.

However, in the case of coefficient modules over second countable groups a different method will generalize the conclusion of Proposition 5.2.1 to all closed subgroups $H < G$ (Proposition 5.8.1).

5.3. Amenable actions : definitions. *We suppose for the rest of the Section 5 that G is locally compact second countable.*

R. Zimmer defines the amenability of a G-action on a measure space S as follows.

DEFINITION 5.3.1. Let S be a standard Borel space endowed with a probability measure and a measure class preserving Borel G-action. The G-action on S is said *amenable* if for every separable Banach space E and every Borel right cocycle $\alpha : S \times G \to \mathrm{Isom}(E)$ the following holds for the dual α^*-twisted action on E^* :

any α^*-invariant Borel field $\{A_s\}_{s \in S}$ of non-empty convex weak-* compact subsets A_s of the closed unit ball in E^* admits an α^*-invariant Borel section.

When the G-action is amenable, we say also that S is an amenable G-space.

This definition and comments can be found in [**141**, chap. 4]. As the reader will have noticed, the G-space S is not supposed regular in the sense of our Definition 2.1.1. Since we shall use regular spaces throughout the sequel, we will have to add this assumption to the general facts collected below.

Intuitively, the above definition is an analogue of the criterion (A_1) for the amenability of groups (more precisely, an analogue of the particular case of (A_1) where the locally convex space is a dual Banach space endowed with weak-* topology). Instead of a single space F, we have a field of spaces fibering over S, and instead of a fixed point we are looking for an invariant section, so that in this analogy the case where S is a point would correspond to the amenability of the group G.

However, a definition analogous to the amenability criterion (A_2) is more appropriate for our purposes. In [**139**], R. Zimmer gives a partial result in this direction for discrete groups. More recently, S. Adams [**2**] has characterized amenable actions in the following terms :

THEOREM 5.3.2 (S. Adams).— *Let S be a regular G-space. The following assertions are equivalent :*

(i) *The G-action on S is amenable.*

(ii) *There is a G-equivariant conditional expectation $L^\infty(G \times S) \to L^\infty(S)$.* □

ON THE PROOF. The notes [**2**] have not been published, but contributed to a joint work [**3**] of Adams-Elliott-Giordano, where the proof can be found : (i)⇒(ii) is Theorem 3.4 in [**3**], while for (ii)⇒(i), according to [**3**], the proof in [**139**] with G discrete holds without change in the continuous case. □

In the above statement, we understood the

DEFINITION 5.3.3. A *conditional expectation* $\mathfrak{m} : L^\infty(G\times S) \to L^\infty(S)$ is a norm one linear continuous map such that

a) $\mathfrak{m}(\mathbf{1}_{G\times S}) = \mathbf{1}_S$,
b) for all $f \in L^\infty(G \times S)$ and each measurable subset $A \subset S$ one has $\mathfrak{m}(f \cdot \mathbf{1}_{G\times A}) = \mathfrak{m}(f) \cdot \mathbf{1}_A$.

The above result can be interpreted as an analogue of the amenability criterion (A_2) if we generalize the concept of an invariant mean. Here is some terminology.

DEFINITION 5.3.4. Let S be a regular G-space and E a coefficient G-module. The canonical G-morphism

$$\begin{aligned} \epsilon:\ E &\longrightarrow L^\infty_{\mathrm{w*}}(S,E) \\ v &\longmapsto v\mathbf{1}_S \end{aligned}$$

as well as its restriction

$$CE \longrightarrow CL^\infty_{\mathrm{w*}}(S,E)$$

will be called the *coefficient inclusion.*

We shall occasionly also use this terminology for the analogous canonical inclusion G-morphism to spaces of continuous functions.

DEFINITION 5.3.5. Let S be a regular G-space and E a coefficient G-module. A *mean* on $L^\infty_{\mathrm{w*}}(S,E)$ is a norm one morphism of Banach spaces

$$\mathfrak{m}:\ L^\infty_{\mathrm{w*}}(S,E) \longrightarrow E$$

which is a left inverse for the coefficient inclusion G-morphism ϵ. Likewise, a mean on $CL^\infty_{\mathrm{w*}}(S,E)$ is a norm one morphism of Banach spaces

$$\mathfrak{m}:\ CL^\infty_{\mathrm{w*}}(S,E) \longrightarrow CE$$

with $\mathfrak{m}\epsilon = Id$.

DEFINITION 5.3.6. A map $\mathfrak{m} : L^\infty(G\times S) \to L^\infty(S)$ is *local* if for any measurable $A \subset S$ and any $f \in L^\infty(G\times S)$ having its essential support in $G\times A$, the function class $\mathfrak{m}(f)$ has essential support in A.

The point-wise multiplication turns $L^\infty(S)$ into a C *- algebra, so that the coefficient inclusion ϵ endows $L^\infty(G \times S)$ with a $L^\infty(S)$-module structure via the identification

$$L^\infty(G \times S) \cong L^\infty_{\mathrm{w*}}(G, L^\infty(S)) \qquad (*)$$

of Corollary 2.3.3.

Finally, we recall that a linear map $\mathfrak{m} : L^\infty(G \times S) \to L^\infty(S)$ is called *positive* if $\mathfrak{m}(f) \geq 0$ for all $f \geq 0$, where the order on $L^\infty(G \times S)$ and $L^\infty(S)$ is given by the canonical Riesz space structure, namely $f \geq 0$ if f is almost everywhere positive[1].

These various notions overlap as follows (for simplicity, we state and prove this lemma for real coefficients).

[1] We share the French view according to which the adjective *positive* does not imply a strict inequality.

LEMMA 5.3.7.— *Let* $\mathfrak{m} : L^\infty(G \times S) \to L^\infty(S)$ *be a continuous linear map with* $\|\mathfrak{m}\| = 1$ *and* $\mathfrak{m}(\mathbf{1}_{G\times S}) = \mathbf{1}_S$ *(and take real-valued* L^∞ *spaces). Then* $\mathfrak{m}$ *is positive, and moreover the following conditions are equivalent :*

(i) $\mathfrak{m}\epsilon = Id$ *(where* ϵ *is defined via* $(*)$ *above).*

(ii) $\mathfrak{m}$ *is a conditional expectation.*

(iii) $\mathfrak{m}$ *is local.*

(iv) $\mathfrak{m}$ *is* $L^\infty(S)$*-linear.*

PROOF. First we show that $\mathfrak{m}$ is positive. Observe that a function class f, both on $G \times S$ or on S, is positive if and only if f is in the closed ball of radius λ centered at $\lambda\mathbf{1}$ for large λ (actually for any $\lambda \geq \|f\|_\infty$). Since $\mathfrak{m}$ sends $\lambda\mathbf{1}_{G\times S}$ to $\lambda\mathbf{1}_S$ and does not increase distances, the criterion is satisfied by $\mathfrak{m}(f)$ as soon as satisfied by f, hence $\mathfrak{m}$ is positive. Now the scheme of the proof is as follows :

$$\begin{array}{ccc} \text{(iv)} & \Longrightarrow & \text{(i)} \\ \Uparrow & & \Downarrow \\ \text{(ii)} & \Longleftarrow & \text{(iii)} \end{array}$$

(i)$\Rightarrow$(iii) : Suppose f has essential support in $G\times A$, where A is a measurable subset of S. One can write f as a finite combination $\sum_{j=1}^n a_j f_j$ with $a_j \in \mathbf{R}$ and f_j with support in $G \times A$ are ranging in $[0,1]$. Thus $0 \leq f_j \leq \mathbf{1}_{G\times A}$. We know already that $\mathfrak{m}$ is positive, so

$$0 \leq \mathfrak{m}(f_j) \leq \mathfrak{m}(\mathbf{1}_{G\times A}) = m(\epsilon\mathbf{1}_A) = \mathbf{1}_A.$$

This shows that each $\mathfrak{m}(f_j)$, hence also $\mathfrak{m}(f)$, is supported in A.

(iii)$\Rightarrow$(ii) : now $\mathfrak{m}$ is supposed local, and we are given a measurable $A \subset S$. Write $B = S \setminus A$, so that

$$\mathfrak{m}(f) = \mathfrak{m}(f \cdot \mathbf{1}_{G\times A}) + \mathfrak{m}(f \cdot \mathbf{1}_{G\times B}).$$

By locality, $\mathfrak{m}(f \cdot \mathbf{1}_{G\times B})$ vanishes on A, so that multiplying the above equality by $\mathbf{1}_A$ we find $\mathfrak{m}(f) \cdot \mathbf{1}_A = \mathfrak{m}(f \cdot \mathbf{1}_{G\times A}) \cdot \mathbf{1}_A$. But locality, again, implies $\mathfrak{m}(f \cdot \mathbf{1}_{G\times A}) \cdot \mathbf{1}_A = \mathfrak{m}(f \cdot \mathbf{1}_{G\times A})$, whence $\mathfrak{m}(f) \cdot \mathbf{1}_A = \mathfrak{m}(f \cdot \mathbf{1}_{G\times A})$.

(ii)$\Rightarrow$(iv) : it is enough to observe the linear span of the set

$$\{\mathbf{1}_A : A \subset S \text{ is measurable}\}$$

is dense in $L^\infty(S)$ for the norm topology (take e.g. dyadic approximations of $f \in L^\infty(S)$).

(iv)$\Rightarrow$(i) : by $L^\infty(S)$-linearity, we have for all $f \in L^\infty(S)$ the relation

$$\mathfrak{m}(\epsilon f) = \mathfrak{m}(\mathbf{1}_{G\times S} \cdot \epsilon f) = \mathfrak{m}(\mathbf{1}_{G\times S}) \cdot f = \mathbf{1}_S \cdot f = f.$$

□

Keeping the identification $(*)$, we may now rephrase the result of S. Adams :

COROLLARY 5.3.8.— *Let S be a regular G-space. The following assertions are equivalent :*

(i) *The G-action on S is amenable.*

(ii) *There is a G-equivariant mean on $L^\infty_{\mathrm{w*}}(G, L^\infty(S))$.*

PROOF. By linearity, the complex case reduces to the real case. Now read Theorem 5.3.2 with Lemma 5.3.7. □

Finally, we mention one further known characterization of amenable actions :

THEOREM 5.3.9 (S. Adams).— *Let S be a regular G-space. The following assertions are equivalent :*

(i) *The G-action on S is amenable.*

(ii) *The equivalence relation on S defined by the G-action is amenable and the stabilizer $\mathrm{Stab}_G(s) < G$ is amenable for almost every $s \in S$.*

ON THE PROOF. Again, the unpublished proof of [**2**] found its way to [**3**]. In the ergodic case, this is (i)⇔(v) in [**3**, Theorem A]. However, the proof given in Sections IV and V of this reference does actually not depend on ergodicity. □

We recall that we have supposed G locally compact second countable throughout this section. Most of the equivalences between the various characterizations of amenable actions are not known to hold, or at any rate are not proved in the literature, without the assumption of second countability. It is not our intention to venture any generalization in this direction ; however, we shall occasionly extend the definition of amenable actions to the following prudent setting :

DEFINITION 5.3.10. Let G be a locally compact group, not necessarily second countable. Let S be a regular G-space, and suppose that there is a compact normal subgroup $K \lhd G$ with G/K second countable such that the G-action on S factors through G/K and that the G/K-action on S is amenable.

Then we shall also call the G-action on S amenable.

This convention is consistent with the previous definitions of amenable actions. Indeed, we know from point (vii) in Examples 5.4.1 that if in the above setting G is second countable, then the amenability of the G-action is equivalent to the amenability of the G/K-action.

SCHOLIUM 5.3.11. Such an extension is convenient in view of a stratagem due to Sh. Kakutani and K. Kodaira [**90**] : if G is a locally compact σ-compact group, then define $\mathfrak{H}$ to be the closed G-invariant subspace generated in $L^2(G)$ by some non zero compactly supported continuous function. Then $\mathfrak{H}$ is separable (Satz 5 in [**90**]) even though $L^2(G)$ is not so in general. The kernel K of the unitary representation $G \to \mathcal{U}(\mathfrak{H})$ induced by the regular representation is compact, and G/K is second countable (Satz 6 in [**90**]). Thus every locally compact σ-compact group admits a second countable quotient by a compact kernel.

5.4. Basic examples. The basic breeding ground for examples of amenable actions is given by the following list (see Examples 2.1.2 for the regularity).

EXAMPLES 5.4.1.

(i) Let $H < G$ be a closed subgroup. Then the G-action on G/H is amenable if and only if H is an amenable group [**141**, Proposition 4.3.2]. In particular G itself is an amenable G-space.

(ii) If S, T are regular G-spaces and the G-action on S is amenable, then the diagonal G-action on $S \times T$ is amenable [**141**, Proposition 4.3.4].

(iii) The group G is amenable if and only if any regular G-space is amenable (follows immediately from Theorem 5.3.9).

(iv) Let S be the Poisson boundary corresponding to a probability measure μ on G. Recall that μ is said *étalée* if μ or at least some of its convolution powers $\mu * \cdots * \mu$ has a non-singular component with respect to the class of the Haar measures of G. If μ is étalée, then the G-action on S is amenable [**140**, Corollary 5.3].

(v) Generalizing (ii), let $p : T \to S$ be a G-equivariant measurable map of regular G-spaces such that the direct image of the measure class on T coincides with the measure class on S. If S is an amenable G-space, then so is T [**3**, Corollary C].

(vi) Suppose we have a finite family S_j of regular G_j-spaces, $j = 1, \ldots, n$. Then $G_1 \times \cdots \times G_n$ acts amenably on $S_1 \times \cdots \times S_n$ if and only if each G_j acts amenably on the corresponding S_j (apply Theorem 5.3.9).

(vii) Let $H \lhd G$ be a closed normal subgroup and S an amenable G/H-space, and consider the G-action on S defined through the canonical map $G \to G/H$. Then S is an amenable G-space if and only if H is amenable (again Theorem 5.3.9).

(viii) Let Γ be a Gromov-hyperbolic group. By a result of S. Adams [**1**], the Γ-action on its geometric boundary at infinity $\partial_\infty \Gamma$ is amenable for any quasi-invariant measure.

Amenability criteria are often stated for ergodic actions, and the way over to the general case does not appear to be a direct consequence of the definitions. Therefore, we quote here the relevant disintegration result of Adams-Elliott-Giordano :

THEOREM 5.4.2 (Adams-Elliott-Giordano).— *Let S be a regular G-space with disintegration $(S_e)_{e \in E}$ into ergodic components, where E is the space of ergodic components with direct image measure.*

Then S is an amenable G-space if and only if S_e is an amenable G-space for almost every $e \in E$.

ON THE PROOF. This is the faithful transliteration of Corollary B page 804 in [**3**]. □

Amongst the basic examples there is also the following, which we prove for completeness.

LEMMA 5.4.3.— *Let $H < G$ be a closed subgroup and S an amenable regular G-space. The restriction to H of the action on S is amenable.*

PROOF. The separability of G (we recall that G is second countable) implies that the canonical projection $p : G \to H\backslash G$ admits a Borel section $\sigma : H\backslash G \to G$ such that the Borel space isomorphism

$$F: G \longrightarrow H \times H\backslash G$$
$$g \longmapsto \Big(g\big(\sigma p(g)\big)^{-1}, p(g)\Big)$$

preserves the measure class. If we endow G and H with the left H-action and $H\backslash G$ with the trivial H-action, then F is moreover H-equivariant. According to Corollary 5.3.8, there is a G-equivariant mean

$$\mathfrak{m}: L^{\infty}_{\mathrm{w}*}\Big(G, L^{\infty}(S)\Big) \longrightarrow L^{\infty}(S).$$

Composing it with the H-isomorphism induced by F^{-1} by a double application of Corollary 2.3.3

$$F_*^{-1}: L^{\infty}_{\mathrm{w}*}\Big(H, L^{\infty}_{\mathrm{w}*}(H\backslash G, L^{\infty}(S))\Big) \longrightarrow L^{\infty}_{\mathrm{w}*}\Big(G, L^{\infty}(S)\Big)$$

and with the H-morphism

$$\epsilon_*: L^{\infty}_{\mathrm{w}*}\Big(H, L^{\infty}(S)\Big) \longrightarrow L^{\infty}_{\mathrm{w}*}\Big(H, L^{\infty}_{\mathrm{w}*}(H\backslash G, L^{\infty}(S))\Big),$$

induced by the coefficient inclusion

$$\epsilon: L^{\infty}(S) \longrightarrow L^{\infty}_{\mathrm{w}*}(H\backslash G, L^{\infty}(S))$$

we obtain an H-morphism of norm one $\mathfrak{m}F_*^{-1}\epsilon_*$ which is a left inverse to the coefficient inclusion of $L^{\infty}(S)$ into $L^{\infty}_{\mathrm{w}*}\big(H, L^{\infty}(S)\big)$. Therefore, by Corollary 5.3.8, the H-action on S is amenable. □

The last example is about measure-theoretic quotients. We recall that if S is a standard Borel G-space with the invariant class of a probability measure and if $\mathcal{A}$ a unital G-invariant weak-* closed sub- C *-algebra of $L^{\infty}(S)$, then there is a standard probability G-space T and a G-equivariant measure class preserving map $p : S \to T$ such that the induced map $p^* : L^{\infty}(T) \to L^{\infty}(S)$ is a G-equivariant C *-algebra isomorphism onto $\mathcal{A}$; the space T is called a point realization of $\mathcal{A}$, see [**101**].

We observe that if moreover S is regular, then so is T. Indeed, $\mathcal{A}$ being weak-* closed in $L^{\infty}(S)$, the duality principle realizes $L^1(T)$ as a quotient of the Banach G-module $L^1(S)$. The latter being continuous by assumption, the former is so too.

An important particular case is the following : for a closed normal subgroup $N \lhd G$, set $\mathcal{A} = L^{\infty}(S)^N$. Then the space T is called the *measure-theoretic quotient* of S by N.

Now the following lemma is an easy consequence of the definitions ; we shall however give an alternative proof of it in Remark 5.7.2 below.

LEMMA 5.4.4.— *Let $N \lhd G$ a normal closed subgroup and S an amenable regular G-space. Let T be the measure-theoretic quotient of S by N ; in other words, T is a point realization of $L^\infty(S)^N$, this latter space being viewed as a G/N-coefficient module.*

Then the G/N-action on T is amenable. □

5.5. Admissibility of the coefficient inclusion. The following observation concerning the Definition 5.3.4 of coefficient inclusion will be of repeated use :

PROPOSITION 5.5.1.— *Let S be a regular G-space and (π, E) a coefficient G-module. Then the coefficient inclusions $E \to L^\infty_{\mathrm{w*}}(S, E)$ and $CE \to CL^\infty_{\mathrm{w*}}(S, E)$ are admissible.*

PROOF. The case of the coefficient inclusion $\epsilon : E \to L^\infty_{\mathrm{w*}}(S, E)$ is immediate : let μ be a probability measure on S as in Definition 2.1.1. Define a map

$$\sigma : \ L^\infty_{\mathrm{w*}}(S, E) \longrightarrow E$$

by the integral

$$\sigma(f) = \int_S^{(\mathrm{GD})} f(s)\, d\mu(s). \qquad \Big(f \in L^\infty_{\mathrm{w*}}(S, E)\Big)$$

This is a linear map of norm at most one and it follows from the definition that $\sigma\epsilon = Id$.

In order to handle the coefficient inclusion $\epsilon : CE \to CL^\infty_{\mathrm{w*}}(S, E)$, we proceed to show that the restriction of σ to $CL^\infty_{\mathrm{w*}}(S, E)$ ranges in CE. Let f be an element of $CL^\infty_{\mathrm{w*}}(S, E)$. Since E is a coefficient module, the operator $\pi(g)$ is adjoint for every $g \in G$. therefore, by Lemma 3.2.1 point (i), we may commute it with the Gelfand-Dunford integral :

$$\pi(g)\sigma(f) = \int_S^{(\mathrm{GD})} \pi(g)\Big(f(s)\Big)\, d\mu(s).$$

The relation $\pi(g)\Big(f(s)\Big) = \Big(\lambda_\pi(g) f\Big)(gs)$ yields thus

$$\pi(g)\sigma(f) = \int_S^{(\mathrm{GD})} \frac{dg\mu}{d\mu}(s)\Big(\lambda_\pi(g) f\Big)(s)\, d\mu(s),$$

where the Radon-Nikodým derivative $dg\mu/d\mu$ is in $L^1(\mu)$ according to Definition 2.1.1. If now $(g_\alpha)_{\alpha\in A}$ is a net converging to $e \in G$, we estimate

$$\begin{aligned}\|\sigma(f) - \pi(g_\alpha)\sigma(f)\|_E &\leq \\ &\leq \left\| \int_S^{(GD)} f(s)\, d\mu(s) - \int_S^{(GD)} (\lambda_\pi(g_\alpha)f)(s)\, d\mu(s) \right\|_E + \\ &+ \left\| \int_S^{(GD)} (\lambda_\pi(g_\alpha)f)(s)\, d\mu(s) - \int_S^{(GD)} \frac{dg_\alpha\mu}{d\mu}(s)(\lambda_\pi(g_\alpha)f)(s)\, d\mu(s) \right\|_E .\end{aligned}$$

Appealing to the second part of Lemma 3.2.1, we bound the first term by $\|f - \lambda_{(g_\alpha)} f\|_\infty$, which converges to zero along $\alpha \in A$ because f is in $CL^\infty_{w*}(S, E)$. The second term is bounded by

$$\|f\|_\infty \cdot \left\| \mathbf{1}_S - \frac{dg_\alpha\mu}{d\mu} \right\|_1,$$

wherein the right hand side factor converges to zero by the assumption on μ made in Definition 2.1.1. Thus, according to the criterion (iii) in Lemma 1.1.1, σ ranges indeed in CE. □

5.6. A result on equivariant means. In this section, we shall link relative injectivity to the existence of equivariant means. Along the way, we compare two notions of equivariance.

Let (π, E) be a coefficient G-module. Given a left Haar measure m on G, the Gelfand-Dunford integral defines a $L^1(G)$-structure on both E and $L^\infty_{w*}(G, E)$ along the lines explained in Section 3. Therefore, it makes sense to ask whether a mean $\mathfrak{m} : L^\infty_{w*}(G, E) \to E$ is $L^1(G)$-equivariant. In the case of trivial coefficients $E = \mathbf{C}$, it is well known and easily seen that for such a mean G-invariance and $L^1(G)$-invariance, the so-called ***topological invariance***, are equivalent (see [**122**, chap. 2] or [**10**]). A related issue is the connection with equivariant means on the space of uniformly continuous functions. All this goes back at least to Greenleaf [**68**].

In the setting we are dealing with, we have to take into account that we have a coefficient space endowed with a possibly non-continuous G-action.

THEOREM 5.6.1.— *Let (π, E) be a coefficient G-module. The following assertions are equivalent :*

(i) *(π, E) is a relatively injective Banach G-module.*

(ii) *There is a G-equivariant mean $\mathfrak{m}_1 : L^\infty_{w*}(G, E) \to E$.*

(iii) *There is a G-equivariant mean $\mathfrak{m}_2 : CL^\infty_{w*}(G, E) \to CE$.*

(iv) *There is an $L^1(G)$-equivariant mean $\mathfrak{m}_3 : L^\infty_{w*}(G, E) \to E$.*

(v) *There is an $L^1(G)$-equivariant mean $\mathfrak{m}_4 : CL^\infty_{w*}(G, E) \to CE$.*

The scheme of the proof is as follows :

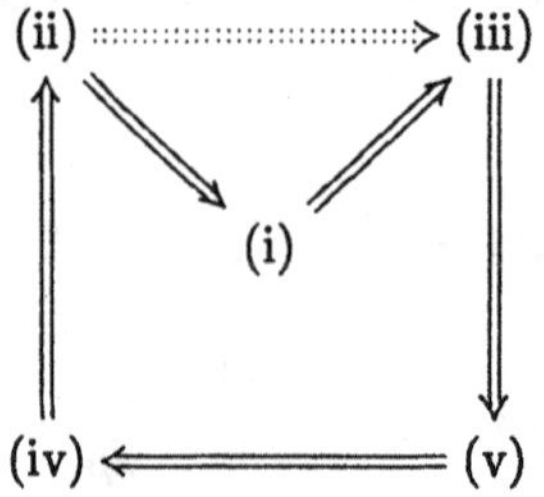

The outer implications establish the equivalence of four notions of equivariant means, which are then linked to the central notion of injectivity (i). The upper arrow, redundant, is only pictured because it represents an obvious implication.

PROOF OF THE THEOREM. The crux of the four outer implications is

(v)⇒(iv) : let $(\varphi_\alpha)_{\alpha\in\mathrm{A}}$ be a strict approximate identity on G (see Definition 3.2.4). Putting together the observations of Lemma 1.2.6 and Proposition 3.2.3 point (i), we see that the composition $\mathfrak{m}_4\lambda_\pi(\varphi_\alpha)$ defines a morphism of Banach spaces

$$\mathfrak{m}_4\lambda_\pi(\varphi_\alpha) : L^\infty_{\mathrm{w*}}(G,E) \longrightarrow \mathcal{C}E \subset E.$$

By Lemma 3.2.1 point (ii), we have $\|\lambda_\pi(\varphi_\alpha)\| \leq 1$ and hence we have a net $\big(\mathfrak{m}_4\lambda_\pi(\varphi_\alpha)\big)_{\alpha\in\mathrm{A}}$ in the closed unit ball of

$$\mathcal{L}\Big(L^\infty_{\mathrm{w*}}(G,E), E\Big) \cong \big(L^\infty_{\mathrm{w*}}(G,E)\widehat{\otimes}E^\flat\big)^*.$$

Applying the theorem of Banach-Alaoğlu [**126**, I.3.15] in order to find a convergent subnet, we conclude that there is another strict approximate identity $(\psi_\beta)_{\beta\in\mathrm{B}}$ such that for the weak-* topology determined by the right hand side of the above identification, $\big(\mathfrak{m}_4\lambda_\pi(\psi_\beta)\big)_{\beta\in\mathrm{B}}$ converges to a linear operator $\mathfrak{m}_3 : L^\infty_{\mathrm{w*}}(G,E) \to E$ of norm $\|\mathfrak{m}_3\|$ at most one. This convergence implies that for all $f \in L^\infty_{\mathrm{w*}}(G,E)$ the net $\mathfrak{m}_4\lambda_\pi(\psi_\beta)f$ converges to $\mathfrak{m}_3$ in E for its weak-* topology. Moreover, if f is essentially constant of essential value $w \in E$, then $\lambda_\pi(\psi_\beta)f$ is constant of value $\pi(\psi_\beta)w \in \mathcal{C}E$. In particular, $\mathfrak{m}_4\lambda_\pi(\psi_\beta)f$ equals $\pi(\psi_\beta)w$, which on the other hand weak-* converges to w as seen in the proof of Proposition 3.2.3. Therefore, $\mathfrak{m}_3 f = w$ and hence $\mathfrak{m}_3$ is indeed a mean $L^\infty_{\mathrm{w*}}(G,E) \to E$.

Thus it remains only to show that $\mathfrak{m}_3$ is $L^1(G)$-equivariant. For $\varphi \in L^1(G)$ and $f \in L^\infty_{\mathrm{w*}}(G,E)$, we have that $\mathfrak{m}_3\lambda_\pi(\varphi)f$ is the weak-* limit of $\mathfrak{m}_4\lambda_\pi(\psi_\beta*\varphi)f$, whilst for all $u \in E^\flat$

$$\begin{aligned}\langle\mathfrak{m}_4\lambda_\pi(\varphi*\psi_\beta)f|u\rangle &= \langle\pi(\varphi)\mathfrak{m}_4\lambda_\pi(\psi_\beta)f|u\rangle\\ &= \langle\mathfrak{m}_4\lambda_\pi(\psi_4)f|\pi^\flat(\varphi^\sim)u\rangle\\ &\xrightarrow{\beta} \langle\mathfrak{m}_3 f|\pi^\flat(\varphi^\sim)u\rangle = \langle\pi(\varphi)\mathfrak{m}_3 f|u\rangle,\end{aligned}$$

where in the very first equality we used the Lemma 3.2.1 point (i) applied to $L^\infty_{\mathbf{w*}}(G,E)$ in order to commute $\mathfrak{m}_4$ with $\lambda_\pi(\varphi)$ according to the hypothesis (v) on $\mathfrak{m}_4$.

This shows that $\mathfrak{m}_3\lambda_\pi(\varphi)f - \pi(\varphi)\mathfrak{m}_3 f$ is the weak-* limit of $\mathfrak{m}_4\lambda_\pi(\psi_\beta * \varphi - \varphi * \psi_\beta)f$. On the other hand, since our strict approximate identity is two-sided, $\psi_\beta*\varphi-\varphi*\psi_\beta$ norm converges to zero in $L^1(G)$. Therefore, by the second point of Lemma 3.2.1, $\lambda_\pi(\psi_\beta*\varphi-\varphi*\psi_\beta)f$ and hence also $\mathfrak{m}_4\lambda_\pi(\psi_\beta*\varphi-\varphi*\psi_\beta)f$ converges to zero in norm. Putting everything together, we conclude that $\mathfrak{m}_3\lambda_\pi(\varphi)f = \pi(\varphi)\mathfrak{m}_3 f$, completing the proof of (v)⇒(iv).

(iv)⇒(ii) : we claim that $\mathfrak{m}_3$ is actually G-equivariant. Indeed, let $f \in L^\infty_{\mathbf{w*}}(G,E)$ and $g \in G$ and fix a strict approximate identity $(\psi_\alpha)_{\alpha\in A}$. Now $\mathfrak{m}_3\lambda_\pi(g)f$ is the weak-* limit of

$$\begin{aligned}\pi(\psi_\alpha^\sim)\mathfrak{m}_3\lambda_\pi(g)f = \mathfrak{m}_3\lambda_\pi(\psi_\alpha^\sim)\lambda_\pi(g)f &= \mathfrak{m}_3\lambda_\pi((\lambda(g^{-1})\psi_\alpha)^\sim)f \\ &= \pi((\lambda(g^{-1})\psi_\alpha)^\sim)\mathfrak{m}_3 f = \pi(\psi_\alpha^\sim)(\pi(g)\mathfrak{m}_3 f),\end{aligned}$$

which converges weak-* to $\pi(g)\mathfrak{m}_3 f$. We may therefore take $\mathfrak{m}_1 = \mathfrak{m}_3$.

(ii)⇒(iii) is a particular case of Lemma 1.2.6.

(iii)⇒(v) : let $f \in \mathcal{C}L^\infty_{\mathbf{w*}}(G,E)$ and $\psi \in L^1(G)$. Since $\mathcal{C}L^\infty_{\mathbf{w*}}(G,E)$ is a continuous Banach G-module, the map

$$\begin{aligned} G &\longrightarrow \mathcal{C}L^\infty_{\mathbf{w*}}(G,E)\\ g &\longmapsto \psi(g)\lambda_\pi(g)f\end{aligned}$$

is m-measurable in Bourbaki's sense, where we recall that m is our fixed left Haar measure on G. Thus the Gelfand-Dunford integral representation of $\lambda_\pi(\psi)$ is actually a Bochner integral, hence can be commuted with $\mathfrak{m}_2$. Therefore we may take $\mathfrak{m}_4 = \mathfrak{m}_2$.

(i)⇒(iii) : considering the coefficient inclusion $\epsilon : \mathcal{C}E \to \mathcal{C}L^\infty_{\mathbf{w*}}(G,E)$, the Proposition 5.5.1 (applied to $S = G$) yields the morphism σ needed to construct the following extension problem for $\mathcal{C}E$:

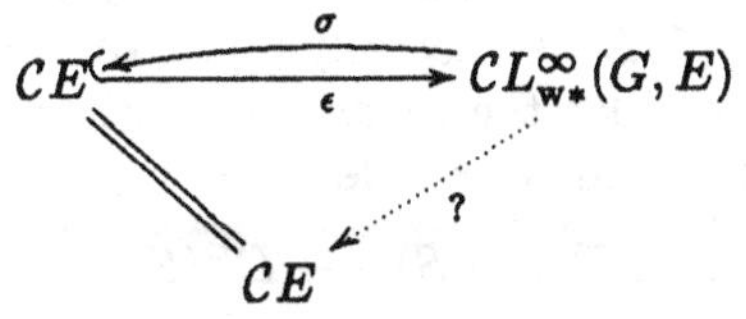

Since $\mathcal{C}E$ is relatively injective (Lemma 4.1.5), there is a solution $\mathfrak{m}_2$ to this problem.

(ii)⇒(i) : by Proposition 4.4.2, the Banach G-module $L^\infty_{\mathbf{w*}}(G,E)$ is relatively injective. Now Proposition 4.3.1 applied to $F = L^\infty_{\mathbf{w*}}(G,E)$ and ϑ the inclusion of coefficients implies that E is relatively injective. This completes the proof of the Theorem 5.6.1. □

5.7. A characterization of amenable actions. Combining the Theorem 5.6.1 with the amenability criterion of Corollary 5.3.8, we obtain a characterization of amenable actions in a homological language.

THEOREM 5.7.1.— *Let S be a regular G-space. The following assertions are equivalent :*

(i) *The G-action on S is amenable.*

(ii) *The coefficient G-module $L^\infty(S)$ is relatively injective.*

(iii) *The coefficient G-module $L^\infty_{\mathbf{w*}}(S^{n+1}, E)$ is relatively injective for every coefficient G-module E and all $n \geq 0$.*

REMARK 5.7.2. Using the above theorem, we can give an immediate proof of Lemma 5.4.4. Suppose S is an amenable regular G-space and $N \lhd G$ a normal closed subgroup. Let T be the point realization of $L^\infty(S)^N$. By the theorem above, $L^\infty(S)$ is G-relatively injective, hence $L^\infty(S)^N$ is G/N-relatively injective by Proposition 4.3.4, and so applying again Theorem 5.7.1 we conclude that T is an amenable regular G/N-space.

EXAMPLE 5.7.3. A trivial case of Theorem 5.7.1 is when S is a point ; then the statement reads :

The following assertions are equivalent :

(i) *The group G is amenable.*

(ii) *The coefficient G-module $\mathbf{C}$ is relatively injective.*

(iii) *Any coefficient G-module is relatively injective.*

PROOF OF THEOREM 5.7.1. (i)$\Rightarrow$(iii) : the second item in Examples 5.4.1 shows that we may as well suppose $n = 1$. We claim that there is a norm one G-morphism

$$\mathfrak{m} : \ L^\infty_{\mathbf{w*}}(G \times S, E) \longrightarrow L^\infty_{\mathbf{w*}}(S, E)$$

providing a left inverse to the canonical inclusion — in view of the Corollary 2.3.3, the claim is equivalent to the existence of an equivariant mean on $L^\infty_{\mathbf{w*}}\big(G, L^\infty_{\mathbf{w*}}(S, E)\big)$. Since by Corollary 4.4.4 the coefficient G-module $L^\infty_{\mathbf{w*}}(G \times S, E)$ is relatively injective, this will allow us to apply the Proposition 4.3.1 in order to conclude that $L^\infty_{\mathbf{w*}}(S, E)$ is relatively injective.

To prove the claim, consider the separable continuous Banach G-module $(\pi^\flat, E^\flat)$ to which E is contragredient and let

$$\mathfrak{n} : \ L^\infty(G \times S) \longrightarrow L^\infty(S)$$

be the G-morphism of norm one granted by Theorem 5.3.2. For every $f \in L^\infty_{\mathbf{w*}}(G \times S, E)$ define a bilinear form $\mathfrak{m}f$ on $L^1(S) \times E^\flat$ by

$$(\mathfrak{m}f)(\psi, v) = \big\langle \mathfrak{n}\langle f(\cdot)|v\rangle \big| \psi \big\rangle. \qquad (\psi \in L^1(S), v \in E^\flat)$$

This definition is well posed even though $\langle f(\cdot)|v\rangle$ is only defined almost everywhere, because it is enough to restrict v to a countable dense subset of $E^\flat$, the latter being separable by definition.

Since we have

$$\big|(\mathfrak{m}f)(\psi, v)\big| \leq \|f\|_\infty \cdot \|v\|_{E^\flat} \cdot \|\psi\|_1,$$

the bilinear form $\mathfrak{m}f$ is continuous of norm at most $\|f\|_\infty$. As $\mathfrak{m}f$ depends linearly on f, we deduce that the corresponding linear map

$$\mathfrak{m}: \ L^\infty_{\mathrm{w*}}(G \times S, E) \longrightarrow \left(L^1(S)\widehat{\otimes}E^\flat\right)^\sharp \cong L^\infty_{\mathrm{w*}}(S, E)$$

is continuous of norm at most one. As for G-equivariance, the relation

$$\langle \lambda_\pi(g) f(\cdot)|v\rangle = \lambda(g)\langle f(\cdot)|\pi^\flat(g)^{-1}v\rangle, \qquad (g \in G, v \in E^\flat)$$

implies for $\psi \in L^1(S)$ that

$$(\mathfrak{m}\lambda_\pi(g)f)(\psi, v) = \big\langle \mathfrak{n}\lambda(g)\langle f(\cdot)|\pi^\flat(g)^{-1}v\rangle\big|\psi\big\rangle$$

and hence by the G-equivariance of $\mathfrak{n}$ this is

$$\begin{aligned}\big\langle\lambda(g)\mathfrak{n}\langle f(\cdot)|\pi^\flat(g)^{-1}v\rangle\big|\psi\big\rangle &= \big\langle\mathfrak{n}\langle f(\cdot)|\pi^\flat(g)^{-1}v\rangle\big|\lambda^\flat(g)^{-1}\psi\big\rangle \\ &= (\mathfrak{m}f)(\lambda^\flat(g)^{-1}\psi, \pi^\flat(g)^{-1}v).\end{aligned}$$

This corresponds precisely to $(\lambda_\pi(g)\mathfrak{m}f)(\psi, v)$ under the identification between continuous bilinear forms on $L^1(S) \times E^\flat$ with $L^\infty_{\mathrm{w*}}(S, E)$. The G-morphism $\mathfrak{m}$ that we have constructed is indeed a left inverse for the inclusion of $L^\infty_{\mathrm{w*}}(S, E)$ in $L^\infty_{\mathrm{w*}}(G \times S, E)$, since the pairing

$$\langle\cdot|\cdot\rangle: \ L^\infty_{\mathrm{w*}}(S, E) \times \left(L^1(S)\widehat{\otimes}E^\flat\right) \longrightarrow \mathbf{C}$$

is obtained by Gelfand-Dunford integration over S of the pairing on $E \times E^\flat$.

(iii)⇒(ii) is obvious.

(ii)⇒(i) : the implication (i)⇒(ii) of Theorem 5.6.1 applied to $E = L^\infty(S)$ implies that there is an equivariant mean on $L^\infty_{\mathrm{w*}}\big(G, L^\infty(S)\big)$. Therefore the Corollary 5.3.8 implies that the G-action on S is amenable. □

5.8. Relative injectivity for subgroups, II. Opening a parenthesis, we remark that for *coefficient modules* we can generalize Proposition 5.2.1 in order to complete the picture of Proposition 4.3.4.

PROPOSITION 5.8.1.— *Let $H < G$ be a closed subgroup of the locally compact second countable G. Then every G-injective coefficient G-module is H-relatively injective.*

The proof parallels the argument for Proposition 5.2.1 in the setting of coefficient modules and L^∞ spaces :

PROOF. Let E be a G-relatively injective coefficient G-module. Because of Lemma 5.4.3, G is an amenable regular H-space. Therefore, the Theorem 5.7.1 implies that the coefficient H-module $L^\infty_{\mathrm{w*}}(G, E)$ is H-relatively injective. Moreover, since E is G-relatively injective, there is by Theorem 5.6.1 a G-equivariant mean $\mathfrak{m}$ on $L^\infty_{\mathrm{w*}}(G, E)$. Now $\mathfrak{m}$ is also a H-morphism

$$L^\infty_{\mathrm{w*}}(G, E) \underset{\epsilon}{\overset{\mathfrak{m}}{\rightleftarrows}} E$$

so that we may apply Proposition 4.3.1 to H with $\vartheta = \epsilon$ and conclude that E is H-relatively injective. □

CHAPTER III

Definition and characterization of continuous bounded cohomology

In the first two chapters, we have collected the material that we need in order to define and benefit from a functorial theory of continuous bounded cohomology adapted to a large class of examples.

However, before proceeding, we intercalate a brief Section 6 in which we give a very crudely concrete definition of the continuous bounded cohomology. The reason for our doing so is that the homological apparatus yields its output up to canonical isomorphisms, and for the sake of definiteness we want to pin down exactly which objects should be characterized by this procedure. We seize the opportunity for a short discussion of the lower degree cohomology, a discussion to be continued more thoroughly later on.

6. A NAIVE DEFINITION

6.1. The canonical complex. Let G be a topological group and E a Banach G-module. We define inductively

$$\mathrm{C}_{\mathrm{b}}^{0}(G,E)=\mathrm{C}_{\mathrm{b}}(G,E),\qquad \mathrm{C}_{\mathrm{b}}^{n}(G,E)=\mathrm{C}_{\mathrm{b}}\Big(G,\mathrm{C}_{\mathrm{b}}^{n-1}(G,E)\Big)\qquad (n\geq 1)$$

and endow $\mathrm{C}_{\mathrm{b}}^{n}(G,E)$ with the Banach G-module structure defined in Section 1.3. The coefficient inclusion G-morphism $\epsilon: E\to \mathrm{C}_{\mathrm{b}}(G,E)$ induces a G-morphism

$$\epsilon_{*}:\ \mathrm{C}_{\mathrm{b}}^{0}(G,E)\longrightarrow \mathrm{C}_{\mathrm{b}}\Big(G,\mathrm{C}_{\mathrm{b}}^{0}(G,E)\Big)\ =\ \mathrm{C}_{\mathrm{b}}^{1}(G,E).$$

In view of $\mathrm{C}_{\mathrm{b}}^{1}(G,E)=\mathrm{C}_{\mathrm{b}}(\mathrm{C}_{\mathrm{b}}^{0}(G,E))$, we write also $\epsilon:\mathrm{C}_{\mathrm{b}}^{0}(G,E)\to \mathrm{C}_{\mathrm{b}}^{1}(G,E)$ for the coefficient inclusion of $\mathrm{C}_{\mathrm{b}}^{0}(G,E)$. Now we define the G-morphism

$$\mathfrak{d}^{1}=\epsilon-\epsilon_{*}:\ \mathrm{C}_{\mathrm{b}}^{0}(G,E)\longrightarrow \mathrm{C}_{\mathrm{b}}^{1}(G,E)$$

and then inductively

$$\mathfrak{d}^{n}=\epsilon-(\mathfrak{d}^{n-1})_{*}:\ \mathrm{C}_{\mathrm{b}}^{n-1}(G,E)\longrightarrow \mathrm{C}_{\mathrm{b}}^{n}(G,E)$$

for all $n\geq 2$. Organizing this data into a sequence of G-morphisms

$$0\to E\xrightarrow{\epsilon}\mathrm{C}_{\mathrm{b}}^{0}(G,E)\xrightarrow{\mathfrak{d}^{1}}\mathrm{C}_{\mathrm{b}}^{1}(G,E)\xrightarrow{\mathfrak{d}^{2}}\mathrm{C}_{\mathrm{b}}^{2}(G,E)\to\cdots$$

one checks readily that $\mathfrak{d}^{n+1}\mathfrak{d}^n = 0$ holds for all $n \geq 1$ (see also the end of Remark 6.1.2 for more explicit formulae). Anticipating a terminology which is to be made systematic in Section 7, we call the above sequence the *canonical augmented resolution* of the Banach G-module E. We are interested in the following sequence of submodules :

$$0 \longrightarrow \mathrm{C}_\mathrm{b}^0(G,E)^G \xrightarrow{\mathfrak{d}^1} \mathrm{C}_\mathrm{b}^1(G,E)^G \xrightarrow{\mathfrak{d}^2} \mathrm{C}_\mathrm{b}^2(G,E)^G \longrightarrow \cdots$$

which we call the *canonical complex* associated to the Banach G-module E. The Banach space of *cocycles* of degree n is

$$\mathrm{Z}^n(G,E) = \mathrm{Ker}\Big(\mathfrak{d}^{n+1} : \mathrm{C}_\mathrm{b}^n(G,E)^G \to \mathrm{C}_\mathrm{b}^{n+1}(G,E)^G\Big)$$

and the linear space of *coboundaries* of degree n is

$$\mathrm{B}^n(G,E) = \mathrm{Im}\Big(\mathfrak{d}^n : \mathrm{C}_\mathrm{b}^{n-1}(G,E)^G \to \mathrm{C}_\mathrm{b}^n(G,E)^G\Big)$$

with the convention $\mathrm{B}^0 = 0$. It follows from the equivariance of $\mathfrak{d}^n$ that the condition $\mathfrak{d}^{n+1}\mathfrak{d}^n = 0$ on the canonical augmented resolution implies $\mathrm{B}^n(G,E) \subset \mathrm{Z}^n(G,E)$. We define the *continuous bounded cohomology* in degree n of G with coefficients in E to be the quotient

$$\mathrm{H}_\mathrm{cb}^n(G,E) \;=\; \mathrm{Z}^n(G,E)\,/\,\mathrm{B}^n(G,E)$$

endowed with the quotient semi-norm.

REMARK 6.1.1. Observe that if E happens to be a coefficient module, the above definition does not depend on the choice of a predual $E^\flat$. In other words, if E is a Banach G-module with two coefficient module structures, say, E_0 and E_1, then the identity map $E_0 \to E_1$ is a priori not adjoint, and it may happen that there exists no adjoint isomorphism at all from E_0 to E_1 ; however, the identity map induces indeed the identity in cohomology.

If Γ is an abstract group with an isometric representation on a Banach space E, we define the *bounded cohomology* of Γ with coefficients in E

$$\mathrm{H}_\mathrm{b}^n(\Gamma,E) = \mathrm{H}_\mathrm{cb}^n(\Gamma_\delta,E)$$

as the continuous bounded cohomology of Γ_δ, where Γ_δ is the topological group resulting from endowing Γ with the discrete topology.

If we consider the trivial coefficients $E = \mathbf{C}$, we write simply $\mathrm{H}_\mathrm{cb}^n(G)$ or $\mathrm{H}_\mathrm{b}^n(\Gamma)$, respectively.

REMARK 6.1.2. The spaces $\mathrm{C}_\mathrm{b}^n(G,E)$ have the disadvantage that the continuity requirement is not symmetric in the $n+1$ variables. Another straightforward object to consider is the sequence of G-morphisms

$$0 \to E \xrightarrow{\epsilon} \mathrm{C}_\mathrm{b}(G,E) \xrightarrow{d^1} \mathrm{C}_\mathrm{b}(G^1,E) \xrightarrow{d^2} \mathrm{C}_\mathrm{b}(G^2,E) \to \cdots$$

where d^n is the *homogeneous differential* defined by the familiar formula

$$(d^n f)(x_0,\ldots,x_n) = \sum_{j=0}^{n}(-1)^j f(x_0,\ldots,\widehat{x_j},\ldots,x_n)$$

(the hat on $\widehat{x_j}$ indicates that this variable is omitted). It is well known and easy to check that $d^{n+1}d^n =$ for all $n \geq 1$. For every $n \geq 0$ there is an isometric, but in general not surjective, G-morphism

$$A^n : \mathrm{C}^n_{\mathrm{b}}(G,E) \longrightarrow \mathrm{C}_{\mathrm{b}}(G^{n+1},E)$$

defined by

$$(A^n f)(x_0,x_1,\ldots,x_n) = \Big(\cdots\big(f(x_0)\big)(x_1)\cdots\Big)(x_n),$$

determining intertwinings $A^n\mathfrak{d}^n = d^n A^{n-1}$ as one readily checks. The relation between this *continuous homogeneous resolution* and bounded cohomology will be investigated in Section 7.4 (see Remark 7.4.9), but is not completely clear if G is not locally compact. The embedding A^n has the advantage to present $\mathfrak{d}^n$ in a notationally more explicit form if we are sparing of brackets :

$$(\mathfrak{d}^n f)(x_0)(x_1)\ldots(x_n) = \sum_{j=0}^{n}(-1)^j f(x_0)\ldots\widehat{(x_j)}\ldots(x_n),$$

for $f \in \mathrm{C}^{n-1}_{\mathrm{b}}(G,E)$.

REMARK 6.1.3. For an abstract group Γ, the situation alluded to in the preceding remark becomes completely transparent. Indeed we have $\mathrm{C}_{\mathrm{b}}(\Gamma_\delta,-) = \ell^\infty(\Gamma,-)$ and therefore the exponential law for ℓ^∞ spaces implies that A^n is an isometric G-isomorphism onto the homogeneous resolution

$$0 \longrightarrow E \longrightarrow \ell^\infty(\Gamma,E) \longrightarrow \ell^\infty(\Gamma^2,E) \longrightarrow \ell^\infty(\Gamma^3,E) \longrightarrow \cdots$$

REMARK 6.1.4. We remark at this occasion that for discrete groups one can use this resolution to define more generally the bounded cohomology of Γ with coefficients in a normed Γ-module, not necessarily a Banach space. Thus for instance the *integral bounded cohomology* $\mathrm{H}^\bullet_{\mathrm{b}}(\Gamma,\mathbf{Z})$ is by definition the cohomology of the complex of invariants

$$0 \longrightarrow \ell^\infty(\Gamma,\mathbf{Z})^\Gamma \longrightarrow \ell^\infty(\Gamma^2,\mathbf{Z})^\Gamma \longrightarrow \ell^\infty(\Gamma^3,\mathbf{Z})^\Gamma \longrightarrow \cdots$$

We shall not study this type of coefficients since (i) we do not see a natural generalization of this to topological groups and (ii) most of the powerful techniques presented in the next sections are impossible to use in this context.

Back to general topological groups, the behaviour of the continuous bounded cohomology with respect to continuity of coefficients is very simple :

PROPOSITION 6.1.5.— *Let G be a topological group and E a Banach G-module. The inclusion map $CE \to E$ induces for all $n \geq 0$ an isometric identification*

$$\mathrm{H}^n_{\mathrm{cb}}(G,CE) \cong \mathrm{H}^n_{\mathrm{cb}}(G,E).$$

This is due to the following basic observation, which we isolate for further use :

LEMMA 6.1.6.— *Let G be a topological group and E a Banach G-module. Then*

$$\mathcal{C}\mathrm{C}^n_\mathrm{b}(G,E) = \mathcal{C}\mathrm{C}_\mathrm{b}\Big(G,\mathcal{C}\mathrm{C}^{n-1}_\mathrm{b}(G,E)\Big) = \mathcal{C}\mathrm{C}^n_\mathrm{b}(G,\mathcal{C}E)$$

holds for all $n \geq 0$, with the convention $\mathrm{C}^{-1}_\mathrm{b}(G,E) = E$.

PROOF OF THE LEMMA. Apply successively Lemma 1.3.1 in an induction over n. □

PROOF OF PROPOSITION 6.1.5. Since for all n we have

$$\mathrm{C}^n_\mathrm{b}(G,E)^G \subset \mathcal{C}\mathrm{C}^n_\mathrm{b}(G,E),$$

the Lemma 6.1.6 implies that the canonical complex

$$0 \to \mathrm{C}^0_\mathrm{b}(G,E)^G \xrightarrow{\partial^1} \mathrm{C}^1_\mathrm{b}(G,E)^G \xrightarrow{\partial^2} \mathrm{C}^2_\mathrm{b}(G,E)^G \to \cdots$$

defining $\mathrm{H}^n_\mathrm{cb}(G,E)$ coincides with the canonical complex

$$0 \to \mathrm{C}^0_\mathrm{b}(G,\mathcal{C}E)^G \xrightarrow{\partial^1} \mathrm{C}^1_\mathrm{b}(G,\mathcal{C}E)^G \xrightarrow{\partial^2} \mathrm{C}^2_\mathrm{b}(G,\mathcal{C}E)^G \to \cdots$$

defining $\mathrm{H}^n_\mathrm{cb}(G,\mathcal{C}E)$. □

We beg the indulgence of the reader and present the following example for the completeness of the exposition :

EXAMPLE 6.1.7. Let $G = 1$ be the trivial group and E a Banach G-module (i.e. a Banach space). Then we have $\mathrm{H}^n_\mathrm{cb}(G,E) = 0$ for all $n \geq 1$ and $\mathrm{H}^0_\mathrm{cb}(G,E) = E$.

Indeed, the space $\mathrm{C}^n_\mathrm{b}(G,E)$ reduces to E for all $n \geq 0$, and the inductive definition of ∂^n shows that for all $n \geq 1$ we have

$$\partial^n = \begin{cases} Id & \text{if } n \text{ is even,} \\ 0 & \text{if } n \text{ is odd.} \end{cases}$$

Therefore, the canonical complex is of the form

$$0 \to E \xrightarrow{0} E \xrightarrow{=} E \xrightarrow{0} E \xrightarrow{=} \cdots$$

and one can read off that its cohomology is as claimed.

We have seen that the very definition of the continuous bounded cohomology presents $\mathrm{H}^n_\mathrm{cb}(G,E)$ as a complete semi-normed space. It may indeed happen that this space is non Hausdorff : T. Soma has shown that if Γ is a non amenable surface group, then $\mathrm{H}^3_\mathrm{b}(\Gamma)$ is non Hausdorff [**132, 133**].

6.2. Low degree, I. There is not much to say in degree zero ; writing down the canonical complex, one can read off

$$\mathrm{H}^0_\mathrm{cb}(G,E) = E^G$$

for every Banach G-module (π, E) and any G.

In order to understand degree one cocycles, we adopt a "semi-inhomogeneous" viewpoint ; inhomogeneous cocycles as such are to be introduced later, at the end of Section 7.4. Denote by A^n the evaluation at $e \in G$ of functions in $\mathrm{C}^n_\mathrm{b}(G,E)$.

Since $\mathrm{C}_{\mathrm{b}}^{n}(G,E)^{G}$ is contained in $\mathcal{C}\mathrm{C}_{\mathrm{b}}^{n}(G,E)$, the Lemma 6.1.6 implies that this defines an isometric morphism

$$A^{n}:\ \mathrm{C}_{\mathrm{b}}^{n}(G,E)^{G}\longrightarrow \mathcal{C}\mathrm{C}_{\mathrm{b}}^{n-1}(G,E)$$

(with the convention $\mathrm{C}_{\mathrm{b}}^{-1}(G,E)=E$). Actually A^{n} is an isomorphism, an inverse being given by the map B^{n} defined for $f\in \mathrm{C}_{\mathrm{b}}^{n-1}(G,E)$ by $(B^{n}f)(x)=\pi^{(n-1)}(x)f$, where $\pi^{(n-1)}$ is the representation on $\mathrm{C}_{\mathrm{b}}^{n-1}(G,E)$. If we consider now the diagram

$$\begin{array}{ccccc}
0 \longrightarrow & \mathrm{C}_{\mathrm{b}}^{0}(G,E)^{G} & \xrightarrow{\ \mathfrak{d}^{1}\ } & \mathrm{C}_{\mathrm{b}}^{1}(G,E)^{G} & \xrightarrow{\ \mathfrak{d}^{2}\ } \cdots \\
 & {\scriptstyle B^{0}}\big\Updownarrow{\scriptstyle A^{0}} & & {\scriptstyle B^{1}}\big\Updownarrow{\scriptstyle A^{1}} & \\
 & \mathcal{C}E & \longrightarrow & \mathcal{C}\mathrm{C}_{\mathrm{b}}(G,E) &
\end{array}$$

we compute that for $f\in\mathcal{C}\mathrm{C}_{\mathrm{b}}(G,E)$ the cocycle equation $\mathfrak{d}^{2}B^{1}f=0$ reads

$$f(xy)=\pi(x)f(y)+f(x) \qquad (\forall x,y\in G)$$

and similarly that f corresponds to a coboundary ($f=A^{1}\mathfrak{d}^{1}B^{0}v$) if and only if there is $v\in\mathcal{C}E$ such that

$$f(x)=\pi(x)v-v. \qquad (\forall x\in G)$$

In particular, for trivial coefficients $E=\mathbf{C}$, we see at once that

$$\mathrm{H}_{\mathrm{cb}}^{1}(G)=0, \qquad (*)$$

irrespective of G. Indeed, in this resolution, a 1-cocycle is a bounded continuous group homomorphism $G\to\mathbf{C}$, which has to be zero since $\mathbf{C}$ has no non-trivial bounded subgroups. The same applies to any trivial Banach G-module $E=E^{G}$.

However, if we consider general coefficients, the first continuous bounded cohomology may become arbitrarily complicated due to the dimension shifting technique — as we shall see later on, see Example 10.3.3.

It turns out that the observation $(*)$ can nevertheless be extended to a considerable generality. As has first been observed in his setting by B.E. Johnson (Theorem 3.4 in [**89**] or Proposition 2.7 in [**74**, chap. III]), bounded cohomology with coefficients in reflexive Banach spaces must vanish in degree one. The reason is that in these spaces bounded closed convex sets are weakly compact, allowing us to apply the Ryll-Nardzewski fixed point theorem.

A much stronger result, going beyond reflexive spaces, will be established in Section 11.4 with the help of double ergodicity.

PROPOSITION 6.2.1.— *Let G be a topological group and (π,E) a reflexive Banach G-module.*

Then $\mathrm{H}_{\mathrm{cb}}^{1}(G,E)=0$.

PROOF. By the discussion preceding the statement of the proposition, a 1-cocycle yields in particular a continuous bounded map $f:G\to E$ satisfying

$$f(xy)=\pi(x)f(y)+f(x). \qquad (x,y\in G)$$

In particular, f defines an isometric affine action $T_f : G \to \mathrm{Isom}(E)$ by

$$T_f(g)v = \pi(g)v + f(g). \qquad (g \in G, v \in E)$$

Let $K \subset E$ be the closure of the convex hull of the orbit of some point in E under this action. Since f is bounded, K is bounded hence compact in the weak topology because E is reflexive. On the other hand, K is invariant under the affine action. This is a situation covered by the Ryll-Nardzewski theorem : see the Corollaire following Théorème 2 in N. Bourbaki's treatise [**30**, IV, Appendice, N° 3]. Therefore the action T_f has a fixed point $v_0 \in E$, that is,

$$v_0 = \pi(g)v_0 + f(g). \qquad (\forall\, g \in G)$$

This realizes $f = A^1 \mathfrak{d}^1 B^0 v_0$ as an "inhomogeneous" coboundary. □

7. The functorial characterization

We attack now the functorial characterization of continuous bounded cohomology with all its apparatus. In order to derive profit from it, we give in Sections 7.4 and 7.5 a number of examples using this technology.

The machinery that we are about to introduce is admittedly heavy ; in particular, we shall pedantically switch from resolutions to continuous subresolutions, and this will introduce an additional complication for contracting homotopies, which will e.g. have to range in exotic spaces like $CL^\infty_{\mathrm{w*}}(-,-)$.

We feel that we should spend a few lines justifying our approach. Why are not all these complications already present in the theory of continuous cohomology ?

The most powerful tools for applications of the theory will be the use of amenable spaces and the corresponding L^∞ spaces, which are not continuous Banach modules. However, continuity of the action (or at least measurability, which also lacks here) is a necessary ingredient of comparison lemmata for relatively injective resolutions.

An easy way out would be to organize the notion of relative injectivity around a weaker continuity, such as weak-* continuity for L^∞ spaces or local uniform convergence for $\mathrm{C_b}$ spaces. As a matter of fact, that is exactly the strategy used in continuous cohomology.

This strategy would indeed be applicable here aswell, but it would yield an easy yet blunt theory, because the various notions of invariant means are not weak-* continuous.

Summing up, the construction presented here is the best combination that we could find to resolve the discord between these contradictory constraints.

7.1. Basic definitions. The first definitions are very straightforward indeed and do not significantly veer off the course of their algebraic counterpart. Slight divergences will appear starting with the additional restriction on contracting homotopies.

A *complex* $(E^\bullet, d^\bullet)$ *of Banach G-modules*, or *complex* for short, is a $\mathbf{Z}$-indexed sequence

$$\cdots \longrightarrow E^{n-1} \xrightarrow{d^n} E^n \xrightarrow{d^{n+1}} E^{n+1} \longrightarrow \cdots \qquad (n \in \mathbf{Z})$$

of Banach G-modules E^n and G-morphisms d^n such that $d^{n+1}d^n = 0$ for all $n \in \mathbf{Z}$. The G-morphisms d^n are called *differentials* or *coboundary maps*, and elements of E^n are referred to as *cochains* of *degree n*. A *right complex* is a complex $(E^\bullet, d^\bullet)$ such that $E^n = 0$ for all $n < 0$, and will also be considered as a $\mathbf{N}$-indexed sequence[1].

NOTATION 7.1.1. We will most of the time omit the superscript and write d for d^n. We will often drop this label altogether and content ourselves with a horizontal arrow in diagrams. Accordingly, we denote the complex $(E^\bullet, d^\bullet)$ simply by $E^\bullet$.

A complex $E^\bullet$ is said continuous, relatively injective etc. whenever all E^n $(n \in \mathbf{Z})$ share the corresponding property. We denote by $CE^\bullet$ the continuous complex

$$\cdots \longrightarrow CE^{n-1} \longrightarrow CE^n \longrightarrow CE^{n+1} \longrightarrow \cdots$$

(see Lemma 1.2.6) and by $E^{\bullet G}$ the complex

$$\cdots \longrightarrow (E^{n-1})^G \longrightarrow (E^n)^G \longrightarrow (E^{n+1})^G \longrightarrow \cdots$$

of invariants. If E is a Banach G-module, we denote by $\underline{E}^\bullet$ the complex

$$\cdots \longrightarrow 0 \longrightarrow 0 \longrightarrow E \longrightarrow 0 \longrightarrow 0 \longrightarrow \cdots$$

where E sits at the degree zero.

A *morphism of complexes* $\alpha^\bullet : E^\bullet \to F^\bullet$ is a sequence α^n $(n \in \mathbf{Z})$ of morphisms $E^n \to F^n$ such that the diagram

$$\begin{array}{ccccccccc} \cdots & \longrightarrow & E^{n-1} & \longrightarrow & E^n & \longrightarrow & E^{n+1} & \longrightarrow & \cdots \\ & & \downarrow{\scriptstyle \alpha^{n-1}} & & \downarrow{\scriptstyle \alpha^n} & & \downarrow{\scriptstyle \alpha^{n+1}} & & \\ \cdots & \longrightarrow & F^{n-1} & \longrightarrow & F^n & \longrightarrow & F^{n+1} & \longrightarrow & \cdots \end{array}$$

commutes. A *G-morphism of complexes* is a morphism of complexes consisting of G-morphisms. The identity and zero morphisms of complexes are simply denoted by Id and 0 respectively.

If $\alpha^\bullet$ and $\beta^\bullet$ are two morphisms of complexes from $(E^\bullet, d^\bullet)$ to $(F^\bullet, \partial^\bullet)$, a *homotopy* $\sigma^\bullet$ from $\alpha^\bullet$ to $\beta^\bullet$ is a sequence of morphisms $\sigma^n : E^n \to F^{n-1}$ $(n \in \mathbf{Z})$ such that one has

$$\partial^n \sigma^n + \sigma^{n+1} d^{n+1} = \beta^n - \alpha^n$$

[1]recall that in our notation $0 \in \mathbf{N}$.

for all $n \in \mathbf{Z}$, as pictured in the familiar diagram

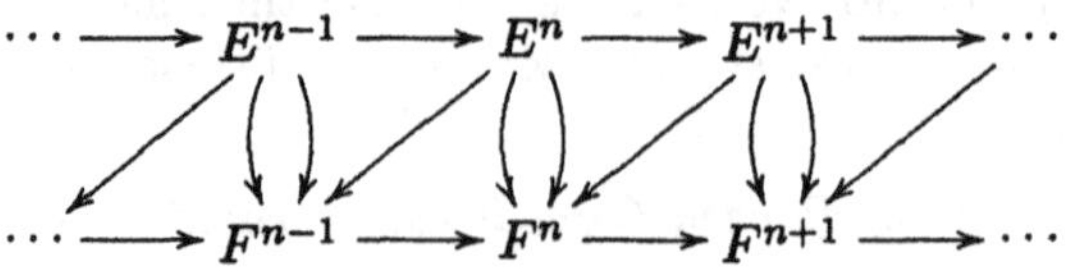

When such a homotopy exists, then $\alpha^\bullet$ is said *homotopic* to $\beta^\bullet$. This defines an equivalence relation since the definition of homotopies is additive. Notice also that by definition $\alpha^\bullet$ is homotopic to $\beta^\bullet$ if and only if the zero morphism of complexes is homotopic to $\beta^\bullet - \alpha^\bullet$. A morphism of complexes is said *null homotopic* if it is homotopic to zero.

A morphism of complexes $\alpha^\bullet : E^\bullet \to F^\bullet$ is called a *homotopy equivalence* if there is a morphism of complexes $\beta^\bullet : F^\bullet \to E^\bullet$ such that $\alpha^\bullet\beta^\bullet$ and $\beta^\bullet\alpha^\bullet$ are homotopic to the identity morphism of the complexes $E^\bullet$ and $F^\bullet$, respectively.

A *G-homotopy* is a homotopy consisting of G-morphisms. One defines accordingly *G-homotopic* G-morphisms of complexes, *null G-homotopic* G-morphisms of complexes and *G-homotopy equivalences*.

A complex $E^\bullet$ is said to admit a *contracting homotopy* $h^\bullet$ if $h^\bullet$ is a homotopy from $0 : E^\bullet \to E^\bullet$ to $Id : E^\bullet \to E^\bullet$ such that $\|h^n\| \leq 1$ for all $n \in \mathbf{Z}$.

LEMMA 7.1.2.— *If the complex* $(E^\bullet, d^\bullet)$ *admits a contracting homotopy, then for all* $n \in \mathbf{Z}$ *the G-morphism* d^n *is admissible and its image coincides with* $\mathrm{Ker}(d^{n+1})$.

PROOF. Let $h^\bullet$ be a contracting homotopy, so that in particular $d^n h^n + h^{n+1}d^{n+1} = Id$. We have

$$d^n h^n d^n = (Id - h^{n+1}d^{n+1})d^n = d^n - h^{n+1}d^{n+1}d^n = d^n,$$

so that the Definition 4.1.1 is satisfied since $\|h^n\| \leq 1$.

The inclusion $\mathrm{Im}(d^n) \subset \mathrm{Ker}(d^{n+1})$ is part of the definition of a complex. On the other hand, if $v \in \mathrm{Ker}(d^{n+1})$, we have

$$v = d^n h^n v + h^{n+1}d^{n+1}v = d^n h^n v,$$

so that $v \in \mathrm{Im}(d^n)$. □

We call a complex $E^\bullet$ *strong* if the complex $\mathcal{C}E^\bullet$ admits a contracting homotopy.

REMARK 7.1.3. Although the notion of a complex with a contracting homotopy does not depend on the group G considered, the concept of strong complex does (because $\mathcal{C}$ depends on the group). We shall however avoid bombastic expressions like "let $E^\bullet$ be a G-strong complex of Banach G-modules".

The n-th cohomology semi-normed space of a complex $(E^\bullet, d^\bullet)$ is by definition the quotient

$$\mathrm{H}^n(E^\bullet) = \mathrm{H}^n(E^\bullet, d^\bullet) = \mathrm{Ker}(d^{n+1})/\mathrm{Im}(d^n)$$

endowed with the quotient semi-norm and the corresponding locally convex not necessarily Hausdorff topology. Any morphism of complexes $\alpha^\bullet : E^\bullet \to F^\bullet$ induces a sequence $\mathrm{H}^n(\alpha^\bullet)$ ($n \in \mathbf{Z}$) of continuous linear maps

$$\mathrm{H}^n(\alpha^\bullet) : \mathrm{H}^n(E^\bullet) \longrightarrow \mathrm{H}^n(F^\bullet).$$

The morphism of complexes $\alpha^\bullet$ is called a ***homologism*** if the induced map $\mathrm{H}^n(\alpha^\bullet)$ is an isomorphism of topological vector spaces for all $n \in \mathbf{Z}$; ***G-homologisms*** are defined accordingly. (We borrow the term ***homologism*** from N. Bourbaki, Définition 3 in [**29**, X §2 N° 2]. It appears that ***quasi-isomorphism*** is a more common synonym.)

It follows from the definition of homotopies that homotopic morphisms of complexes induce identical maps in cohomology, so that in particular any homotopy equivalence is a homologism.

REMARK 7.1.4. Obseve that G-morphisms of complexes $E^\bullet \to F^\bullet$ as well as G-homotopies restrict to the continuous subcomplexes and restrict further to morphisms of complexes and homotopies $E^{\bullet G} \to F^{\bullet G}$ of the subcomplex of invariants. In particular, a G-homologism $E^\bullet \to F^\bullet$ induces a homologism $E^{\bullet G} \to F^{\bullet G}$.

DEFINITION 7.1.5. Let E be a Banach G-module. A ***resolution*** of E is a pair $(\mathfrak{a}, E^\bullet)$ where $E^\bullet$ is a right complex and $\mathfrak{a}$ a G-homologism $\mathfrak{a} : \underline{E}^\bullet \to E^\bullet$.

If any confusion is possible, we use the terminology ***G-resolution***.

Since $\mathfrak{a}$ is concentrated in degree zero, we abuse notation and denote also by $\mathfrak{a}$ the G-morphism $E \to E^0$. This G-morphism is called the ***augmentation***. We say that $(\mathfrak{a}, E^\bullet)$ is a resolution of E by continuous, relatively injective etc. Banach G-modules if the complex $E^\bullet$ has the corresponding property.

REMARK 7.1.6. The condition that $\mathfrak{a}$ be a G-homologism

$$\begin{array}{ccccccccc}
\cdots \longrightarrow & 0 & \longrightarrow & E & \cdots\cdots\to & 0 & \cdots\cdots\to & 0 & \cdots\cdots\to \cdots \\
& \vdots & & \downarrow{\scriptstyle \mathfrak{a}} & & \vdots & & \vdots & \\
\cdots\cdots\cdots\to & 0 & \cdots\cdots\to & E^0 & \longrightarrow & E^1 & \longrightarrow & E^2 & \longrightarrow \cdots
\end{array}$$

amounts to say that $E^\bullet$ is exact in all degrees $n \neq 0$ and that $\mathfrak{a}$ is a topological G-isomorphism

$$\mathfrak{a} : E \longrightarrow \mathrm{Ker}(d^1).$$

Therefore, if we "unfold" the above diagram by forgetting the dotted lines, we get a new complex which is exact everywhere. This is the object of the following definition.

Given a resolution $(\mathfrak{a}, E^\bullet)$ of the Banach G-module E, the ***augmented resolution*** is the complex

$$\cdots \longrightarrow 0 \longrightarrow E \xrightarrow{\mathfrak{a}} E^0 \xrightarrow{d^1} E^1 \xrightarrow{d^2} E^2 \longrightarrow \cdots$$

which begins at degree -1. We say that $(\mathfrak{a}, E^\bullet)$ is a ***strong resolution*** of E if the corresponding augmented resolution is a strong complex. This complex is then said to be the corresponding ***strong augmented resolution***.

If $(\mathfrak{a}, E^\bullet)$ is a strong resolution of E, then by restricting $\mathfrak{a}$ we obtain a strong resolution $(\mathfrak{a}, \mathcal{C}E^\bullet)$ of $\mathcal{C}E$.

7.2. Statement of the functorial characterization. We can now state precisely what it means that the language of relatively injective Banach G-modules, strong resolutions, etc. characterizes functorially the continuous bounded cohomology of a topological group with coefficient in a Banach G-module :

THEOREM 7.2.1.— *Let G be a topological group and E a banach G-module.*

(i) *There exists a strong resolution of E by relatively injective Banach G-modules.*

(ii) *For any strong resolution $(\mathfrak{a}, E^\bullet)$ of E by relatively injective Banach G-modules, the cohomology $\mathrm{H}^n(E^{\bullet G})$ of the complex $E^{\bullet G}$ of invariants is canonically isomorphic, as a topological vector space, to the continuous bounded cohomology $\mathrm{H}^n_{\mathrm{cb}}(G, E)$ for all $n \geq 0$.*

REMARK 7.2.2. The point (ii) of Theorem 7.2.1 states the existence of a canonical isomorphism of topological vector spaces, and not a canonical isometry of semi-normed spaces. At first sight, this may seem somewhat disappointing in view of the part c) of our desiderata listed at the beginning of Section 4. Technically, the reason of this apparent shortcoming is that in the Lemma 7.2.4 below we construct by induction morphisms α^n for which one can a priori only show

$$\|\alpha^n\| \leq \|\partial^1\| \cdot \cdots \cdot \|\partial^n\| \cdot \|\alpha\|,$$

whilst even for the canonical resolution we have in general $\|\mathfrak{d}^n\| = n + 1$. Moreover, one could influence the semi-norm of the cohomology spaces associated to a resolution by multiplying its successive terms by increasing factors.

However, all this is only seemingly a defect of the functorial approach. It turns out that the semi-norm on $\mathrm{H}^n_{\mathrm{cb}}(G, E)$ is indeed canonical and even minimal, as will be shown in Section 7.3. Even more, we shall show in Section 7.5 that the resolutions in which we are especially interested, namely resolutions on amenable G-measure spaces and proper G-topological spaces, share this minimality property.

The proof of Theorem 7.2.1 is based on the three following lemmata. The definition of the preceding section have been balanced in such a way that the proofs of Lemma 7.2.4 and 7.2.6 are formally almost identical to their classical counterpart.

LEMMA 7.2.3.— *The canonical resolution $(\epsilon, \mathrm{C}^\bullet_{\mathrm{b}}(G, E))$ is a strong resolution of E by relatively injective Banach G-modules.*

LEMMA 7.2.4.— *Let A be a continuous Banach G-module and $(\mathfrak{a}, A^\bullet)$ be a strong resolution of A by a continuous complex $(A^\bullet, d^\bullet)$. Let*

$$0 \longrightarrow B \xrightarrow{\partial^0} B^0 \xrightarrow{\partial^1} B^1 \xrightarrow{\partial^2} B^2 \xrightarrow{\partial^3} \cdots$$

be any complex beginning at degree -1 and let $\alpha : A \to B$ be a G-morphism. If B^n is relatively injective for all $n \geq 0$, then there exists a G-morphism of complexes $\alpha^\bullet$ from the augmented resolution

$$\begin{array}{ccccccccccc} 0 & \longrightarrow & A & \xrightarrow{a} & A^0 & \xrightarrow{d^1} & A^1 & \xrightarrow{d^2} & A^2 & \xrightarrow{d^3} & \cdots \\ & & \Big\downarrow{\scriptstyle \alpha^{-1}=\alpha} & & \Big\downarrow{\scriptstyle \alpha^0} & & \Big\downarrow{\scriptstyle \alpha^1} & & \Big\downarrow{\scriptstyle \alpha^2} & & \\ 0 & \longrightarrow & B & \xrightarrow[\partial^0]{} & B^0 & \xrightarrow[\partial^1]{} & B^1 & \xrightarrow[\partial^2]{} & B^2 & \xrightarrow[\partial^3]{} & \cdots \end{array}$$

such that $\alpha^{-1} = \alpha$.

DEFINITION 7.2.5. In the situation of Lemma 7.2.4, one says that the G-morphism α *extends* to a G-morphism of complexes, and $\alpha^\bullet$ is called an *extension* of α.

LEMMA 7.2.6.— *Keep the notation of Lemma 7.2.4 and Definition 7.2.5. Then any two extensions of α are G-homotopic.*

PROOF OF LEMMA 7.2.3. Let U^n be the evaluation at $e \in G$ map

$$U^n : C\mathrm{C}_\mathrm{b}^n(G,E) \longrightarrow C\mathrm{C}_\mathrm{b}^{n-1}(G,E)$$

with $\mathrm{C}_\mathrm{b}^{-1}(G,E) = E$, as at the beginning of Section 6.2 ; U^n ranges in the maximal continuous submodule because of Lemma 6.1.6. On one hand, U^n is a left inverse for the coefficient inclusion

$$\epsilon : C\mathrm{C}_\mathrm{b}^{n-1}(G,E) \longrightarrow C\mathrm{C}_\mathrm{b}^n(G,E),$$

so that in particular $U^0\epsilon = Id$. On the other hand, we have $U^1\epsilon_* = \epsilon U^0$ and more generally $U^{n+1}(\partial^n)_* = \partial^n U^n$ for all $n \geq 1$, so that

$$\epsilon U^0 + U^1\partial^1 = \epsilon U^0 + U^1(\epsilon - \epsilon_*) = U^1\epsilon = Id$$

and more generally for $n \geq 1$

$$\partial^n U^n + U^{n+1}\partial^{n+1} = \partial^n U^n + U^{n+1}(\epsilon - (\partial^n)_*) = U^{n+1}\epsilon = Id.$$

This shows that the augmented resolution

$$0 \to CE \xrightarrow{\epsilon} C\mathrm{C}_\mathrm{b}^0(G,E) \xrightarrow{\partial^1} C\mathrm{C}_\mathrm{b}^1(G,E) \xrightarrow{\partial^2} C\mathrm{C}_\mathrm{b}^2(G,E) \to \cdots$$

admits a contracting homotopy $U^\bullet$, so that the rewsolution of the statement is strong. On the other hand, the inductive definition of $\mathrm{C}_\mathrm{b}^n(G,E)$ presents this Banach module as a space of continuous functions on G, so that it is relatively injective by the case $n = 0$ of Proposition 4.4.1. □

PROOF OF LEMMA 7.2.4. The G-morphism α^n is constructed inductively on n, setting $\alpha^{-1} = \alpha$. For uniform notation, we denote by $d^\bullet$ the differential of the augmented resolution, so that $d^0 = a$ (in spite of the ambiguity arising from the fact that in the complex $A^\bullet$ itself $d^0 = 0$). Since the augmented resolution is a strong continuous complex, we may fix a contracting homotopy $(h^n)_{n\in\mathbf{Z}}$ for it (so $h^n = 0$ for $n \leq -1$). In particular,

$$h^0 d^0 = d^{-1}h^{-1} + h^0 d^0 = Id,$$

so that we have an extension problem

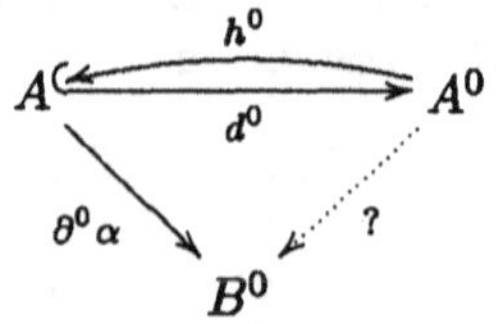

By the relative injectivity of B^0, there is a solution α^0 such that

$$\alpha^0 d^0 = \partial^0\alpha = \partial^0\alpha^{-1},$$

settling the case $n = 0$ of the induction.

Let now $n \geq 1$ and suppose that we have a collection of G-morphisms $(\alpha^k)_{k=-1}^{n-1}$ such that $\alpha^k d^k = \partial^k\alpha^{k-1}$ holds for all $0 \leq k \leq n-1$. We claim that

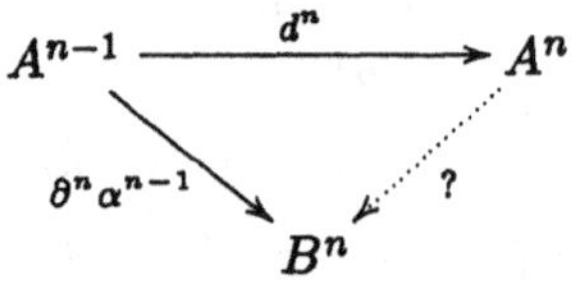

is a generalized extension problem (see Definition 4.3.3). To begin with, d^n is indeed admissible by Lemma 7.1.2. Thus we proceed to check the kernel condition of Definition 4.3.3. Let $v \in \operatorname{Ker}(d^n)$. We have

$$v = d^{n-1}h^{n-1}v + h^n d^n v = d^{n-1}h^{n-1}v,$$

so that

$$\partial^n\alpha^{n-1}v = \partial^n\alpha^{n-1}d^{n-1}h^{n-1}v$$

which by the induction hypothesis is

$$\partial^n\partial^{n-1}\alpha^{n-2}h^{n-1}v = 0,$$

so that $\operatorname{Ker}(d^n) \subset \operatorname{Ker}(\partial^n\alpha^{n-1})$ holds. We may therefore fix a solution α^n to the generalized induction problem, so that $\alpha^n d^n = \partial^n\alpha^{n-1}$, completing the proof of Lemma 7.2.4. □

Proof of Lemma 7.2.6. We keep the notation introduced in the beginning of the proof of Lemma 7.2.4. Considering the difference of any two extensions, we see that it is sufficient to prove that any extension $\alpha^\bullet$ of the zero G-morphism $\alpha = 0 = \alpha^{-1}$ is null G-homotopic. To this end, we construct inductively a sequence $(\sigma^n)_{n=-1}^{\infty}$ of G-morphisms such that

$$\partial^n\sigma^n + \sigma^{n+1}d^{n+1} = \alpha^n \qquad (*)$$

holds for all $n \geq -1$. If we set σ^{-1} and σ^0 to zero, the equation $(*)$ holds indeed for $n = -1$.

For the induction step, we let $n \geq 0$ and assume that we have a sequence $(\sigma^k)_{k=-1}^{n}$ such that $(*)$ holds for all $-1 \leq k \leq n-1$. We have to find σ^{n+1}

satisfying (*) for n, which we depict as

$$\begin{tikzcd} & A^{n-1} \arrow[r, "d^n"] \arrow[dl, "\sigma^{n-1}"'] \arrow[d, "\alpha^{n-1}"] & A^n \arrow[r, "d^{n+1}"] \arrow[dl, "\sigma^n"'] \arrow[d, "\alpha^n"] & A^{n+1} \arrow[dl, dotted, "\sigma^{n+1}"] \\ B^{n-2} \arrow[r, "\partial^{n-1}"'] & B^{n-1} \arrow[r, "\partial^n"'] & B^n & \end{tikzcd}$$

(where $B^{-1} = B$ and $B^{-2} = 0$ is understood). We claim that

$$\begin{tikzcd} A^n \arrow[rr, "d^{n+1}"] \arrow[dr, "\alpha^n - \partial^n\sigma^n"'] & & A^{n+1} \arrow[dl, dotted, "?"] \\ & B^n & \end{tikzcd}$$

is a generalized extension problem, the admissibility of d^{n+1} being granted as before by Lemma 7.1.2. Let $v \in \operatorname{Ker}(d^{n+1})$, so that $v = d^n h^n v$, and compute

$$(\alpha^n - \partial^n\sigma^n)v = (\alpha^n - \partial^n\sigma^n)d^n h^n v = \partial^n\alpha^{n-1}h^n v - \partial^n\sigma^n d^n h^n v$$

(because $\alpha^n d^n = \partial^n\alpha^{n-1}$). Since by induction $\sigma^n d^n = \alpha^{n-1} - \partial^{n-1}\sigma^{n-1}$, the second term above is

$$\partial^n(\alpha^{n-1} - \partial^{n-1}\sigma^{n-1})h^n v = \partial^n\alpha^{n-1}h^n v,$$

so that $(\alpha^n - \partial^n\sigma^n)v = 0$ and the condition of Definition 4.3.3 is satisfied. Since $n \geq 0$, the Banach G-module B^n is injective, so that there is a solution σ^{n+1} to the generalized extension problem. This completes the induction step. □

As far as the proof of Theorem 7.2.1 is concerned, we need only retain the following from Lemma 7.2.4 and Lemma 7.2.6 :

COROLLARY 7.2.7.— *Let $(\mathfrak{a}, E^\bullet)$ and $(\mathfrak{b}, F^\bullet)$ be two strong resolutions of a Banach G-module E by relatively injective Banach G-modules. Then there is a G-homotopy equivalence $CE^\bullet \to CF^\bullet$ which induces a canonical isomorphism of topological vector spaces*

$$\mathrm{H}^n(E^{\bullet G}) \cong \mathrm{H}^n(F^{\bullet G})$$

for all $n \geq 0$.

PROOF. By Lemma 7.2.4, there is a G-morphism of complexes $\alpha^\bullet$ extending the identity morphism $CE \to CE$

$$\begin{tikzcd} 0 \arrow[r] & CE \arrow[r, "\mathfrak{a}"] \arrow[d, equal, "\alpha^{-1}=Id"'] & CE^0 \arrow[r] \arrow[d, "\alpha^0"'] & CE^1 \arrow[r] \arrow[d, "\alpha^1"'] & CE^2 \arrow[r] \arrow[d, "\alpha^2"'] & \cdots \\ 0 \arrow[r] & CE \arrow[r, "\mathfrak{b}"] & CF^0 \arrow[r] & CF^1 \arrow[r] & CF^2 \arrow[r] & \cdots \end{tikzcd}$$

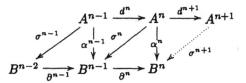

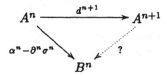

Likewise, there is a G-morphism of complexes $\beta^\bullet$ extending the identity morphism

$$\begin{array}{ccccccccc} 0 & \longrightarrow & CE & \xrightarrow{\mathfrak{b}} & CF^0 & \longrightarrow & CF^1 & \longrightarrow & CF^2 & \longrightarrow \cdots \\ & & \Big\| {\scriptstyle \beta^{-1}=Id} & & \Big\downarrow {\scriptstyle \beta^0} & & \Big\downarrow {\scriptstyle \beta^1} & & \Big\downarrow {\scriptstyle \beta^2} & \\ 0 & \longrightarrow & CE & \xrightarrow{\mathfrak{a}} & CE^0 & \longrightarrow & CE^1 & \longrightarrow & CE^2 & \longrightarrow \cdots \end{array}$$

Since both the identity morphism of complexes and $\beta^\bullet\alpha^\bullet$ extend the identity $CE \to CE$ to the augmented resolution of CE by $(\mathfrak{a}, CE^\bullet)$, the Lemma 7.2.6 implies that there are G-homotopic. In particular, $\beta^\bullet\alpha^\bullet$ restricted to the (non-augmented) complex of invariants $E^{\bullet G}$ is G-homotopic to the identity morphism of complexes and hence induces the identity $\mathrm{H}^\bullet(E^{\bullet G}) \to \mathrm{H}^\bullet(E^{\bullet G})$. Likewise, $\alpha^\bullet\beta^\bullet$ is G-homotopic to the identity so that $\alpha^\bullet$ and $\beta^\bullet$ are G-homotopy equivalences.

In particular, $\alpha^\bullet$ and $\beta^\bullet$ restrict to homotopy equivalences between $E^{\bullet G}$ and $F^{\bullet G}$ and thus induce topological isomorphisms $\mathrm{H}^n(E^{\bullet G}) \cong \mathrm{H}^n(F^{\bullet G})$. These isomorphisms are canonical because by Lemma 7.2.6 another choice of extensions $\alpha^\bullet$ and $\beta^\bullet$ would induce the same maps in cohomology. □

End of proof of Theorem 7.2.1. The Lemma 7.2.3 establishes point (i). The Corollary 7.2.7 establishes point (ii), since by Lemma 7.2.3 we may take for $(\mathfrak{b}, F^\bullet)$ the resolution $(\epsilon, \mathrm{C}^\bullet_{\mathrm{b}}(G, E))$. □

7.3. The canonical semi-norm. The following result complements the part (ii) of Theorem 7.2.1. In the case of the trivial coefficients $\mathbf{R}$ for discrete groups an analogue is due to N. Ivanov [**86**, Theorem 3.6].

Theorem 7.3.1.— *Let G be a topological group and E a banach G-module. For every strong resolution $(\mathfrak{a}, E^\bullet)$ of E by relatively injective Banach G-modules and all $n \geq 0$, the canonical isomorphism*

$$\mathrm{H}^n(E^{\bullet G}) \longrightarrow \mathrm{H}^n_{\mathrm{cb}}(G, E)$$

granted by Theorem 7.2.1 does not increase the norm.

In particular, the semi-norm on $\mathrm{H}^n_{\mathrm{cb}}(G, E)$ is canonical in asmuch as it can be recovered as the infimum of all semi-norms induced by the identifications with the cohomology of the invariants of all strong resolutions of E by relatively injective Banach G-modules.

In the next sections, we shall see other resolutions which realize this infimum.

Proof of the theorem. It is sufficient to strengthen the Lemma 7.2.4 in the following way :

Let A be a continuous Banach G-module and $(\mathfrak{a}, A^\bullet)$ a strong resolution of A by a continuous complex $(A^\bullet, d^\bullet)$. Let $\alpha : A \to B$ be a G-morphism to a Banach G-module B. Then we claim that α extends to a morphism of complexes $\alpha^\bullet$ from the augmented resolution corresponding to $(\mathfrak{a}, A^\bullet)$ to the canonical augmented resolution of B such that $\|\alpha^n\| \leq \|\alpha\|$ for all $n \geq 0$.

The only difference with the proof of Lemma 7.2.4 is in the inductive definition of α^n for $n \geq 1$. We consider the induction step

$$\begin{array}{ccccccc} \cdots \longrightarrow & A^{n-1} & \xrightarrow{d^n} & A^n & \xrightarrow{d^{n+1}} \cdots \\ & \downarrow{\scriptstyle \alpha^{n-1}} & & \vdots\, ? & \\ \cdots \longrightarrow & \mathrm{C}_{\mathrm{b}}^{n-1}(G,B) & \xrightarrow{\partial^n} & \mathrm{C}_{\mathrm{b}}^{n}(G,B) & \xrightarrow{\partial^{n+1}} \cdots \end{array}$$

Fix a contracting homotopy $h^\bullet$ for augmented resolution of A. Now for any $v \in A^n$ we define $\alpha^n v$ by

$$\alpha^n v(x) = \alpha^{n-1}\big(xh^n(x^{-1}v)\big). \qquad (x \in G)$$

Writing the left hand side as

$$\Big(\epsilon\alpha^{n-1}\big(xh^n(x^{-1}v)\big)\Big)(x_0),$$

it is apparent that this formula defines a G-morphism $\alpha^n : A^n \to \mathrm{C}_{\mathrm{b}}^n(G,B)$ exactly as did the definition of β in the proof of Proposition 4.4.1 wherein n would be zero and $E = \mathrm{C}_{\mathrm{b}}^{n-1}(G,B)$. The particularity of the present case is that

$$\begin{array}{ccc} A^{n-1} & \underset{d^n}{\overset{h^n}{\rightleftarrows}} & A^n \\ & {\scriptstyle \epsilon\alpha^{n-1}} \searrow \quad \swarrow {\scriptstyle ?} & \\ & \mathrm{C}_{\mathrm{b}}^n(G,B) & \end{array}$$

is *not* a generalized extension problem in general, since $\epsilon\alpha^{n-1}$ need not be zero on the kernel of d^n. However, due to the special nature of the canonical resolution, we can show by induction on n that $\partial^n\alpha^{n-1} = \alpha^n d^n$ holds (for $n = 0$, the definition of α^n does not differ from Lemma 7.2.4). For all $m \geq 1$, we write out ∂^m as $\partial^m = \sum_{j=0}^{m}(-1)^j\partial_j^m$, where ∂_j^m omits the jth variable as in Remark 6.1.2 (so $\partial_0^m = \epsilon$). Now for $u \in A^{n-1}$ and $x \in G$ we compute

$$\begin{aligned} \partial^n\alpha^{n-1}u(x) &= \alpha^{n-1}u + \sum_{j=1}^{n}(-1)^j\partial_j^n\alpha^{n-1}u(x) \\ &= \alpha^{n-1}u - \sum_{j=0}^{n-1}(-1)^j\partial_{j+1}^n\alpha^{n-1}u(x). \end{aligned}$$

Since for all $0 \leq j \leq n-1$ we have $\partial_{j+1}^n = (\partial_j^{n-1})_*$, the inductive definition of α^{n-1} implies

$$\partial_{j+1}^n\alpha^{n-1}u(x) = \partial_j^{n-1}\alpha^{n-2}\big(xh^{n-1}(x^{-1}u)\big),$$

so that

$$\partial^n\alpha^{n-1}u(x) = \alpha^{n-1}u - \partial^{n-1}\alpha^{n-2}\big(xh^{n-1}(x^{-1}u)\big)$$

which by the induction hypothesis is

$$\alpha^{n-1}u - \alpha^{n-1}d^{n-1}\big(xh^{n-1}(x^{-1}u)\big).$$

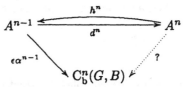

Since d^{n-1} is equivariant, we rewrite this as

$$\alpha^{n-1}(x\,Id\,x^{-1}u) - \alpha^{n-1}\big(xd^{n-1}h^{n-1}(x^{-1}u)\big) = \\ = \alpha^{n-1}\big(x(Id - d^{n-1}h^{n-1})x^{-1}(u)\big).$$

Since $d^{n-1}h^{n-1} + h^n d^n = Id$, this is

$$\alpha^{n-1}\big(xh^n d^n(x^{-1}u)\big) = \alpha^{n-1}\big(xh^n(x^{-1}d^n u)\big)$$

which is $\alpha^n d^n u$ by definition. Last, $\|\epsilon\| = 1$ implies that $\|\alpha^n\| \leq \|\alpha^{n-1}\|$. □

7.4. Examples of resolutions of Banach modules. Before we dwell further upon functoriality, we give now a few examples of strong resolutions, mostly for locally compact groups. These examples are basically a harvest of Chapter II and Sections 7.1 to 7.3.

We have divided the examples in two series : in this section we consider still general Banach modules, whilst in the next Section 7.5 we focus on coefficient modules. The latter admit much more interesting resolutions because of the connections that we have established with amenable actions.

The first and easiest example regards injective coefficients :

PROPOSITION 7.4.1.— *Let G be a topological group and E a relatively injective Banach G-module. Then*

$$\mathrm{H}^n_{\mathrm{cb}}(G, E) = 0$$

for every $n \geq 1$.

PROOF. We consider the following complex $E^\bullet$ starting at degree -1 :

$$\cdots \longrightarrow 0 \longrightarrow E \xrightarrow{=} E \longrightarrow 0 \longrightarrow 0 \longrightarrow \cdots$$

This is an augmented resolution of E by relatively injective Banach G-modules. We claim that it is strong. Indeed, the sequence $h^\bullet$ of maps

$$h^n = \begin{cases} 0 & \text{if } n < 0 \\ Id & \text{if } n = 0 \\ 0 & \text{if } n > 0 \end{cases}$$

is a contracting homotopy for $E^\bullet$; since $h^\bullet$ restricts to $\mathcal{C}E^\bullet$, the resolution $E^\bullet$ is strong. Thus Theorem 7.2.1 implies that $\mathrm{H}^\bullet_{\mathrm{cb}}(G, E)$ is realized on the non augmented complex of invariants

$$0 \longrightarrow E^G \longrightarrow 0 \longrightarrow 0 \longrightarrow \cdots$$

Now we read off that the cohomology vanishes except in degree zero. □

COROLLARY 7.4.2.— *Let G be a compact topological group. Then*

$$\mathrm{H}^n_{\mathrm{cb}}(G, E) = 0$$

for any Banach G-module E and every $n \geq 1$.

PROOF. Every Banach G-module is relatively injective by Proposition 5.1.5. □

Since almost all examples below and in the next section are variations on the concept of homogeneous cocycles, we specify once and for all the following general notation, generalizing what has been introduced in Remark 6.1.2.

NOTATION 7.4.3. If f is a map or map class of any kind (continuous, measurable, alternating) in n variables ranging over some set (topological space, measure space) then $d^n f$ denotes the map in $n+1$ variables defined by

$$(d^n f)(x_0, \ldots, x_n) = \sum_{j=0}^{n} (-1)^j f(x_0, \ldots, \widehat{x_j}, \ldots, x_n),$$

where the variable under the hat is omitted. The map d^n is called the ***homogeneous differential*** and will often be denoted simply by d or by a horizontal arrow, in accordance with Notation 7.1.1. One has always $d^{n+1}d^n = 0$. In certain computations, we shall denote by d^n_j the map corresponding to the term

$$(d^n_j f)(x_0, \ldots, x_n) = f(x_0, \ldots, \widehat{x_j}, \ldots, x_n)$$

for $0 \leq j \leq n$.

All resolutions arising in this setting with differential d will be called ***homogeneous resolutions***.

Although it is a merely formal issue, it will be useful on certain occasions to have at our disposal the possibility of considering alternating cochains. Therefore, we fix the relevant notations at the same level of generality than in Notation 7.4.3.

NOTATION 7.4.4. For $n \geq 1$, we denote by $\mathcal{S}_n$ the symmetric group on n elements and by $\text{sign} : \mathcal{S}_n \to \{\pm 1\}$ the signature character. If X is some set (topological space, measure space) then we endow tacitly X^n with the $\mathcal{S}_n$-action by factor permutation, so that the induced action on a function (or function class) f is given for $\sigma \in \mathcal{S}_n$ by

$$(\sigma^* f)(x_1, \ldots, x_n) = f(x_{\sigma^{-1}(1)}, \ldots, x_{\sigma^{-1}(n)}).$$

Now f is said *alternating* if it is equivariant for the signature character, that is, $\sigma^* f = \text{sign}(\sigma) \cdot f$ for all $\sigma \in \mathcal{S}_n$. We denote by Alt_{n-1} the familiar alternation idempotent operator defined by

$$\text{Alt}_{n-1} f = \frac{1}{n!} \sum_{\sigma \in \mathcal{S}_n} \text{sign}(\sigma) \cdot \sigma^* f.$$

It is a well known formal verification to check that the homogeneous differential takes alternating maps to alternating maps and that $\text{Alt}_n d^n = d^n \text{Alt}_{n-1}$ holds. Moreover, if X is endowed with an action of some group G, then Alt_{n-1} is G-equivariant for the diagonal action on X^n.

THEOREM 7.4.5.— *Let G be a locally compact group, X a locally compact topological space with continuous proper G-action such that $G \backslash X^{n+1}$ is paracompact for all $n \geq 0$. Let E be any Banach G-module. Then*

$$0 \longrightarrow E \stackrel{\epsilon}{\longrightarrow} \mathrm{C_b}(X, E) \longrightarrow \mathrm{C_b}(X^2, E) \longrightarrow \mathrm{C_b}(X^3, E) \longrightarrow \cdots$$

is a strong augmented resolution of E by relatively injective G-modules.

Moreover, the cohomology of the complex

$$0 \longrightarrow \mathrm{C_b}(X,E)^G \longrightarrow \mathrm{C_b}(X^2,E)^G \longrightarrow \mathrm{C_b}(X^3,E)^G \longrightarrow \cdots$$

is canonically isometrically isomorphic to $\mathrm{H}^\bullet_{\mathrm{cb}}(G,E)$.

The corresponding statements hold for the sub-resolution and subcomplex of alternating cochains.

SCHOLIUM 7.4.6. The following comment on Theorem 7.4.5 does also concern, mutatis mutandis, most statements of this Section 7.4 and the next one. Once the above statements about not necessarily alternating cochains are proved, the presence of the surjective norm one G-morphism of complexes $\mathrm{Alt}_\bullet$ is actually sufficient to deduce that the cohomology of the complex of alternating invariant cochains realizes isometrically $\mathrm{H}^\bullet_{\mathrm{cb}}(G,E)$. We repeat however the full statement for alternating cochains because it gives us a more natural and more complete statement without really adding to the length of the proofs.

PROOF OF THEOREM 7.4.5. We begin with the non alternating case. The object of Theorem 4.5.2 was to show that the Banach G-modules in the above resolution are relatively injective. Therefore, in order to conclude with the functorial characterization of Theorem 7.2.1, there are two point yet to be established : the fact that the resolution is indeed strong, and the fact that the canonical isomorphism given by point (ii) in Theorem 7.2.1 is isometric in the present situation.

We proceed first to construct a contracting homotopy $\sigma^\bullet$ for the subcomplex

$$0 \longrightarrow \mathcal{C}E \xrightarrow{\ \epsilon\ } \mathcal{C}\mathrm{C_b}(X,E) \longrightarrow \mathcal{C}\mathrm{C_b}(X^2,E) \longrightarrow \mathcal{C}\mathrm{C_b}(X^3,E) \longrightarrow \cdots$$

of maximal continuous submodules in the augmented resolution. Fix a left Haar measure m on G and let h be a generalized Bruhat function corresponding to m, as granted by Lemma 4.5.4. Fix $x_0 \in X$ and let $n \geq 0$. Given $f \in \mathcal{C}\mathrm{C_b}(X^{n+1},E)$ and $x_1,\ldots,x_n$ in X, we consider the Bochner integral

$$(\sigma^n f)(x_1,\ldots,x_n) = \int_G^{(\mathrm{B})} h(g^{-1}x_0) f(gx_0,x_1,\ldots,x_n)\, dm(g).$$

Lemma 4.5.5 applied to $Y = X^n$ implies that $\sigma^n f$ is continuous. In particular, since $\int_G h(g^{-1}x_0)\, dm(g) = 1$, we have obtained a linear map

$$\sigma^n : \mathcal{C}\mathrm{C_b}(X^{n+1},E) \longrightarrow \mathrm{C_b}(X^n,E)$$

of norm at most one. Now we claim that σ^n ranges in $\mathcal{C}\mathrm{C_b}(X^n,E)$, with the convention $\mathrm{C_b}(X^0,E) = E$. To see this, let $f \in \mathcal{C}\mathrm{C_b}(X^{n+1},E)$ and consider a net $(g_\alpha)_{\alpha\in\mathrm{A}}$ converging to $e \in G$. We have

$$\|\lambda_\pi(g_\alpha)\sigma^n f - \sigma^n f\|_\infty \leq \big\|\lambda_\pi(g_\alpha)\sigma^n f - \sigma^n \lambda_\pi(g_\alpha) f\big\|_\infty + \\ + \big\|\sigma^n\big(\lambda_\pi(g_\alpha) f - f\big)\big\|_\infty,$$

where the second term converges to zero because f is in the maximal continuous submodule. Since $\pi(g_\alpha)$ can be commuted with the Bochner integral defining

σ^n and is isometric, the first term is just

$$\begin{aligned}\left\|\lambda(g_\alpha)\sigma^n f - \sigma^n\lambda(g_\alpha)f\right\|_\infty &= \\ = \sup_{x_1,\dots,x_n\in X}\Bigg\| &\int_G^{(\mathrm{B})} h(g^{-1}x_0)f(gx_0, g_\alpha^{-1}x_1,\dots,g_\alpha^{-1}x_n)\,dm(g) \\ &- \int_G^{(\mathrm{B})} h(g^{-1}x_0)f(g_\alpha^{-1}gx_0, g_\alpha^{-1}x_1,\dots,g_\alpha^{-1}x_n)\,dm(g)\Bigg\|_\infty .\end{aligned}$$

By left invariance, the second integral is

$$\int_G^{(\mathrm{B})} h(g^{-1}g_\alpha^{-1}x_0)f(gx_0, g_\alpha^{-1}x_1,\dots,g_\alpha^{-1}x_n)\,dm(g),$$

so that we are reduced to consider

$$\|f\|_\infty \int_G \Big(h(g^{-1}x_0) - h(g^{-1}g_\alpha^{-1}x_0)\Big)\,dm(G).$$

This converges to zero because $g_\alpha^{-1}x_0$ converges to x (apply e.g. the case Y trivial and F constant of Lemma 4.5.5).

Thus it remains only to check the homotopy relations for $\sigma^\bullet$. By abuse of notation we write $d^0 = \epsilon$. On one hand, the first condition of Lemma 4.5.4 implies $\sigma^n d^n_0 = Id$ for all $n \geq 0$. In particular, we have already $\sigma^0 d^0 = Id$. On the other hand, for all $n \geq 1$ and $1 \leq j \leq n$ the relation

$$\sigma^n d^n_j = d^{n-1}_{j-1}\sigma^{n-1}$$

holds. This implies readily $d^n\sigma^n + \sigma^{n+1}d^{n+1} = Id$ for all $n \geq 0$.

We consider now our second task, the isometry question. All we have to do is to find some G-morphism of complexes $\alpha^\bullet$

$$\begin{array}{ccccccccccc} 0 & \longrightarrow & CE & \xrightarrow{\epsilon} & C\mathrm{C}^0_{\mathrm{b}}(G,E) & \xrightarrow{\partial^1} & C\mathrm{C}^1_{\mathrm{b}}(G,E) & \xrightarrow{\partial^2} & C\mathrm{C}^2_{\mathrm{b}}(G,E) & \xrightarrow{\partial^3} & \cdots \\ & & {\scriptstyle \alpha^{-1}=Id}\big\| & & \big\downarrow{\scriptstyle \alpha^0} & & \big\downarrow{\scriptstyle \alpha^1} & & \big\downarrow{\scriptstyle \alpha^2} & & \\ 0 & \longrightarrow & CE & \xrightarrow[\epsilon]{} & C\mathrm{C}_{\mathrm{b}}(X,E) & \xrightarrow[d^1]{} & C\mathrm{C}_{\mathrm{b}}(X^2,E) & \xrightarrow[d^2]{} & C\mathrm{C}_{\mathrm{b}}(X^3,E) & \xrightarrow[d^3]{} & \cdots \end{array}$$

extending the identity with $\|\alpha^n\| \leq 1$ for all n. Indeed, the map

$$\mathrm{H}^n_{\mathrm{cb}}(G,E) \longrightarrow \mathrm{H}^n\Big(\mathrm{C}_{\mathrm{b}}(X^{\bullet+1},E)^G\Big)$$

induced in cohomology by $\alpha^\bullet$ is then norm non-decreasing, and by Lemma 7.2.6 it is an inverse for the canonical norm non-decreasing isomorphism of Theorem 7.3.1.

We construct $\alpha^\bullet$ as follows. First we consider the isometric (but not necessarily surjective) G-morphism of complexes $A^\bullet$

$$\begin{array}{ccccccccccc}
0 & \longrightarrow & E & \xrightarrow{\epsilon} & \mathrm{C}^0_{\mathrm{b}}(G,E) & \xrightarrow{\partial^1} & \mathrm{C}^1_{\mathrm{b}}(G,E) & \xrightarrow{\partial^2} & \mathrm{C}^2_{\mathrm{b}}(G,E) & \xrightarrow{\partial^3} & \cdots \\
 & & \Big\| {\scriptstyle A^{-1}=Id} & & \Big\downarrow {\scriptstyle A^0} & & \Big\downarrow {\scriptstyle A^1} & & \Big\downarrow {\scriptstyle A^2} & & \\
0 & \longrightarrow & E & \xrightarrow[\epsilon]{} & \mathrm{C_b}(G,E) & \xrightarrow[d^1]{} & \mathrm{C_b}(G^2,E) & \xrightarrow[d^2]{} & \mathrm{C_b}(G^3,E) & \xrightarrow[d^3]{} & \cdots
\end{array}$$

introduced in Remark 6.1.2 (and restrict it to the subcomplex of maximal continuous submodules). We compose it then with the G-morphism of complexes $B^\bullet$ defined as follows. We retain the left Haar measure m and the generalized Bruhat function h chosen above. For $n \geq 0$, $x_0, \dots, x_n$ in X and f in $C\mathrm{C_b}(G^{n+1}, E)$ define the element $B^n f(x_0, \dots, x_n)$ of E by the Bochner integral

$$\int^{(\mathrm{B})}_{G^{n+1}} h(g_0^{-1}x_0) \cdots h(g_n^{-1}x_n) \cdot f(g_0, \dots, g_n)\, dm^{\otimes(n+1)}(g_0, \dots, g_n).$$

Applying the Lemma 4.5.5 successively to the $n+1$ integrations, we deduce as usual that $B^n f$ is continuous, and thus we get a linear map

$$B^n : C\mathrm{C_b}(G^{n+1}, E) \longrightarrow \mathrm{C_b}(X^{n+1}, E)$$

which is of norm at most one because of Lemma 4.5.4 point (i). The left invariance of m implies (using that π commutes to Bochner integration) that B^n is G-equivariant. It follows from the definition of the collection $(B^n)_{n\geq 0}$ that it is compatible with the homogeneous differentials and extends the identity.

In order to handle the alternating case, it is enough to remark that $\sigma^\bullet$ preserves the property of being alternated and that the presence of Alt_n insures, by Proposition 4.3.1, that the Banach G-module of alternating elements in $\mathrm{C_b}(X^{n+1}, E)$ is relatively injective. □

The Theorem 7.4.5 has the following important particular case :

COROLLARY 7.4.7.— *Let G be a locally compact group and E a Banach G-module. Then*

$$0 \longrightarrow E \xrightarrow{\epsilon} \mathrm{C_b}(G,E) \longrightarrow \mathrm{C_b}(G^2,E) \longrightarrow \mathrm{C_b}(G^3,E) \longrightarrow \cdots$$

is a strong augmented resolution of E by relatively injective G-modules.

Moreover, the cohomology of the complex

$$0 \longrightarrow \mathrm{C_b}(G,E)^G \longrightarrow \mathrm{C_b}(G^2,E)^G \longrightarrow \mathrm{C_b}(G^3,E)^G \longrightarrow \cdots$$

is canonically isometrically isomorphic to $\mathrm{H}^\bullet_{\mathrm{cb}}(G,E)$.

The corresponding statements hold for the sub-resolution and subcomplex of alternating cochains.

DEFINITION 7.4.8. We call the above resolution the (alternating) ***continuous homogeneous resolution.***

REMARK 7.4.9. Coming back to the Remark 6.1.2, we observe that the continuous homogeneous resolution $(\epsilon, \mathrm{C_b}(G^{\bullet+1}, E))$ is a resolution of E by relatively injective Banach G-modules for any topological group G and any Banach G-modules E. Indeed, the injectivity is proved in Proposition 4.4.1, and a contracting homotopy $V^\bullet$ of the associated augmented resolution is given by the map

$$V^n : \mathrm{C_b}(G^{n+1}, E) \longrightarrow \mathrm{C_b}(G^n, E)$$

defined by

$$V^n f(x_1, \ldots, x_n) = f(e, x_1, \ldots, x_n).$$

However, we can prove that the resolution is *strong* only for locally compact groups.

PROOF OF COROLLARY 7.4.7. This is indeed a particular case of Theorem 7.4.5. Recall from the proof of Corollary 4.5.6 that the paracompactness of the quotient $\Delta(G)\backslash G^{n+1}$ follows e.g. from the results of [**27**, III §4]. □

There is the following more general form :

COROLLARY 7.4.10.— *Let G be a locally compact group, $H < G$ a closed subgroup and $K < G$ a compact subgroup. Let (π, E) be any Banach H-module. Then*

$$0 \longrightarrow E \xrightarrow{\epsilon} \mathrm{C_b}(G/K, E) \longrightarrow \mathrm{C_b}((G/K)^2, E) \longrightarrow \mathrm{C_b}((G/K)^3, E) \longrightarrow \cdots$$

is a strong augmented resolution of E by relatively injective H-modules.

Moreover, the cohomology of the complex

$$0 \longrightarrow \mathrm{C_b}(G/K, E)^H \longrightarrow \mathrm{C_b}((G/K)^2, E)^H \longrightarrow \mathrm{C_b}((G/K)^3, E)^H \longrightarrow \cdots$$

is canonically isometrically isomorphic to $\mathrm{H}^\bullet_{\mathrm{cb}}(H, E)$.

The corresponding statements hold for the sub-resolution and subcomplex of alternating cochains.

PROOF. Again, this is a case of Theorem 7.4.5. The paracompactness issue is addressed in the proof of Corollary 4.5.7. □

We end this section by mentioning the *inhomogeneous complexes*.

DEFINITION 7.4.11. Let G be a topological group and (π, E) a Banach G-module. For all $n \geq 1$, define the *inhomogeneous differential*

$$\delta_\pi = \delta_\pi^n : \mathrm{C_b}(G^{n-1}, \mathcal{C}E) \longrightarrow \mathrm{C_b}(G^n, \mathcal{C}E)$$

for $f \in \mathrm{C_b}(G^{n-1}, \mathcal{C}E)$ by the formula

$$\begin{aligned}\delta_\pi^n f(x_1, \ldots, x_n) = \pi(x_1) f(x_2, \ldots, x_n) + \\ + \sum_{j=1}^{n-1} (-1)^j f(x_1, \ldots, x_j x_{j+1}, \ldots, x_n) + \\ + (-1)^n f(x_1, \ldots, x_{n-1})\end{aligned}$$

($\mathrm{C_b}(G^0, \mathcal{C}E) = \mathcal{C}E$ is understood). One checks readily $\delta_\pi^{n+1}\delta_\pi^n = 0$. The complex

$$0 \to \mathcal{C}E \xrightarrow{\delta_\pi} \mathrm{C_b}(G, \mathcal{C}E) \xrightarrow{\delta_\pi} \mathrm{C_b}(G^2, \mathcal{C}E) \xrightarrow{\delta_\pi} \mathrm{C_b}(G^3, \mathcal{C}E) \to \cdots$$

is called the *continuous inhomogeneous complex*.

PROPOSITION 7.4.12.— *Let G be a locally compact group and (π, E) a Banach G-module.*

Then the cohomology of the continuous inhomogeneous complex is isometrically isomorphic to $\mathrm{H}^\bullet_{\mathrm{cb}}(G, E)$.

PROOF. Let $n \geq 0$. The isometric embedding

$$V^n : \mathrm{C_b}(G^{n+1}, E)^G \longrightarrow \mathrm{C_b}(G^n, E)$$

defined for $f \in \mathrm{C_b}(G^{n+1}, E)^G$ by

$$V^n f(x_1, \ldots, x_n) = f(e, x_1, x_1 x_2, \ldots, x_1 \cdots x_n)$$

ranges in $\mathrm{C_b}(G^n, \mathcal{C}E)$ by Lemma 1.3.1. On the latter space one can define an inverse

$$U^n : \mathrm{C_b}(G^n, \mathcal{C}E) \longrightarrow \mathrm{C_b}(G^{n+1}, E)^G$$

by letting for $f \in \mathrm{C_b}(G^n, \mathcal{C}E)$

$$U^n f(x_0, \ldots, x_n) = \pi(x_0) f(x_0^{-1} x_1, x_1^{-1} x_2, \ldots, x_{n-1}^{-1} x_n) ;$$

in the case $n = 0$ this reduces for $f \in \mathcal{C}\mathrm{C_b}(G^0, \mathcal{C}E) = \mathcal{C}E$ to $U^0 f(x_0) = \pi(x_0) f$. We obtain thus isometric isomorphisms intertwining δ_π with the homogeneous differential d. This entails the statement of the theorem since by the Corollary 7.4.7 the cohomology of the complex

$$0 \to \mathrm{C_b}(G, E)^G \xrightarrow{d} \mathrm{C_b}(G^2, E)^G \xrightarrow{d} \mathrm{C_b}(G^3, E)^G \to \cdots$$

is isometrically isomorphic to $\mathrm{H}^\bullet_{\mathrm{cb}}(G, E)$. □

7.5. Examples of resolutions of coefficient modules. From our viewpoint, the most important class of resolutions is the family of homogeneous resolutions on amenable regular G-spaces.

We keep the notations introduced in Notation 7.4.3 for the homogeneous differentials. In case S is a regular G-space and E a coefficient G-module, we denote by $L^\infty_{\mathrm{w*,alt}}(S^n, E)$ the Banach G-module of alternating elements of $L^\infty_{\mathrm{w*}}(S^n, E)$ (so $L^\infty_{\mathrm{w*,alt}}(S, E) = L^\infty_{\mathrm{w*}}(S, E)$). As the S_n-action is adjoint, $L^\infty_{\mathrm{w*,alt}}(S^n, E)$ is weak-* closed in $L^\infty_{\mathrm{w*}}(S^n, E)$, hence is again a coefficient G-module. (If $E^\flat$ is the chosen predual of E, the duality of Proposition 2.3.1 identifies $L^\infty_{\mathrm{w*,alt}}(S^n, E)$ as the dual of $\bigwedge^n L^1(S) \widehat{\otimes} E^\flat$, where $\bigwedge^n$ denotes the exterior product of Banach spaces, on which informations can be found e.g. in [**43**].)

The first example is the following.

PROPOSITION 7.5.1.— *Let G be a locally compact second countable group and E a coefficient G-module. Then*

$$0 \longrightarrow E \xrightarrow{\epsilon} L^\infty_{\mathrm{w*}}(G, E) \longrightarrow L^\infty_{\mathrm{w*}}(G^2, E) \longrightarrow L^\infty_{\mathrm{w*}}(G^3, E) \longrightarrow \cdots$$

is a strong augmented resolution of E by relatively injective G-modules.

Moreover, the cohomology of the complex

$$0 \longrightarrow L^\infty_{\mathrm{w*}}(G,E)^G \longrightarrow L^\infty_{\mathrm{w*}}(G^2,E)^G \longrightarrow L^\infty_{\mathrm{w*}}(G^3,E)^G \longrightarrow \cdots$$

is canonically isometrically isomorphic to $\mathrm{H}^\bullet_{\mathrm{cb}}(G,E)$.

The corresponding statements hold for the sub-resolution and subcomplex of alternating cochains.

DEFINITION 7.5.2. We call the above resolution the (alternating) L^∞ *homogeneous resolution*.

More generally, we shall prove the following theorem, which is the main result of the section :

THEOREM 7.5.3.— *Let G be a locally compact second countable group, S an amenable regular G-space and E a coefficient G-module. Then*

$$0 \longrightarrow E \xrightarrow{\epsilon} L^\infty_{\mathrm{w*}}(S,E) \longrightarrow L^\infty_{\mathrm{w*}}(S^2,E) \longrightarrow L^\infty_{\mathrm{w*}}(S^3,E) \longrightarrow \cdots$$

is a strong augmented resolution of E by relatively injective G-modules.

Moreover, the cohomology of the complex

$$0 \longrightarrow L^\infty_{\mathrm{w*}}(S,E)^G \longrightarrow L^\infty_{\mathrm{w*}}(S^2,E)^G \longrightarrow L^\infty_{\mathrm{w*}}(S^3,E)^G \longrightarrow \cdots$$

is canonically isometrically isomorphic to $\mathrm{H}^\bullet_{\mathrm{cb}}(G,E)$.

The corresponding statements hold for the sub-resolution and subcomplex of alternating cochains.

REMARK 7.5.4. The differentials in the above resolution are adjoint maps (equivalently, they are weak-* continuous). Indeed, recall that the differential

$$d^n : L^\infty_{\mathrm{w*}}(S^n,E) \longrightarrow L^\infty_{\mathrm{w*}}(S^{n+1},E)$$

is given by $d^n = \sum_{j=0}^n (-1)^j d^n_j$ (Notation 7.4.3). If μ is a measure on S as in Definition 2.1.1, $E^\flat$ the fixed predual of E and $\langle\cdot|\cdot\rangle$ the corresponding pairing on $E \times E^\flat$, we recall (Proposition 2.3.1) that a predual of $L^\infty_{\mathrm{w*}}(S^n,E)$ is given by

$$L^1(\mu)\widehat{\otimes}\cdots\widehat{\otimes}L^1(\mu)\widehat{\otimes}E^\flat$$

with pairing $\langle\cdot|\cdot\rangle_n$ given by integration of $\langle\cdot|\cdot\rangle$ against $\mu\otimes\cdots\otimes\mu$. Therefore, the function

$$D^n_j : L^1(\mu)^{\otimes(n+1)} \otimes E^\flat \longrightarrow L^1(\mu)^{\otimes n} \otimes E^\flat$$

given by

$$D^n_j(\mu(\psi_j)\cdot\psi_0\otimes\cdots\otimes\psi_n\otimes u) = \mu(\psi_j)\cdot\psi_0\otimes\cdots\widehat{\psi_j}\cdots\otimes\psi_n\otimes u$$

satisfies for all $f \in L^\infty_{\mathrm{w*}}(S^n,E)$ and $\psi \in L^1(\mu)^{\otimes(n+1)} \otimes E^\flat$ the relation

$$\langle d^n_j f|\psi\rangle_{n+1} = \langle f|D^n_j(\psi)\rangle_n.$$

Since moreover D^n_j is continuous (of norm one), it extends to the completion for the projective norm and hence yields a predual for d^n_j.

The following fact is needed for the proof of the above proposition and theorem :

LEMMA 7.5.5.— *Let G be a locally compact second countable group, S a regular G-space and E a coefficient G-module. Then the complexes*

$$0 \to E \xrightarrow{\epsilon} L^\infty_{\mathrm{w*}}(S,E) \xrightarrow{d} L^\infty_{\mathrm{w*}}(S^2,E) \xrightarrow{d} L^\infty_{\mathrm{w*}}(S^3,E) \xrightarrow{d} \cdots$$

and

$$0 \to E \xrightarrow{\epsilon} L^\infty_{\mathrm{w*,alt}}(S,E) \xrightarrow{d} L^\infty_{\mathrm{w*,alt}}(S^2,E) \xrightarrow{d} L^\infty_{\mathrm{w*,alt}}(S^3,E) \xrightarrow{d} \cdots$$

are strong.

PROOF OF THE LEMMA. For each $n \geq 0$, a successive application of Corollary 2.3.3 implies that $L^\infty_{\mathrm{w*}}(S^{n+1},E)$ identifies canonically with the coefficient G-module L^n_E defined inductively by

$$L^n_E = L^\infty_{\mathrm{w*}}(S, L^{n-1}_E)$$

starting with $L^{-1}_E = E$. Under this identification, the term d^n_0 corresponds to the coefficient inclusion

$$\epsilon:\ L^{n-1}_E \longrightarrow L^n_E.$$

Now the Proposition 5.5.1 applied to L^{n-1}_E instead of E states that there is a morphism of Banach spaces

$$\sigma^n :\ CL^n_E \longrightarrow CL^{n-1}_E$$

of norm at most one such that $\sigma^n\epsilon = Id$. We abuse notation and denote still by σ^n the corresponding morphism

$$\sigma^n :\ CL^\infty_{\mathrm{w*}}(S^{n+1},E) \longrightarrow CL^\infty_{\mathrm{w*}}(S^n,E)$$

for $n \geq 1$ and

$$\sigma^0 :\ CL^\infty_{\mathrm{w*}}(S,E) \longrightarrow CE.$$

Now we have the relation

$$\sigma^n d^n_j = d^{n-1}_{j-1}\sigma^{n-1}$$

for all $n \geq 1$ and $1 \leq j \leq n$, which together with $\sigma^n d^n_0 = Id$ implies readily $d^n\sigma^n + \sigma^{n+1}d^{n+1} = Id$ for all $n \geq 0$. Thus σ^n is a contracting homotopy for the complex

$$0 \to CE \xrightarrow{\epsilon} CL^\infty_{\mathrm{w*}}(S,E) \longrightarrow CL^\infty_{\mathrm{w*}}(S^2,E) \longrightarrow CL^\infty_{\mathrm{w*}}(S^3,E) \longrightarrow \cdots,$$

so that the first complex considered in the lemma is indeed strong.

As $\sigma^\bullet$ preserves the property of being alternated, we conclude that the second complex of the statement is also strong. □

PROOF OF PROPOSITION 7.5.1. The augmented resolution

$$0 \longrightarrow E \xrightarrow{\epsilon} L^\infty_{\mathrm{w*}}(G,E) \longrightarrow L^\infty_{\mathrm{w*}}(G^2,E) \longrightarrow L^\infty_{\mathrm{w*}}(G^3,E) \longrightarrow \cdots$$

is strong by Lemma 7.5.5, and the Banach G-modules $L^\infty_{\mathrm{w*}}(G^{n+1},E)$ are relatively injective for all $n \geq 0$ by the Corollary 4.4.5. Therefore, by Theorem 7.2.1, there is a canonical topological isomorphism

$$\mathrm{H}^n(E^{\bullet G}) \longrightarrow \mathrm{H}^n_{\mathrm{cb}}(G,E)$$

and by Theorem 7.3.1 this isomorphism does not increase the semi-norm. It is therefore sufficient to find any G-morphism of complexes $\alpha^\bullet$

$$\begin{array}{ccccccccc} 0 \longrightarrow & E & \xrightarrow{\epsilon} & \mathrm{C}_\mathrm{b}^0(G,E) & \xrightarrow{\partial^1} & \mathrm{C}_\mathrm{b}^1(G,E) & \xrightarrow{\partial^2} & \mathrm{C}_\mathrm{b}^2(G,E) & \xrightarrow{\partial^3} \cdots \\ & \| {\scriptstyle \alpha^{-1}=Id} & & \downarrow {\scriptstyle \alpha^0} & & \downarrow {\scriptstyle \alpha^1} & & \downarrow {\scriptstyle \alpha^2} & \\ 0 \longrightarrow & E & \xrightarrow[\epsilon]{} & L^\infty_{\mathrm{w}*}(G,E) & \xrightarrow[d^1]{} & L^\infty_{\mathrm{w}*}(G^2,E) & \xrightarrow[d^2]{} & L^\infty_{\mathrm{w}*}(G^3,E) & \xrightarrow[d^3]{} \cdots \end{array}$$

extending the identity and satisfying $\|\alpha^n\| \leq 1$ for all n in order to conclude that the above isomorphism is isometric (exactly as in the proof of Theorem 7.4.5). Such a morphism is given by the composition of the G-equivariant isometric embeddings

$$A^n : \mathrm{C}_\mathrm{b}^n(G,E) \longrightarrow \mathrm{C}_\mathrm{b}(G^{n+1},E)$$

of Remark 6.1.2 with the mere inclusions

$$\mathrm{C}_\mathrm{b}(G^{n+1},E) \longrightarrow L^\infty_{\mathrm{w}*}(G^{n+1},E).$$

As for alternating cochains, the coefficient G-modules $L^\infty_{\mathrm{w}*,\mathrm{alt}}(G^{n+1},E)$ are relatively injective by an application of Proposition 4.3.1, so that $\mathrm{Alt}_\bullet$ is a norm one G-homotopy equivalence. □

In order to handle the isometry statement of the above theorem, we prove a stronger version of the Lemma 7.2.3 valid for resolutions on regular G-spaces :

LEMMA 7.5.6.— *Let G be a locally compact second countable group, T and S regular G-spaces and E a coefficient G-module.*

If there is a norm one G-morphism $\mathfrak{m} : L^\infty(T) \to L^\infty(S)$ such that $\mathfrak{m}(\mathbf{1}_T) = \mathbf{1}_S$, then there is a G-morphism of complexes $\mathfrak{m}_E^\bullet$ extending the identity

$$\begin{array}{ccccccccc} 0 \longrightarrow & E & \longrightarrow & L^\infty_{\mathrm{w}*}(T,E) & \longrightarrow & L^\infty_{\mathrm{w}*}(T^2,E) & \longrightarrow & L^\infty_{\mathrm{w}*}(T^3,E) & \longrightarrow \cdots \\ & \| {\scriptstyle \mathfrak{m}_E^{-1}=Id} & & \downarrow {\scriptstyle \mathfrak{m}_E^0} & & \downarrow {\scriptstyle \mathfrak{m}_E^1} & & \downarrow {\scriptstyle \mathfrak{m}_E^2} & \\ 0 \longrightarrow & E & \longrightarrow & L^\infty_{\mathrm{w}*}(S,E) & \longrightarrow & L^\infty_{\mathrm{w}*}(S^2,E) & \longrightarrow & L^\infty_{\mathrm{w}*}(S^3,E) & \longrightarrow \cdots \end{array}$$

with all $\mathfrak{m}_E^n$ of norm at most one.

PROOF OF THE LEMMA. Choose measures μ, ν on T, S as in Definition 2.1.1 and consider the corresponding canonical isometric G-equivariant identifications

$$\mathcal{L}(L^\infty(T), L^\infty(S)) \cong \left(L^\infty(T)\widehat{\otimes}L^1(\nu)\right)^\sharp \cong \mathcal{L}(L^1(\nu), L^\infty(T)^\sharp).$$

Denote by $\underline{\mathfrak{m}}$ the invariant element of the closed unit ball in the right hand side which corresponds to $\mathfrak{m}$. One can fix a directed set A such that for each $\varphi \in L^1(\nu)$ there is a net $(M^0_\alpha(\varphi))_{\alpha\in\mathrm{A}}$ in $L^1(\mu)$ converging weak-* in its bidual $L^\infty(T)^\sharp$ to $\underline{\mathfrak{m}}^0(\varphi)$. Moreover, we may suppose $\mu(M^0_\alpha(\varphi)) = \nu(\varphi)$ since $\mathfrak{m}^0(\mathbf{1}_T) = \mathbf{1}_S$. Let $n \geq 0$ and write $C_{n,E}$ for the closed unit ball of $L^\infty_{\mathrm{w}*}(T^{n+1},E)$ endowed with the weak-* topology corresponding to integration in $E^\flat$. The product space

$$C = \prod_{n=0}^{\infty} \prod_{\substack{\varphi_j \in L^1(\nu) \\ 0\leq j\leq n}} \prod_{v\in E^\flat} C_{n,E}$$

is compact by the theorems of Banach-Alaoğlu and Tychonoff. We define a net $(M^n_{E,\alpha})_{\alpha\in A}$ in C by assigning to $M^n_{E,\alpha}(\varphi_0,\dots,\varphi_n;v)$ the image of

$$M^0_\alpha(\varphi_0)\otimes\cdots\otimes M^0_\alpha(\varphi_n)\otimes v\ \in L^1(\mu^{\otimes(n+1)},E^\flat)$$

under the canonical embedding into the bidual. By compactness of C, there is an accumulation point $(\underline{\mathfrak{m}}^n_E)_{n=0}^\infty$, which must be linear in v and the φ_j. Therefore, we view it as simultaneous weak-* accumulation points $\underline{\mathfrak{m}}^n_E$ of nets $(M^n_{E,\alpha})_{\alpha\in A}$ in

$$\mathcal{L}\Big(L^1(\nu)\widehat{\otimes}\cdots\widehat{\otimes}L^1(\nu)\widehat{\otimes}E^\flat, L^\infty_{\mathrm{w*}}(T^{n+1},E)^\sharp\Big).$$

We claim that the maps $\mathfrak{m}^n_E$ corresponding to $\underline{\mathfrak{m}}^n_E$ under the identification of the latter space with

$$\mathcal{L}\Big(L^\infty_{\mathrm{w*}}(T^{n+1},E), L^\infty_{\mathrm{w*}}(S^{n+1},E)\Big)$$

have all required properties. The only point that is not an immediate consequence of the weak-* continuity of the G-module structures is that the coboundaries intertwine $\mathfrak{m}^n_E$ with $\mathfrak{m}^{n-1}_E$. We shall actually show that each summand d^n_j of the coboundary d^n (see Notation 7.4.3) intertwines them. Under the above identification, this reduces to show that for every $\varphi\in L^1(\nu)$, $\psi\in L^1(\nu^{\otimes n},E^\flat)$ and $\chi\in L^\infty_{\mathrm{w*}}(T^n,E)$ the relation

$$\underline{\mathfrak{m}}^n_E(\varphi\otimes\psi)(\mathbf{1}_T\otimes\chi)=\langle\mathbf{1}_S|\varphi\rangle\underline{\mathfrak{m}}^{n-1}_E(\psi)(\chi)\tag{$*$}$$

holds. Indeed, the standard coboundary map is but an alternating sum of various tensorisations against $\mathbf{1}$, and our definition of $\mathfrak{m}^n_E$ is compatible with permutation of the factors. We conclude the proof with the remark that $(*)$ follows from

$$M^0_\alpha(\varphi)(\mathbf{1}_T)=\mu(M^0_\alpha(\varphi))=\nu(\varphi)=\langle\mathbf{1}_S|\varphi\rangle.$$

□

PROOF OF THEOREM 7.5.3. By Lemma 7.5.5, the augmented resolution

$$0\longrightarrow E\overset{\epsilon}{\longrightarrow}L^\infty_{\mathrm{w*}}(S,E)\longrightarrow L^\infty_{\mathrm{w*}}(S^2,E)\longrightarrow L^\infty_{\mathrm{w*}}(S^3,E)\longrightarrow\cdots$$

is strong. Moreover, the characterization of amenable actions given in Theorem 5.7.1 implies that the Banach G-modules $L^\infty_{\mathrm{w*}}(S^{n+1},E)$ are relatively injective for all $n\geq 0$. In particular, there is a canonical topological isomorphism

$$\mathrm{H}^n\Big(L^\infty_{\mathrm{w*}}(S^{\bullet+1},E)^G\Big)\cong\mathrm{H}^n\Big(L^\infty_{\mathrm{w*}}(G^{\bullet+1},E)^G\Big)\cong\mathrm{H}^n_{\mathrm{cb}}(G,E)$$

We claim now that Lemma 7.5.6 applies to $T=G$. Indeed, denoting by $\mathfrak{m}$ the composition of the canonical inclusion

$$L^\infty(G)\longrightarrow L^\infty(G\times S)$$

with a G-morphism of norm one as in the criterion of S. Adams stated in Theorem 5.3.2

$$L^\infty(G\times S)\longrightarrow L^\infty(S),$$

we have $\mathfrak{m}(\mathbf{1}_G)=\mathbf{1}_S$ as required. Therefore, the Lemma 7.5.6 implies that for all $n\geq 0$ the canonical topological isomorphism

$$\mathrm{H}^n\Big(L^\infty_{\mathrm{w*}}(G^{\bullet+1},E)^G\Big)\longrightarrow\mathrm{H}^n\Big(L^\infty_{\mathrm{w*}}(S^{\bullet+1},E)^G\Big)$$

does not increase the norm. But we know from Proposition 7.5.1 that the left hand side carries already the canonical semi-norm, which is minimal by Theorem 7.3.1. Therefore the semi-norm on the right hand side is also the minimal, hence canonical, semi-norm.

As for alternating cochains, $\mathrm{Alt}_\bullet$ is again a norm one G-homotopy equivalence as in the previous proofs. □

EXAMPLE 7.5.7. Let Γ be a Gromov-hyperbolic group and endow its boundary at infinity $\partial_\infty\Gamma$ with the class of a quasi-invariant measure. Then, applying S. Adams' result [**1**] (see the last point in Examples 5.4.1 above), the Γ-action on $\partial_\infty\Gamma$ is amenable. Thus we deduce from Theorem 7.5.3 that for any coefficient Γ-module E we have a strong augmented resolution of E by relatively injective Γ-modules

$$0 \longrightarrow E \stackrel{\epsilon}{\longrightarrow} L^\infty_{\mathrm{w*}}(\partial_\infty\Gamma, E) \longrightarrow L^\infty_{\mathrm{w*}}((\partial_\infty\Gamma)^2, E) \longrightarrow L^\infty_{\mathrm{w*}}((\partial_\infty\Gamma)^3, E) \longrightarrow \cdots$$

and that the cohomology of the complex

$$0 \longrightarrow L^\infty_{\mathrm{w*}}(\partial_\infty\Gamma, E)^\Gamma \longrightarrow L^\infty_{\mathrm{w*}}((\partial_\infty\Gamma)^2, E)^\Gamma \longrightarrow L^\infty_{\mathrm{w*}}((\partial_\infty\Gamma)^3, E)^\Gamma \longrightarrow \cdots$$

is canonically isometrically isomorphic to $\mathrm{H}^\bullet_{\mathrm{cb}}(\Gamma, E)$. One may aswell consider alternating cochains.

EXAMPLE 7.5.8. Let G be a locally compact second countable group and μ an étalée measure on G (see point (iv) in Examples 5.4.1 above). If B is the corresponding Poisson boundary, then Theorem 7.5.3 applies and in particular the complex

$$0 \longrightarrow L^\infty_{\mathrm{w*}}(B, E)^G \longrightarrow L^\infty_{\mathrm{w*}}(B^2, E)^G \longrightarrow L^\infty_{\mathrm{w*}}(B^3, E)^G \longrightarrow \cdots$$

realises canonically isometrically $\mathrm{H}^\bullet_{\mathrm{cb}}(G, E)$. The Poisson transform yields equivariant isometric isomorphisms

$$L^\infty_{\mathrm{w*}}(B^{n+1}, E) \cong L^\infty_{\mathrm{w*}}(G^{n+1}, E)_\mu,$$

wherein the right hand side is the space of bounded μ-pluriharmonic functions. Moreover, the Poisson transform intertwines the differential d. Thus the complex

$$0 \longrightarrow L^\infty_{\mathrm{w*}}(G, E)^G_\mu \longrightarrow L^\infty_{\mathrm{w*}}(G^2, E)^G_\mu \longrightarrow L^\infty_{\mathrm{w*}}(G^3, E)^G_\mu \longrightarrow \cdots$$

also realises canonically isometrically $\mathrm{H}^\bullet_{\mathrm{cb}}(G, E)$. Again, one may aswell consider alternating cochains.

The Theorem 7.5.3 has the following important particular case.

COROLLARY 7.5.9.— *Let G be a locally compact second countable group, let $G', H < G$ be closed subgroups and E a coefficient G'-module. If H is amenable, then*

$$0 \longrightarrow E \stackrel{\epsilon}{\longrightarrow} L^\infty_{\mathrm{w*}}(G/H, E) \longrightarrow L^\infty_{\mathrm{w*}}((G/H)^2, E) \longrightarrow L^\infty_{\mathrm{w*}}((G/H)^3, E) \longrightarrow \cdots$$

is a strong augmented resolution of E by relatively injective G'-modules and the cohomology of the complex

$$0 \longrightarrow L^\infty_{\mathrm{w*}}(G/H, E)^{G'} \longrightarrow L^\infty_{\mathrm{w*}}((G/H)^2, E)^{G'} \longrightarrow L^\infty_{\mathrm{w*}}((G/H)^3, E)^{G'} \longrightarrow \cdots$$

is canonically isometrically isomorphic to $\mathrm{H}^\bullet_{\mathrm{cb}}(G', E)$.

The corresponding statements hold for the subresolution and subcomplex of alternating cochains.

PROOF. According to the first item of the Examples 5.4.1, the G-action on G/H is amenable. Hence the G'-action is amenable aswell (Lemma 5.4.3). Therefore we may apply Theorem 7.5.3 to G'. □

In case the amenable subgroup is normal, we deduce

COROLLARY 7.5.10.— *Let G be a locally compact second countable group and $N \lhd G$ a closed normal subgroup. If N is amenable, then there is a canonical isometric isomorphism*

$$\mathrm{H}^n_{\mathrm{cb}}(G, E) \cong \mathrm{H}^n_{\mathrm{cb}}(G/N, E^N)$$

for every coefficient G-module E and all $n \geq 0$.

This canonical isometric isomorphism is an instance of the *inflation* map which is to be introduced in Section 8.5 (Definition 8.5.1).

PROOF. In the case $H = N$ normal in G, the complex of invariants of the Corollary 7.5.9 reduces to the terms

$$L^\infty_{\mathrm{w*}}\big((G/N)^{n+1}, E\big)^G = L^\infty_{\mathrm{w*}}\big((G/N)^{n+1}, E^N\big)^G.$$

Since the N-action on $L^\infty_{\mathrm{w*}}\big((G/N)^{n+1}, E^N\big)$ is trivial, one can also write the above invariants as $L^\infty_{\mathrm{w*}}\big((G/N)^{n+1}, E^N\big)^{G/N}$, obtaining thus the complex of Proposition 7.5.1 for $\mathrm{H}^n_{\mathrm{cb}}(G/N, E^N)$. □

The extreme case is of course when G is itself amenable ; then the above Corolarry 7.5.10 reduces $\mathrm{H}^n_{\mathrm{cb}}(G, E)$ to the cohomology of the trivial group (with coefficients in E^G), which has been considered in Example 6.1.7. Therefore, we have the

COROLLARY 7.5.11.— *Let G be a locally compact second countable amenable group. Then*

$$\mathrm{H}^n_{\mathrm{cb}}(G, E) = 0$$

for every coefficient G-module E and all $n \geq 1$. □

Another argument leading to the above statement is the following. We know from Example 5.7.3 that any coefficient G-module is relatively injective if G is amenable ; on the other hand, the cohomology vanishes in degree $n \geq 1$ for relatively injective coefficients by Proposition 7.4.1.

For the same reason, we have

COROLLARY 7.5.12.— *Let G be a amenable topological group. Then*

$$\mathrm{H}^n_{\mathrm{cb}}(G) = 0$$

holds for all $n \geq 1$.

PROOF. The Banach G-module $\mathbf{C}$ is relatively injective by Corollary 5.1.4, so that we can apply Proposition 7.4.1. □

REMARKS 7.5.13.

(i) Our definition of coefficient modules imposes a separable pre-dual for technical reasons ; this asumption can however be dropped in the statement of Corollary 7.5.11, which then regards all modules contragredient to a continuous Banach module. The latter statement is already in B.E. Johnson's memoir [**89**], together with a converse.

(ii) For Banach G-modules that are not contragredient, however, the statement of Corollary 7.5.11 does not necessarily hold. Indeed, G.A. Noskov considers in [**117**] for every $\varrho > 0$ and $\mu \in \mathbf{R}$ the Banach $\mathbf{Z}$-module $\mathcal{A}^{\mu}_{\varrho}$ of 2π-periodic functions that are analytic in the strip $|\mathrm{im}(z)| < \varrho$ and continuous in the closure of the strip, endowed with the translation by multiples of $2\pi\mu$ and sup-norm. He shows that results of V.I. Arnold imply $\dim \mathrm{H}^1_{\mathrm{b}}(\mathbf{Z}, \mathcal{A}^{\mu}_{\varrho}) = \infty$ for $2^{\aleph_0}$ many $\mu \in \mathbf{R}$ (we read Arnold's relevant results in the translation [**4**], chap. 3 §12 ; there is an English version [**5**]).

Let us come back to homogeneous L^∞ resolutions such as given by Proposition 7.5.1 or Corollary 7.5.9. In certain situations, it happens that one can give very simple and concrete formulae for some morphism of complexes inducing the canonical isometric isomorphisms :

EXAMPLE 7.5.14. Let G be a locally compact second countable group, $\Gamma < G$ a discrete subgroup and (π, E) a coefficient Γ-module. Fix a Borelian left Γ-equivariant map $\sigma : G \to \Gamma$ (this amounts to choosing a Borelian fundamental domain for the left Γ-action on G). Precomposition by Σ gives a Γ-morphism of complexes

$$\begin{array}{ccccccccc}
0 \longrightarrow & E & \longrightarrow & \ell^\infty(\Gamma, E) & \longrightarrow & \ell^\infty(\Gamma^2, E) & \longrightarrow & \ell^\infty(\Gamma^3, E) & \longrightarrow \cdots \\
& \Big\| & & \Big\downarrow {\scriptstyle \sigma^*} & & \Big\downarrow {\scriptstyle \sigma^*} & & \Big\downarrow {\scriptstyle \sigma^*} & \\
0 \longrightarrow & E & \longrightarrow & L^\infty_{\mathrm{w*}}(G, E) & \longrightarrow & L^\infty_{\mathrm{w*}}(G^2, E) & \longrightarrow & L^\infty_{\mathrm{w*}}(G^3, E) & \longrightarrow \cdots
\end{array}$$

defined for $f \in \ell^\infty(\Gamma^{n+1}, E)$ by

$$(\sigma^* f)(g_0, \ldots, g_n) \;=\; f\big(\sigma(g_0), \ldots, \sigma(g_n)\big).$$

This is measurable and makes sense almost everywhere because Γ is discrete and countable by the second countability of G. The second line is the canonical augmented resolution for $\mathrm{H}^\bullet_{\mathrm{b}}(\Gamma, E)$ as observed in Remark 6.1.3. Now σ^* is clearly a norm one Γ-morphism of complexes and thus realizes the canonical isomorphism granted by, say, Corollary 7.5.9. We observe that in this particular case the morphism is isometric already at the cochain level

For coefficient modules, there is a L^∞ analogue of the continuous inhomogeneous complex. Indeed, if G is a locally compact second countable group and

(π, E) a coefficient G-module, then the very formula of Definition 7.4.11, namely

$$\delta_\pi^n f(x_1, \dots, x_n) = \pi(x_1) f(x_2, \dots, x_n) + \\ + \sum_{j=1}^{n-1} (-1)^j f(x_1, \dots, x_j x_{j+1}, \dots, x_n) + \\ + (-1)^n f(x_1, \dots, x_{n-1})$$

defines a map

$$\delta_\pi = \delta_\pi^n : \; L^\infty_{\mathrm{w*}}(G^{n-1}, E) \longrightarrow L^\infty_{\mathrm{w*}}(G^n, E)$$

since π is weak-* continuous. The corresponding complex

$$0 \to E \xrightarrow{\delta_\pi} L^\infty_{\mathrm{w*}}(G, E) \xrightarrow{\delta_\pi} L^\infty_{\mathrm{w*}}(G^2, E) \xrightarrow{\delta_\pi} L^\infty_{\mathrm{w*}}(G^3, E) \to \cdots$$

is called the L^∞ *inhomogeneous complex.*

PROPOSITION 7.5.15.— *Let G be a locally compact second countable group and (π, E) a coefficient G-module.*

Then the cohomology of the L^∞ inhomogeneous complex is isometrically isomorphic to $\mathrm{H}^\bullet_{\mathrm{cb}}(G, E)$.

PROOF. The proof begins in similarity with the proof of Proposition 7.4.12 ; that is, we consider the maps

$$U^n : L^\infty_{\mathrm{w*}}(G^n, E) \longrightarrow L^\infty_{\mathrm{w*}}(G^{n+1}, E)^G$$

defined for $f \in L^\infty_{\mathrm{w*}}(G^n, E)$ almost everywhere by

$$U^n f(x_0, \dots, x_n) = \pi(x_0) f(x_0^{-1} x_1, x_1^{-1} x_2, \dots, x_{n-1}^{-1} x_n),$$

where for $n = 0$ and $f \in L^\infty_{\mathrm{w*}}(S^0, E) = E$ one sets $U^0 f(x_0) = \pi(x_0) f$.

The difference is that the definition we gave for V^n does not, as it is, make sense for function *classes.* However, for $f \in L^\infty_{\mathrm{w*}}(G^{n+1}, E)^G$ we may define almost everywhere

$$W^n f(x_0, x_1, \dots, x_n) = f(x_0, x_0 x_1, \dots, x_0 x_1 \cdots x_n).$$

This yields an element $W^n f$ of $L^\infty_{\mathrm{w*}}(G^{n+1}, E)$ which is invariant under the translation action of G on the first variable, hence by Fubini's theorem we can consider $W^n F$ as an element $V^n f$ of $L^\infty_{\mathrm{w*}}(G^n, E)$. This gives an inverse for the map U^n, and we get an isometric isomorphism with the complex

$$0 \to L^\infty_{\mathrm{w*}}(G, E)^G \xrightarrow{d} L^\infty_{\mathrm{w*}}(G^2, E)^G \xrightarrow{d} L^\infty_{\mathrm{w*}}(G^3, E)^G \to \cdots$$

whose cohomology is isometrically isomorphic to $\mathrm{H}^\bullet_{\mathrm{cb}}(G, E)$ by Proposition 7.5.1. □

8. Functoriality

When it comes to functoriality, the precise book-keeping of biunivoque notational conventions can generate exaggerate loads of decorations. On the other hand, these notations are to denote very straightforward and most familiar concepts. Therefore, we adopt the following frame of abuse of notation :

NOTATION 8.0.1. We use α_* as a generic notation for a map covariantly induced in some natural and obvious way by a map α. The notation α^* refers to a contravariant situation. These shorthands do not exclude occasional use of more precise notations when required by the context.

For the sake of definiteness, we proceed for the functoriality maps as for the cohomology spaces $\mathrm{H}^\bullet_{\mathrm{cb}}$ themselves : we begin by a concrete definition and then characterize it functorially.

8.1. Covariance. Let G be a topological group and $\alpha : A \to B$ a G-morphism of Banach G-modules. By post-composition, the G-morphism α induces a G-morphism of Banach G-modules

$$\alpha_* : \mathrm{C_b}(G, A) \longrightarrow \mathrm{C_b}(G, B)$$

with same norm. Since $\mathrm{C}^n_{\mathrm{b}}(G, -) = \mathrm{C_b}(G, \mathrm{C}^{n-1}_{\mathrm{b}}(G, -))$, we obtain inductively a G-morphism of complexes

$$\begin{array}{ccccccccccc}
0 & \longrightarrow & A & \longrightarrow & \mathrm{C}^0_{\mathrm{b}}(G,A) & \longrightarrow & \mathrm{C}^1_{\mathrm{b}}(G,A) & \longrightarrow & \mathrm{C}^2_{\mathrm{b}}(G,A) & \longrightarrow & \cdots \\
 & & \downarrow{\scriptstyle \alpha} & & \downarrow & & \downarrow & & \downarrow & & \\
0 & \longrightarrow & B & \longrightarrow & \mathrm{C}^0_{\mathrm{b}}(G,B) & \longrightarrow & \mathrm{C}^1_{\mathrm{b}}(G,B) & \longrightarrow & \mathrm{C}^2_{\mathrm{b}}(G,B) & \longrightarrow & \cdots
\end{array}$$

consisting of G-morphisms of norm $\|\alpha\|$. Therefore, α induces for all $n \geq 0$ a continuous linear map

$$\mathrm{H}^n_{\mathrm{cb}}(G, \alpha) : \mathrm{H}^n_{\mathrm{cb}}(G, A) \longrightarrow \mathrm{H}^n_{\mathrm{cb}}(G, B)$$

of semi-norm at most $\|\alpha\|$. (The semi-norm of a continuous linear map between semi-normed spaces is defined exactly as the operator norm in the Hausdorff case.)

With this definition, $\mathrm{H}^n_{\mathrm{cb}}(G, \alpha)$ depends functorially on α, that is, one has obviously $\mathrm{H}^n_{\mathrm{cb}}(G, \alpha\beta) = \mathrm{H}^n_{\mathrm{cb}}(G, \alpha) \circ \mathrm{H}^n_{\mathrm{cb}}(G, \beta)$ and $\mathrm{H}^n_{\mathrm{cb}}(G, Id) = Id$. Moreover, $\mathrm{H}^n_{\mathrm{cb}}(G, \alpha)$ is additive with respect to α.

We claim now that the definition is compatible with the functorial characterization of $\mathrm{H}^n_{\mathrm{cb}}(G, -)$ in the following sense :

PROPOSITION 8.1.1.— *If $(\mathfrak{a}, A^\bullet)$ and $(\mathfrak{b}, B^\bullet)$ are strong resolutions of A respectively B by relatively injective Banach G-modules and if $\alpha : A \to B$ is a G-morphism, then there is an extension of α to a G-morphism of complexes of the corresponding augmented resolutions.*

Moreover, if $\alpha^\bullet$ is any such extension, then the topological isomorphisms

$$\mathrm{H}^n_{\mathrm{cb}}(G, A) \cong \mathrm{H}^n(A^{\bullet G}) \quad \text{and} \quad \mathrm{H}^n_{\mathrm{cb}}(G, B) \cong \mathrm{H}^n(B^{\bullet G})$$

of Theorem 7.2.1 conjugate the corresponding maps

$$\alpha_*^\bullet : \mathrm{H}^n(A^{\bullet G}) \longrightarrow \mathrm{H}^n(B^{\bullet G})$$

to $\mathrm{H}^n_{\mathrm{cb}}(G, \alpha)$.

PROOF. The existence of the extension is granted by the Lemma 7.2.4 and the conjugation statement follows from Lemma 7.2.6. □

In view of Proposition 8.1.1 and the remarks preceding it, we may consider $\mathrm{H}^\bullet_{\mathrm{cb}}(G, -)$ as a sequence of additive functors taking Banach G-modules to the inhospitable category of semi-normed spaces. The next natural question is now how much of the "triangulated" approach to usual cohomology can survive in the present setting. More specifically, we want to investigate to which extent a short exact sequence $0 \to A \to B \to C \to 0$ of Banach G-modules induces a long exact sequence in cohomology.

8.2. The long exact sequence. It turns out that the situation is the following. We can establish in Proposition 8.2.7 the fundamental long exact sequence for all topological groups G and all *weakly admissible* short exact sequences of continuous Banach G-modules (see Definition 4.2.5). This restriction is rather natural from the viewpoint of relative homological agebra.

However, thanks to an important result in functional analysis, *we can go beyond this restriction* provided the group G is locally compact (e.g. discrete). We begin with this setting, since its proof illustrates in detail the relevant techniques.

PROPOSITION 8.2.1.— *Let G be a locally compact group and let*

$$0 \to A \xrightarrow{\alpha} B \xrightarrow{\beta} C \to 0$$

be a short exact sequence of G-morphisms of Banach G-modules. Suppose either

(i) *A, B, C are continuous Banach G-modules*

or (ii) *G is second countable and $0 \to A \to B \to C \to 0$ is an adjoint sequence of coefficient G-modules.*

Then there is family of continuous maps (τ^n) so that the infinite sequence

$$\cdots \xrightarrow{\tau^n} \mathrm{H}^n_{\mathrm{cb}}(G, A) \longrightarrow \mathrm{H}^n_{\mathrm{cb}}(G, B) \longrightarrow \mathrm{H}^n_{\mathrm{cb}}(G, C) \xrightarrow{\tau^{n+1}} \mathrm{H}^{n+1}_{\mathrm{cb}}(G, A) \longrightarrow \cdots$$

is exact.

Moreover, this infinite sequence is natural in the coefficients in asmuch as the maps induced by a commutative diagram

$$\begin{array}{ccccccccc} 0 & \longrightarrow & A & \longrightarrow & B & \longrightarrow & C & \longrightarrow & 0 \\ & & \downarrow & & \downarrow & & \downarrow & & \\ 0 & \longrightarrow & A' & \longrightarrow & B' & \longrightarrow & C' & \longrightarrow & 0 \end{array}$$

with both lines as in (i) *or* (ii) *make the following infinite diagram commutative*

$$\begin{array}{ccccccccc}
\xrightarrow{\tau^n} & \mathrm{H}^n_{\mathrm{cb}}(G,A) & \longrightarrow & \mathrm{H}^n_{\mathrm{cb}}(G,B) & \longrightarrow & \mathrm{H}^n_{\mathrm{cb}}(G,C) & \xrightarrow{\tau^{n+1}} & \mathrm{H}^{n+1}_{\mathrm{cb}}(G,A) \cdots \\
& \downarrow & & \downarrow & & \downarrow & & \downarrow \\
\xrightarrow{\tau^n} & \mathrm{H}^n_{\mathrm{cb}}(G,A') & \longrightarrow & \mathrm{H}^n_{\mathrm{cb}}(G,B') & \longrightarrow & \mathrm{H}^n_{\mathrm{cb}}(G,C') & \xrightarrow{\tau^{n+1}} & \mathrm{H}^{n+1}_{\mathrm{cb}}(G,A') \cdots
\end{array}$$

DEFINITION 8.2.2. The above maps τ^n are called *transgression maps*.

REMARKS 8.2.3.

a) For the sake of readability, we have omitted to label the arrows corresponding to $\mathrm{H}^n_{\mathrm{cb}}(G,\alpha)$ and $\mathrm{H}^n_{\mathrm{cb}}(G,\beta)$. It is moreover understood that the above sequence begins with

$$0 \longrightarrow \mathrm{H}^0_{\mathrm{cb}}(G,A) \longrightarrow \mathrm{H}^0_{\mathrm{cb}}(G,B) \longrightarrow \cdots$$

b) It is an easy exercise in duality using the closed range theorem to show that an adjoint sequence $0 \to A \to B \to C \to 0$ of coefficient modules is exact if and only if the pre-dual sequence $0 \to C^\flat \to B^\flat \to A^\flat \to 0$ is exact.

The ingredient from functional analysis which is relevant to case (i) is the following application of Michael's selection theorem. Recall that if $\beta : B \to C$ is a surjective continuous linear map of Banach spaces, then β admits a continuous linear section if and only if its kernel happens to be complemented in B (see Proposition 4.2.1). However, giving up the linearity, a theorem of R.G. Bartle and L.M. Graves [**8**, Theorem 4] implies that there is always a continuous section $\sigma : C \to B$. Now Michael's selection theorem implies [**109**, Proposition 7.2] that one can moreover for any $\lambda > 1$ choose σ so as to satisfy

$$\|\sigma(c)\| \leq \lambda \inf\{\|b\| : \beta(b) = c\}$$

for all $c \in C$. Observe that contrary to the linear case, the mere existence of such a λ is not a consequence of the continuity of σ.

The selection theorem enters the proof of Proposition 8.2.1 under the following form :

LEMMA 8.2.4.— *Let* $0 \to A \xrightarrow{\alpha} B \xrightarrow{\beta} C \to 0$ *be a short exact sequence of G-morphisms of continuous Banach G-modules. Then the induced sequence*

$$0 \longrightarrow \mathrm{C_b}(G^{n+1},A)^G \xrightarrow{\alpha_*} \mathrm{C_b}(G^{n+1},B)^G \xrightarrow{\beta_*} \mathrm{C_b}(G^{n+1},C)^G \longrightarrow 0$$

is also exact for all $n \geq 0$.

PROOF OF THE LEMMA. The only non-trivial point is the surjectivity of β_*. The isomorphism

$$V = U^{-1} : \ \mathrm{C_b}(G^{n+1},-) \cong \mathrm{C_b}(G^n,\mathrm{C}-)$$

given in the proof of Proposition 7.4.12 identifies the sequence of the present lemma with

$$0 \longrightarrow \mathrm{C_b}(G^n, A) \xrightarrow{\alpha_*} \mathrm{C_b}(G^n, B) \xrightarrow{\beta_*} \mathrm{C_b}(G^n, C) \longrightarrow 0$$

because A, B, C are continuous. We still label the arrows above by α_* and β_* because the corresponding maps $V\alpha_*U$ and $V\beta_*U$ are still induced by post-composition with α and β, respectively. (For the convenience of the reader, we check this assertion, say, for β_*. Let f be in $\mathrm{C_b}(G^n, B)$. Then we compute

$$\begin{aligned}(V\beta_*Uf)(x_1, \dots, x_n) &= \beta_*Uf(e, x_1, x_1x_2, \dots, x_1 \cdots x_n)\\ &= \beta\Big(Uf(e, x_1, x_1x_2, \dots, x_1 \cdots x_n)\Big)\\ &= \beta\Big(e.f(x_1, \dots, x_n)\Big)\\ &= (\beta \circ f)(x_1, \dots, x_n),\end{aligned}$$

as claimed.)

Applying now the above quoted consequence of Michael's selection theorem, we choose some $\lambda > 1$ and fix a continuous, yet in general non-linear, map $\sigma : C \to B$ such that $\beta\sigma = Id$ and

$$\|\sigma(c)\| \leq \lambda \inf\{\|b\| : \beta(b) = c\}$$

for all $c \in C$. By the open mapping theorem [**126**, I, Corollary 2.12], the induced morphism $B/\mathrm{Ker}\beta \to C$ admits an inverse morphism of Banach spaces $\overline{\beta}^{-1}$. Now for all $f \in \mathrm{C_b}(G, C)$ the map σf is continuous, bounded by $\lambda\|\overline{\beta}^{-1}\| \cdot \|f\|_\infty$ and is a pre-image of f under β_*. □

In the setting of (ii), Michael's theorem is replaced with a specific property of L^∞ spaces in order to get the counterpart of Lemma 8.2.4 :

LEMMA 8.2.5.— *Let G be a locally compact second countable group and let $0 \to A \xrightarrow{\alpha} B \xrightarrow{\beta} C \to 0$ be an adjoint short exact sequence of G-morphisms of coefficient G-modules. Then the induced sequence*

$$0 \longrightarrow L^\infty_{\mathrm{w*}}(G^{n+1}, A)^G \xrightarrow{\alpha_*} L^\infty_{\mathrm{w*}}(G^{n+1}, B)^G \xrightarrow{\beta_*} L^\infty_{\mathrm{w*}}(G^{n+1}, C)^G \longrightarrow 0$$

is also exact for all $n \geq 0$.

PROOF. Using the isomorphisms

$$L^\infty_{\mathrm{w*}}(G^{n+1}, -)^G \cong L^\infty_{\mathrm{w*}}(G^n, -)$$

of Proposition 7.5.15, we are reduced to consider

$$0 \longrightarrow L^\infty_{\mathrm{w*}}(G^n, A) \xrightarrow{\alpha_*} L^\infty_{\mathrm{w*}}(G^n, B) \xrightarrow{\beta_*} L^\infty_{\mathrm{w*}}(G, C) \longrightarrow 0.$$

Because of the open mapping theorem, there is no loss in generality in supposing that α is actually the inclusion map of a submodule $A \subset B$, which will again be a coefficient module because by the closed range theorem [**126**, I 4.14], the space A is weak-* closed in B. For the same reason, there is no loss of generality in assuming that $C = B/A$ and that β is the quotient map. Now the exactness in the middle follows from the open mapping theorem and hence the only point left is the surjectivity of β_*. Denoting $\beta^\flat : C^\flat \to B^\flat$ the map of preduals to which β is adjoint, our assumption $C = B/A$ implies that $\beta^\flat$ is an inclusion map, so that we conclude by the injectivity of the map of Bochner-Lebesgue L^1 spaces

$$\beta^\flat_* : L^1(G^n, C^\flat) \longrightarrow L^1(G^n, B^\flat).$$

□

REMARKS 8.2.6.

(i) One one hand, the Lemma 8.2.5 established above enters the proof of Proposition 8.2.1. We shall see that on the other hand this proposition allows us in turn to generalize very easily the Lemma 8.2.5 to amenable regular G-spaces (Corollary 8.2.11).

(ii) In view of Grothendieck's isomorphism $L^1(G^n, C^\flat) \cong L^1(G)\widehat{\otimes}C^\flat$, the argument above is actually a very special case of characterization of L^1 spaces due to A. Grothendieck (Théorème 1 and Proposition 1(1) pp. 553–554 in [**73**]). He proves namely that a Banach space L is isometrically isomorphic to some $L^1(\mu)$ (where μ is a Radon measure on a locally compact space) if and only if for any closed subspace F of any Banach space E the canonical map $L\widehat{\otimes}F \to L\widehat{\otimes}E$ is an isometric embedding. In particular, if (S, μ) is a standard probability space, we deduce that the map

$$L^\infty_{\mathrm{w*}}(S, E^*) \longrightarrow L^\infty_{\mathrm{w*}}(S, F^*)$$

induced by the adjoint map $E^* \to F^*$ is surjective.

We can now complete the proof of Proposition 8.2.1.

PROOF OF PROPOSITION 8.2.1. All we have to do is to observe that the Lemmata 8.2.4 and 8.2.5 put us in position to apply a "snake lemma" to any extension of α and β, and that the thus obtained transgression map is continuous. Once this is done, the remaining verifications follow verbatim as in the usual homological setting, and therefore we will not spell them out.

So let

$$\begin{array}{ccc}
0 & & 0 \\
\downarrow & & \downarrow \\
A_1 & \xrightarrow{d_A} & A_2 \\
\downarrow{\scriptstyle \alpha_1} & & \downarrow{\scriptstyle \alpha_2} \\
B_1 & \xrightarrow{d_B} & B_2 \\
\downarrow{\scriptstyle \beta_1} & & \downarrow{\scriptstyle \beta_2} \\
C_1 & \xrightarrow{d_C} & C_2 \\
\downarrow & & \downarrow \\
0 & & 0
\end{array}$$

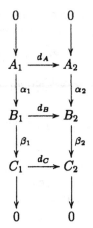

be a commutative diagram of G-morphisms of Banach G-modules with exact columns. Recall that for the standard snake lemma one starts with $c \in C_1$ such that $d_C c = 0$ and picks $b \in B_1$ with $\beta_1 b = c$. Since $\beta_2 d_B b = d_C \beta_1 b = d_C c = 0$, there is $a \in A_2$ with $\alpha_2 a = d_B b$. Then one verifies that this correspondance $c \mapsto a$ gives a well posed characterization of the transgression map.

In the setting of Proposition 8.2.1 case (i), we may compute $\mathrm{H}^\bullet_{\mathrm{cb}}(G, -)$ with the continuous homogeneous resolutions (by Corollary 7.4.7). Thus we take for the above diagram the following "slice"

$$
\begin{array}{ccc}
0 & & 0 \\
\downarrow & & \downarrow \\
\mathrm{C_b}(G^{n+1}, A)^G & \longrightarrow & \mathrm{C_b}(G^{n+2}, A)^G \\
\downarrow{\scriptstyle \alpha_*} & & \downarrow{\scriptstyle \alpha_*} \\
\mathrm{C_b}(G^{n+1}, B)^G & \longrightarrow & \mathrm{C_b}(G^{n+2}, B)^G \\
\downarrow{\scriptstyle \beta_*} & & \downarrow{\scriptstyle \beta_*} \\
\mathrm{C_b}(G^{n+1}, C)^G & \longrightarrow & \mathrm{C_b}(G^{n+2}, C)^G \\
\downarrow & & \downarrow \\
0 & & 0
\end{array}
$$

where $A_1 = \mathrm{C_b}(G^{n+1}, A)^G$, $A_2 = \mathrm{C_b}(G^{n+2}, A)^G$ etc. It is therefore the Lemma 8.2.4 that allows us to pick b with $\beta_1 b = c$. In case (ii), we may use the L^∞ homogeneous resolutions (see Proposition 7.5.1), where $A_1 = L^\infty_{\mathrm{w*}}(G^{n+1}, A)^G$ etc. so that we apply Lemma 8.2.5.

Moreover, in both cases, the open mapping theorem implies that for any $\epsilon > 0$, the element b can be chosen with

$$\|b\| \leq (\|\overline{\beta_1}^{-1}\| + \epsilon)\|c\|,$$

where $\overline{\beta_1} : B_1/\mathrm{Ker}\beta_1 \to C_1$ is the induced map. Since $\mathrm{Im}\alpha_2 = \mathrm{Ker}\beta_2$ is closed, we apply again the open mapping theorem, so that we may choose a such that

$$\|a\| \leq (\|(\alpha_2|^{\mathrm{Im}\alpha_2})^{-1}\| + \epsilon)\|b\|,$$

where $\alpha_2|^{\mathrm{Im}\alpha_2} : A_2 \to \mathrm{Im}\alpha_2$ is the co-restriction. Thus the transgression map is continuous, of seminorm at most

$$\|\overline{\beta_1}^{-1}\| \cdot \|d_B\| \cdot \|(\alpha_2|^{\mathrm{Im}\alpha_2})^{-1}\|.$$

□

As announced, the long exact sequence of Proposition 8.2.1 can be generalized to all topological groups provided the short exact sequence under consideration is weakly admissible :

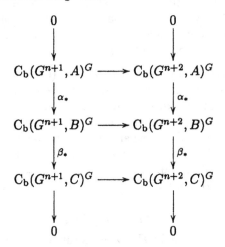

PROPOSITION 8.2.7.— *Let G be a topological group and let*

$$0 \to A \xrightarrow{\alpha} B \xrightarrow{\beta} C \to 0$$

be a weakly admissible short exact sequence of continuous Banach G-modules. Then there is family of continuous maps (τ^n) so that the infinite sequence

$$\cdots \xrightarrow{\tau^n} \mathrm{H}^n_{\mathrm{cb}}(G,A) \longrightarrow \mathrm{H}^n_{\mathrm{cb}}(G,B) \longrightarrow \mathrm{H}^n_{\mathrm{cb}}(G,C) \xrightarrow{\tau^{n+1}} \mathrm{H}^{n+1}_{\mathrm{cb}}(G,A) \longrightarrow \cdots$$

is exact.

Moreover, this infinite sequence is natural in the coefficients.

Again, the maps τ^n are called *transgression maps*, and it is understood that the above sequence begins with

$$0 \longrightarrow \mathrm{H}^0_{\mathrm{cb}}(G,A) \longrightarrow \mathrm{H}^0_{\mathrm{cb}}(G,B) \longrightarrow \cdots$$

The weak admissibility of the short exact sequence is exploited as follows :

LEMMA 8.2.8.— *Let G be a topological group and let*

$$0 \to A \xrightarrow{\alpha} B \xrightarrow{\beta} C \to 0$$

be a weakly admissible short exact sequence of continuous Banach G-modules. Then the sequence

$$0 \to \mathrm{C}^n_{\mathrm{b}}(G,A)^G \xrightarrow{\alpha_*} \mathrm{C}^n_{\mathrm{b}}(G,B)^G \xrightarrow{\beta_*} \mathrm{C}^n_{\mathrm{b}}(G,C)^G \to 0$$

is also exact for every $n \geq 0$.

PROOF. The only non-trivial point is the surjectivity of β_*. According to the definition of weakly admissible sequences, we may fix a morphism of Banach spaces $\sigma^0 : C \to B$ with $\beta\sigma^0 = Id$. Recall from the first lines of Section 6.2 that the evaluation at $e \in G$ yields an isometric isomorphism

$$U^n : \mathrm{C}^n_{\mathrm{b}}(G,-)^G \xrightarrow{\cong} C\mathrm{C}^{n-1}_{\mathrm{b}}(G,-).$$

Thus U^n identifies the sequence of the lemma with the sequence

$$0 \to C\mathrm{C}^{n-1}_{\mathrm{b}}(G,A) \xrightarrow{\alpha_*} C\mathrm{C}^{n-1}_{\mathrm{b}}(G,B) \xrightarrow{\beta_*} C\mathrm{C}^{n-1}_{\mathrm{b}}(G,C) \to 0,$$

where the maps α_*, β_* are still defined by post-composition. We shall show by induction on n that there is a morphism of Banach spaces

$$\sigma^n : C\mathrm{C}^{n-1}_{\mathrm{b}}(G,C) \longrightarrow C\mathrm{C}^{n-1}_{\mathrm{b}}(G,B)$$

such that $\beta_*\sigma^n = Id$. The case $n = 0$ is void since then $C\mathrm{C}^{n-1}_{\mathrm{b}}(G,C) = CC = C$ and similarily for B, so we suppose $n \geq 1$. The inductive relation $C\mathrm{C}^{n-1}_{\mathrm{b}}(G,-) = C\mathrm{C}_{\mathrm{b}}\Big(G, C\mathrm{C}^{n-2}_{\mathrm{b}}(G,E)\Big)$ of Lemma 6.1.6 shows that it is enough to handle the case $n = 1$. Hence, let f be an element of $C\mathrm{C}_{\mathrm{b}}(G,C)$ and define $\sigma^1 f : G \to B$ by

$$\sigma^1 f(x) \;=\; \pi_B(x)\sigma^0\pi_C(x)^{-1}f(x), \qquad (x \in G)$$

where π_B, π_C are the G-representation on B and C, respectively. This defines a continuous linear map

$$\sigma^1 : C\mathrm{C}_{\mathrm{b}}(G,C) \longrightarrow \mathrm{C}_{\mathrm{b}}(G,B)$$

of norm $\|\sigma^1\| = \|\sigma^0\|$ and satisfying $\beta_*\sigma^1 = Id$, so that it remains only to show that σ^1 does indeed range in the maximal continuous submodule $\mathcal{C}\mathrm{C_b}(G,B)$. To this end, we pick a net $(g_\iota)_{\iota\in I}$ converging to $e \in G$ and check the criterion (iii) of Lemma 1.1.1 :

$$\begin{aligned}
\left\|\lambda_{\pi_B}(g_\iota)\sigma^1 f - \sigma^1 f\right\|_\infty &= \sup_{x\in G} \left\|\pi_B(g_\iota)\sigma^1 f(g_\iota^{-1}x) - \sigma^1 f(x)\right\|_B \\
&= \sup_{x\in G} \left\|\pi_B(g_\iota)\pi_B(g_\iota^{-1}x)\sigma^0\pi_C(g_\iota^{-1}x)^{-1} f(g_\iota^{-1}x) - \pi_B(x)\sigma^0\pi_C(x)^{-1}f(x)\right\|_B \\
&= \sup_{x\in G} \left\|\sigma^0\Big(\pi_C(x)^{-1}\pi_C(g_\iota)f(g_\iota^{-1}x) - \pi_C(x)^{-1}f(x)\Big)\right\|_B \\
&\leq \|\sigma^0\| \sup_{x\in G} \left\|\pi_C(g_\iota)f(g_\iota^{-1}x) - f(x)\Big)\right\|_B \\
&= \|\sigma^0\| \cdot \left\|\lambda_{\pi_C}(g_\iota)f - f\right\|_\infty,
\end{aligned}$$

which converges to zero. □

Proof of Proposition 8.2.7. With Lemma 8.2.8 at hand instead of Lemmata 8.2.4 and 8.2.5, the remaining part of the proof of Proposition 8.2.7 goes exactly like the proof of Proposition 8.2.1. This time the snake lemma will be applied to "slices"

$$\begin{array}{ccc}
0 & & 0 \\
\downarrow & & \downarrow \\
\mathrm{C}_\mathrm{b}^n(G,A)^G & \longrightarrow & \mathrm{C}_\mathrm{b}^{n+1}(G,A)^G \\
\downarrow{\scriptstyle \alpha_*} & & \downarrow{\scriptstyle \alpha_*} \\
\mathrm{C}_\mathrm{b}^n(G,B)^G & \longrightarrow & \mathrm{C}_\mathrm{b}^{n+1}(G,B)^G \\
\downarrow{\scriptstyle \beta_*} & & \downarrow{\scriptstyle \beta_*} \\
\mathrm{C}_\mathrm{b}^n(G,C)^G & \longrightarrow & \mathrm{C}_\mathrm{b}^{n+1}(G,C)^G \\
\downarrow & & \downarrow \\
0 & & 0
\end{array}$$

of the canonical complexes. □

As is to be expected, the long exact sequence degenerates in the special case where the short exact sequence is G-split (see Definition 4.2.7).

Proposition 8.2.9.— *Let G be a topological group and let*

$$0 \to A \xrightarrow{\alpha} B \xrightarrow{\beta} C \to 0$$

be a G-split short exact sequence of Banach G-modules. Then one has for all $n \geq 0$ the short exact sequence

$$0 \longrightarrow \mathrm{H}_\mathrm{cb}^n(G,A) \longrightarrow \mathrm{H}_\mathrm{cb}^n(G,B) \longrightarrow \mathrm{H}_\mathrm{cb}^n(G,C) \longrightarrow 0.$$

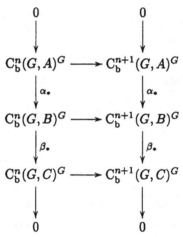

(Here A, B, C are not supposed continuous).

PROOF. Recall that according to Proposition 6.1.5 we have $\mathrm{H}^\bullet_{\mathrm{cb}}(G,-) = \mathrm{H}^\bullet_{\mathrm{cb}}(G,\mathcal{C}-)$. Since the original sequence is G-split, the sequence

$$0 \to \mathcal{C}A \xrightarrow{\alpha} \mathcal{C}B \xrightarrow{\beta} \mathcal{C}C \to 0$$

is also exact and G-split (Lemma 4.2.8). Therefore, there is no loss of generality in supposing that A, B, C are continuous Banach G-modules. We may thus apply Proposition 8.2.7 and all that is left to do is to show that the transgression maps vanish. To this end, we fix left respectively right inverse G-morphisms τ, σ for α respectively β. By post-composition we obtain commutative diagrams

$$\begin{array}{ccc}
\mathrm{C}^n_{\mathrm{b}}(G,A)^G & \longrightarrow & \mathrm{C}^{n+1}_{\mathrm{b}}(G,A)^G \\
\uparrow{\scriptstyle \tau_*} & & \uparrow{\scriptstyle \tau_*} \\
\mathrm{C}^n_{\mathrm{b}}(G,B)^G & \longrightarrow & \mathrm{C}^{n+1}_{\mathrm{b}}(G,B)^G \\
\uparrow{\scriptstyle \sigma_*} & & \uparrow{\scriptstyle \sigma_*} \\
\mathrm{C}^n_{\mathrm{b}}(G,C)^G & \longrightarrow & \mathrm{C}^{n+1}_{\mathrm{b}}(G,C)^G
\end{array}$$

Now the element $a \in \mathrm{C}^{n+1}_{\mathrm{b}}(G,A)^G$ that the snake lemma associates to a cocycle $c \in \mathrm{C}^n_{\mathrm{b}}(G,C)^G$ is trivialized by $\tau_*\sigma_*c$. □

With the convention of Notation 1.2.12, we deduce

COROLLARY 8.2.10.— *Let G be a topological group and let E_1, E_2 be two Banach G-modules. Then the canonical maps $E_j \hookrightarrow E$ and $E \twoheadrightarrow E_j$ $(j = 1, 2)$ induce for each $n \geq 0$ a topological isomorphism of semi-normed spaces*

$$\mathrm{H}^n_{\mathrm{cb}}(G, E_1 \oplus E_2) \cong \mathrm{H}^n_{\mathrm{cb}}(G, E_1) \oplus \mathrm{H}^n_{\mathrm{cb}}(G, E_2). \qquad \square$$

As announced in point (i) of Remarks 8.2.6, we can generalize the Lemma 8.2.5 as follows :

COROLLARY 8.2.11.— *Let G be a locally compact second countable group, S an amenable regular G-space. Let $0 \to A \xrightarrow{\alpha} B \xrightarrow{\beta} C \to 0$ be an adjoint short exact sequence of G-morphisms of coefficient G-modules. Then the induced sequence*

$$0 \longrightarrow L^\infty_{\mathrm{w*}}(S,A)^G \xrightarrow{\alpha_*} L^\infty_{\mathrm{w*}}(S,B)^G \xrightarrow{\beta_*} L^\infty_{\mathrm{w*}}(S,C)^G \longrightarrow 0$$

is also exact.

PROOF. In the lines preceding Lemma 8.2.5, we have explained how Grothendieck's result [**73**] implies that the sequence

$$0 \longrightarrow L^\infty_{\mathrm{w*}}(S,A) \xrightarrow{\alpha_*} L^\infty_{\mathrm{w*}}(S,B) \xrightarrow{\beta_*} L^\infty_{\mathrm{w*}}(S,C) \longrightarrow 0$$

is exact, so we have only to pass to the G-invariants. Since the sequence above satisfies the condition (ii) of Proposition 8.2.1, we have an associated long exact sequence. Recalling that $\mathrm{H}^0_{\mathrm{cb}}(G,-)$ consists in taking G-invariants, the long exact sequence begins with

$$0 \longrightarrow L^\infty_{\mathrm{w*}}(S,A)^G \xrightarrow{\beta_*} L^\infty_{\mathrm{w*}}(S,B)^G \xrightarrow{\beta_*} L^\infty_{\mathrm{w*}}(S,C)^G \xrightarrow{\tau^1} \\ \longrightarrow \mathrm{H}^1_{\mathrm{cb}}\Big(G, L^\infty_{\mathrm{w*}}(S,A)\Big) \longrightarrow \cdots$$

According to Theorem 5.7.1, the coefficient G-module $L^\infty_{\mathrm{w*}}(S,A)$ is relatively injective, so that by Proposition 7.4.1 we have

$$\mathrm{H}^1_{\mathrm{cb}}\Big(G, L^\infty_{\mathrm{w*}}(S,A)\Big) = 0,$$

whence the statement. □

More applications of the long exact sequence will be given in Section 10. We end our treatment of the long exact sequence by mentioning a related result about the integral bounded cohomology of discrete groups (as defined in Remark 6.1.4).

PROPOSITION 8.2.12 (S. Gersten).— *Let Γ be a discrete group. Then the inclusion $\mathbf{Z} \subset \mathbf{R}$ induces a long exact sequence*

$$\cdots \to \mathrm{H}^n_{\mathrm{b}}(\Gamma,\mathbf{Z}) \to \mathrm{H}^n_{\mathrm{b}}(\Gamma,\mathbf{R}) \to \mathrm{H}^n(\Gamma,\mathbf{R}/\mathbf{Z}) \to \mathrm{H}^{n+1}_{\mathrm{cb}}(\Gamma,\mathbf{Z}) \to \cdots$$

starting with $0 \to \mathrm{H}^1(\Gamma,\mathbf{R}/\mathbf{Z}) \to \mathrm{H}^2_{\mathrm{cb}}(\Gamma,\mathbf{Z}) \to \cdots$.

ON THE PROOF. This statement, due to S. Gersten (Proposition 1.1 in [**62**]), can be checked directly on the cochains ; the point is of course that usual cocycles ranging in $\mathbf{R}/\mathbf{Z}$ can only be bounded, and that there is a bounded fundamental domain for $\mathbf{Z}$ in $\mathbf{R}$. □

8.3. The coefficient representation. We conclude our overhaul of covariance with the following observation, which is also a transition to the next section. Suppose $H \lhd G$ is a normal subgroup of the topological group G, and let (π, E) be a Banach G-module. View it also as a Banach H-module, denote by $Z_G(H)$ the centralizer of H in G and by $Z(H)$ the center of H. For all $z \in Z_G(H)$, the morphism

$$\pi(z):\ E \longrightarrow E$$

is a H-morphism. Therefore, the functor $\mathrm{H}^n_{\mathrm{cb}}(H,\pi(-))$ yields for all $n \geq 0$ an isometric representation of $Z_G(H)$ on $\mathrm{H}^n_{\mathrm{cb}}(H,E)$. We call this representation the *coefficient representation*. It can be checked easily that this representation factors through the quotient

$$Z_G(H) \twoheadrightarrow Z_G(H)/Z(H).$$

We do not verify this right now because we shall identify this action later on in a setting where this factorization will be a consequence of a more general result (see Corollary 8.7.6 below).

8.4. Contravariance. We turn now to the contravariant aspect of $\mathrm{H}^\bullet_{\mathrm{cb}}$, letting the group vary.

Let $\psi : H \to G$ be a morphism of topological groups, that is, a continuous group homomorphism. To any Banach G-module (π, E) we associate the Banach H-module $(\pi\psi, E)$. We shall occasionly write $\psi^* E$ instead of E if any confusion is possible (e.g. when $H = G$).

By successive pre-compositions, ψ induces for all $n \geq 0$ a norm one map

$$\psi^* : \ \mathrm{C}^n_{\mathrm{b}}(G, E) \longrightarrow \mathrm{C}^n_{\mathrm{b}}(H, E)$$

which is equivariant with respect to ψ since indeed for every $h \in H$ one has $\psi^* \lambda_\pi(\psi(h)) = \lambda_{\pi\psi}(h)\psi^*$. Hence ψ^* is an H-morphism when we view $\mathrm{C}^n_{\mathrm{b}}(G, E)$ as a Banach H-module according to the above convention. We obtain in this way an extension of the identity map $(\pi, E) \to (\pi\psi, E)$ to an H-morphism of complexes. Therefore, since

$$\mathrm{C}^n_{\mathrm{b}}(G, E)^G \subset \mathrm{C}^n_{\mathrm{b}}(G, E)^H,$$

we compose the morphisms of complexes

$$\begin{array}{ccccccccc}
0 & \longrightarrow & \mathrm{C}^0_{\mathrm{b}}(G,E)^G & \longrightarrow & \mathrm{C}^1_{\mathrm{b}}(G,E)^G & \longrightarrow & \mathrm{C}^2_{\mathrm{b}}(G,E)^G & \longrightarrow & \cdots \\
 & & \downarrow & & \downarrow & & \downarrow & & \\
0 & \longrightarrow & \mathrm{C}^0_{\mathrm{b}}(G,E)^H & \longrightarrow & \mathrm{C}^1_{\mathrm{b}}(G,E)^H & \longrightarrow & \mathrm{C}^2_{\mathrm{b}}(G,E)^H & \longrightarrow & \cdots \\
 & & \downarrow{\scriptstyle \psi^*} & & \downarrow{\scriptstyle \psi^*} & & \downarrow{\scriptstyle \psi^*} & & \\
0 & \longrightarrow & \mathrm{C}^0_{\mathrm{b}}(H,E)^H & \longrightarrow & \mathrm{C}^1_{\mathrm{b}}(H,E)^H & \longrightarrow & \mathrm{C}^2_{\mathrm{b}}(H,E)^H & \longrightarrow & \cdots
\end{array}$$

and thus we end up with a sequence of continuous linear maps

$$\mathrm{H}^n_{\mathrm{cb}}(\psi, E) : \ \mathrm{H}^n_{\mathrm{cb}}(G, E) \longrightarrow \mathrm{H}^n_{\mathrm{cb}}(H, E) \qquad (n \geq 0)$$

of semi-norm at most one. We emphasize once again that the right hand side is more exactly denoted by $\mathrm{H}^n_{\mathrm{cb}}(H, \psi^* E)$.

Remark 8.4.1. The spaces $\mathrm{C}^n_{\mathrm{b}}(G, E)^H$ appearing in the definition of $\mathrm{H}^n_{\mathrm{cb}}(\psi, E)$ are of course identical with $\mathrm{C}^n_{\mathrm{b}}(G, E)^{\psi(H)}$, where in the latter case $\mathrm{C}^n_{\mathrm{b}}(G, E)$ is again thought of as a Banach G-module. The obviousness of this fact should not let us forget that in general the functors $\mathcal{C}_{\pi\psi}$ and $\mathcal{C}_{\pi|_{\psi(H)}}$ do not coincide. However, the morphism of topological groups $\psi : H \to G$ determines for every Banach G-module (ϱ, E) the following inclusions :

$$\mathcal{C}_{\varrho|_{\psi(H)}} F \subset \mathcal{C}_{\varrho\psi} F \subset \mathcal{C}_{\varrho} F.$$

It follows from the definitions above that if we are given moreover a G-morphism $\alpha : E \to F$ of Banach G-modules, then the diagram

$$\begin{array}{ccc} \mathrm{H}^n_{\mathrm{cb}}(G,E) & \longrightarrow & \mathrm{H}^n_{\mathrm{cb}}(H,\psi^*E) \\ \downarrow & & \downarrow \\ \mathrm{H}^n_{\mathrm{cb}}(G,F) & \longrightarrow & \mathrm{H}^n_{\mathrm{cb}}(H,\psi^*F) \end{array}$$

commutes. We denote the common composition by $\mathrm{H}^n_{\mathrm{cb}}(\psi,\alpha)$.

It follows also that $\mathrm{H}^n_{\mathrm{cb}}(\psi,E)$ depends functorially on ψ. It is moreover compatible with the functorial characterization of $\mathrm{H}^n_{\mathrm{cb}}(-,E)$ in the following sense :

PROPOSITION 8.4.2.— *Let $(\mathfrak{a},A^\bullet)$ be a strong resolution of (π,E) by G-relatively injective Banach G-modules and let $(\mathfrak{b},B^\bullet)$ be a strong resolution of $(\pi\psi,E)$ by H-relatively injective Banach H-modules. Consider the former as a complex of H-modules.*

Then the identity map on E extends to an H-morphism of the augmented complexes and moreover for all $n \geq 0$ the map

$$\iota^\bullet_* : \mathrm{H}^n(A^{\bullet G}) \longrightarrow \mathrm{H}^n(B^{\bullet H})$$

induced by any such extension $\iota^\bullet$ is conjugated to $\mathrm{H}^n_{\mathrm{cb}}(\psi,E)$ by the topological isomorphisms

$$\mathrm{H}^n_{\mathrm{cb}}(G,E) \cong \mathrm{H}^n(A^{\bullet G}) \quad \textit{and} \quad \mathrm{H}^n_{\mathrm{cb}}(H,E) \cong \mathrm{H}^n(B^{\bullet H})$$

given by Theorem 7.2.1.

PROOF. The existence of an extension of the identity which is conjugated to $\mathrm{H}^n_{\mathrm{cb}}(\psi,E)$ is a consequence of Lemma 7.2.4. First one applies the lemma over G to the two augmented complexes $\mathcal{C}_\pi A^\bullet$ and $\mathrm{C}^\bullet_{\mathrm{b}}(G,E)$, then one composes with ψ^* and finally one applies the lemma over H to the complexes $\mathrm{C}^\bullet_{\mathrm{b}}(H,E)$ and $\mathcal{C}_{\pi|H}B^\bullet$. The symbol $\mathcal{C}$ stands for G-continuity in the first application and for H-continuity in the second.

For the uniqueness, we have to take into account the Remark 8.4.1 (and Remark 7.1.3) : we may in general not assume that $(\mathfrak{a},A^\bullet)$ is a strong resolution of E when we endow it with its H-structure. However, the augmented resolution

$$0 \longrightarrow \mathcal{C}_\pi E \longrightarrow \mathcal{C}_\pi A^0 \longrightarrow \mathcal{C}_\pi A^1 \longrightarrow \mathcal{C}_\pi A^2 \longrightarrow \cdots$$

of maximal G-continuous submodules is both G-strong and H-strong because of the inclusions of Remark 8.4.1. Therefore, given any extension $\iota^\bullet$ as in the statement, we restrict it to the latter augmented resolution and may then apply the Lemma 7.2.6 (over H) to conclude that $\iota^\bullet$ induces the same map in cohomology than the map constructed in the first part of the proof. □

EXAMPLE 8.4.3. Let G be a topological group and G' the same underlying group endowed with a finer topology (yet still compatible with the group structure). The the identity map $G' \to G$ is a morphism of topological groups and hence induces a natural map $\mathrm{H}^\bullet_{\mathrm{cb}}(G,E) \to \mathrm{H}^\bullet_{\mathrm{cb}}(G',E)$ for every Banach G-module E.

In particular, if G is a topological group and $\underline{G}$ is the underlying group, then there is a natural map $\mathrm{H}^\bullet_{\mathrm{cb}}(G,E) \to \mathrm{H}^\bullet_{\mathrm{b}}(\underline{G},E)$ (recall that the right hand side is *defined* as $\mathrm{H}^\bullet_{\mathrm{cb}}(G_\delta,E)$, see Section 6.1).

Two further interesting particular cases of the contravariance are the *inflation* and *restriction.*

8.5. Inflation.

DEFINITION 8.5.1. Let $N \lhd G$ be a normal subgroup of the topological group G and let $p : G \to G/N$ be the quotient map. Let F be a Banach G/N-module. Then the natural map

$$\inf = \mathrm{H}^\bullet_{\mathrm{cb}}(p,F) : \ \mathrm{H}^\bullet_{\mathrm{cb}}(G/N,F) \longrightarrow \mathrm{H}^\bullet_{\mathrm{cb}}(G,F)$$

is called the *inflation*.

Observe that if we start with a Banach G-module E, then E^N is both a Banach G/N- and G-module. Denoting by $\iota : E^N \to E$ the inclusion map, we can consider the natural map

$$\mathrm{H}^\bullet_{\mathrm{cb}}(p,\iota) : \ \mathrm{H}^\bullet_{\mathrm{cb}}(G/N,E^N) \longrightarrow \mathrm{H}^\bullet_{\mathrm{cb}}(G,E).$$

With this terminology, the Corollary 7.5.10 reads

COROLLARY 8.5.2.— *Let G be a locally compact second countable group and $N \lhd G$ a closed normal subgroup. If N is amenable, then the natural map*

$$\mathrm{H}^\bullet_{\mathrm{cb}}(p,\iota) : \ \mathrm{H}^n_{\mathrm{cb}}(G/N,E^N) \longrightarrow \mathrm{H}^n_{\mathrm{cb}}(G,E)$$

is an isometric isomorphism for every coefficient G-module E and all $n \geq 0$. In particular, for every coefficient G/N-module E, the inflation

$$\inf : \ \mathrm{H}^n_{\mathrm{cb}}(G/N,E) \longrightarrow \mathrm{H}^n_{\mathrm{cb}}(G,E)$$

is an isometric isomorphism for all $n \geq 0$.

PROOF. The proof of Corollary 7.5.10 realizes canonically both $\mathrm{H}^n_{\mathrm{cb}}(G/N,E^N)$ and $\mathrm{H}^n_{\mathrm{cb}}(G,E)$ on L^∞ homogeneous complexes over G/N, so that the above Propositions 8.1.1 and 8.4.2 imply that the obtained isomorphism coincides with the map $\mathrm{H}^\bullet_{\mathrm{cb}}(G,\iota) \circ \mathrm{H}^\bullet_{\mathrm{cb}}(p,E^N) = \mathrm{H}^\bullet_{\mathrm{cb}}(p,\iota)$. □

As a by-product, we get the

COROLLARY 8.5.3.— *Let G be a locally compact second countable group and E a coefficient G-module. If $N \lhd G$ is an amenable closed normal subgroup, then the natural map*

$$\mathrm{H}^n_{\mathrm{cb}}(G,E^N) \longrightarrow \mathrm{H}^n_{\mathrm{cb}}(G,E)$$

induced by the inclusion $\iota : E^N \to E$ is an isometric isomorphism for all $n \geq 0$.

PROOF. The Corollary 8.5.2 yields an isometric isomorphism

$$\mathrm{H}^\bullet_{\mathrm{cb}}(p,\iota) : \ \mathrm{H}^\bullet_{\mathrm{cb}}(G/N,E^N) \longrightarrow \mathrm{H}^\bullet_{\mathrm{cb}}(G,E).$$

However, applying the same corollary to the Banach G-module E^N instead of E, we see that

$$\inf : \ \mathrm{H}^\bullet_{\mathrm{cb}}(G/N,E^N) \longrightarrow \mathrm{H}^\bullet_{\mathrm{cb}}(G,E^N)$$

is also an isometric isomorphism. We have by definition

$$\mathrm{H}^\bullet_{\mathrm{cb}}(p,\iota) = \mathrm{H}^\bullet_{\mathrm{cb}}(G,\iota)\circ \inf,$$

whence the claim. □

REMARK 8.5.4. The statement of Corollary 8.5.2 is well known in the special case of discrete groups. In this setting, our technical assumption of second countability would amount to countability, but since in the discrete case there are no measurability issues whatsoever, one can directly obtain the following general statement :

Let Γ *be a group and* $\mathrm{A} \lhd \Gamma$ *a normal amenable subgroup. Then the for every contragredient Banach* Γ*-module* E *the natural map*

$$\mathrm{H}^\bullet_{\mathrm{b}}(p,\iota):\ \mathrm{H}^n_{\mathrm{b}}(\Gamma/\mathrm{A}, E^{\mathrm{A}}) \longrightarrow \mathrm{H}^n_{\mathrm{b}}(\Gamma, E)$$

is an isometric isomorphism for all $n \geq 0$*. In particular, for every contragredient Banach* Γ/A*-module* E *the inflation*

$$\inf:\ \mathrm{H}^n_{\mathrm{b}}(\Gamma/\mathrm{A}, E) \longrightarrow \mathrm{H}^n_{\mathrm{b}}(\Gamma, E)$$

is an isometric isomorphism for all $n \geq 0$*.*

The proof consists simply in averaging cochains over A, and can be found as Theorem 1 in [**117**] ; the case with trivial coefficients, attributed to Trauber, is carefully proved by N. Ivanov in [**86**], Theorem 3.8.4.

Here is an elementary concrete application of Corollary 8.5.2 :

COROLLARY 8.5.5.— *For all* $k, n \in \mathbf{N}$ *the natural inclusion*

$$\mathrm{SL}_k(\mathbf{C}) \longrightarrow \mathrm{GL}_k(\mathbf{C})$$

induces by contravariance an isometric isomorphism

$$\mathrm{H}^n_{\mathrm{cb}}(\mathrm{GL}_k(\mathbf{C})) \xrightarrow{\cong} \mathrm{H}^n_{\mathrm{cb}}(\mathrm{SL}_k(\mathbf{C})).$$

PROOF. Let us take the the quotients by the centres ; since $\mathbf{C}$ is algebraically closed, we obtain one and the same group $\mathrm{PSL}_k(\mathbf{C}) = \mathrm{PGL}_k(\mathbf{C})$. Thus, by functoriality of the contravariance, we have a commutative diagram

$$\begin{array}{ccc}
\mathrm{H}^n_{\mathrm{cb}}(\mathrm{PGL}_k(\mathbf{C})) & = & \mathrm{H}^n_{\mathrm{cb}}(\mathrm{PSL}_k(\mathbf{C})) \\
{\scriptstyle\inf}\downarrow & & \downarrow{\scriptstyle\inf} \\
\mathrm{H}^n_{\mathrm{cb}}(\mathrm{GL}_k(\mathbf{C})) & \longrightarrow & \mathrm{H}^n_{\mathrm{cb}}(\mathrm{SL}_k(\mathbf{C}))
\end{array}$$

in which both lateral arrows are isometric isomorphisms in virtue of Corollary 8.5.2. Therefore the lower arrow is an isometric isomorphism aswell. □

As usual, there is a statement similar to Corollary 8.5.2 for more general Banach modules provided we replace amenability by compactness :

PROPOSITION 8.5.6.— *Let G be a locally compact group, $K \lhd G$ a compact normal subgroup and E a Banach G-module. The natural map*

$$\mathrm{H}^{\bullet}_{\mathrm{cb}}(p,\iota): \ \mathrm{H}^{n}_{\mathrm{cb}}(G/K, E^{K}) \longrightarrow \mathrm{H}^{n}_{\mathrm{cb}}(G,E)$$

is an isometric isomorphism for every Banach G-module E and all $n \geq 0$. In particular, for every Banach $/K$-module E, the inflation

$$\inf: \ \mathrm{H}^{n}_{\mathrm{cb}}(G/K, E) \longrightarrow \mathrm{H}^{n}_{\mathrm{cb}}(G,E)$$

is an isometric isomorphism for all $n \geq 0$.

PROOF. If we apply Corollary 7.4.10 with $H = G$, we see that $\mathrm{H}^{n}_{\mathrm{cb}}(G,E)$ is realized by homogeneous cochains on G/K. One concludes then as in the proof of Corollary 8.5.2 above. □

We have again the by-product

COROLLARY 8.5.7.— *Let G be a locally compact group and E a Banach G-module. If $K \lhd G$ is a compact normal subgroup, then the natural map*

$$\mathrm{H}^{n}_{\mathrm{cb}}(G, E^{K}) \longrightarrow \mathrm{H}^{n}_{\mathrm{cb}}(G,E)$$

induced by the inclusion $\iota : E^{K} \to E$ is an isometric isomorphism for all $n \geq 0$.
□

8.6. Restriction. We turn now to second particular case we wanted to consider.

DEFINITION 8.6.1. Let H be a subgroup of the topological group G and let $\iota : H \to G$ be the inclusion map. Let E be a Banach G-module. Then the natural map

$$\mathrm{res} = \mathrm{H}^{\bullet}_{\mathrm{cb}}(\iota, E): \ \mathrm{H}^{\bullet}_{\mathrm{cb}}(G,E) \longrightarrow \mathrm{H}^{\bullet}_{\mathrm{cb}}(H,E)$$

is called the *restriction.*

In usual continuous cohomology of locally compact groups (with, say, Banach coefficients), it is well known that for *co-compact* subgroups such that the quotient admits a finite invariant measure, the restriction is injective. This encompasses notably the case of uniform lattices and finite index subgroups. The standard argument uses the existence of a *transfer* map which gives a left inverse to the restriction. The transfer map is obtained by integration, so that it is crucial that the subgroup be co-compact (see Examples 8.6.3).

In bounded cohomology, the situation can be appreciably improved. A first and immediate remark is that the same transfer principle goes through also without the co-compactness assumption (Proposition 8.6.2). Then, replacing in Proposition 8.6.6 integrals by invariant means, we observe that for coefficient modules it is enough to suppose the quotient amenable in the sense of P. Eymard [**54**], namely that the quotient admits an invariant mean (this property is sometimes called *co-Følner*, see [**85**]).

More information about the image of the restriction maps will be given in Section 8.8 (Corollary 8.8.6). We defer also to that section the non-locally compact case, about which we have anyway nothing more to say than the obvious statement for finite index subgroups (Corollary 8.8.5).

PROPOSITION 8.6.2.— *Let H be a closed subgroup of the locally compact group G and let (π, E) be a continuous Banach G-module. If the quotient G/H admits a finite invariant measure, then the restriction*

$$\mathrm{res}: \mathrm{H}^n_{\mathrm{cb}}(G, E) \longrightarrow \mathrm{H}^n_{\mathrm{cb}}(H, E)$$

is isometrically injective for all $n \geq 0$.

EXAMPLES 8.6.3. In order to illustrate the above statement and the remarks preceding it, we observe that if $\Gamma < G$ is a lattice in the locally compact group G, then the restriction $\mathrm{H}^n_{\mathrm{cb}}(G) \to \mathrm{H}^n_{\mathrm{b}}(\Gamma)$ is always injective, whilst the restriction $\mathrm{H}^n_{\mathrm{c}}(G) \to \mathrm{H}^n(\Gamma)$ need not be so. As first example, take $\Gamma = \mathrm{SL}_2(\mathbf{Z})$. Then Γ is a lattice in $G = \mathrm{SL}_2(\mathbf{R})$, but whilst $\mathrm{H}^2_{\mathrm{c}}(G)$ is one-dimensional (generated by the Euler class), the cohomology group $\mathrm{H}^2(\Gamma)$ is trivial. The same holds of course for finite index subgroups of Γ, among which we find free groups.

Thanks to a result of A. Borel and J.-P. Serre [**18**], this example can be considerably generalized as follows. Let $\mathbf{G}$ be a connected $\mathbf{Q}$-semi-simple $\mathbf{Q}$-group and Γ a torsion-free arithmetic subgroup of $\mathbf{G}(\mathbf{Q})$. Let $G = \mathbf{G}(\mathbf{R})$, write d for the dimension of the associated symmetric space and l for the $\mathbf{Q}$-rank of $\mathbf{G}$. Then A. Borel and J.-P. Serre prove that the cohomological dimension of Γ equals $d - l$ (Corollary 11.4.3 in [**18**], or in N° 98 in the *Œuvres* [**16**]).

Thus for example with complex coefficients we see that $\mathrm{H}^k(\mathrm{SL}_n(\mathbf{Z}))$ vanishes at least when k exceeds $n(n-1)/2$, whilst the volume form of the associated symmetric space yields a non-trivial class in $\mathrm{H}^d_{\mathrm{c}}(\mathrm{SL}_n(\mathbf{R}))$ for $d = n(n+1)/2 - 1$. We observe moreover that by Savage's result [**127**], this class happens to be in the image of $\mathrm{H}^d_{\mathrm{cb}}(\mathrm{SL}_n(\mathbf{R}))$ so that in view of the above Proposition 8.6.2 it restricts to a non-trivial class of $\mathrm{H}^d_{\mathrm{b}}(\mathrm{SL}_n(\mathbf{Z}))$.

In contrast to this, we mention that for $\Gamma, \mathbf{G}, G = \mathbf{G}(\mathbf{R})$ as above a result of A. Borel (Theorem 7.5 in [**17**] or in N° 100 in [**16**]) ensures that the restriction $\mathrm{H}^k_{\mathrm{c}}(G) \to \mathrm{H}^k(\Gamma)$ is injective if $k < \mathrm{rank}_{\mathbf{Q}}(\mathbf{G})/2$ and even bijective if moreover $k < \mathrm{rank}_{\mathbf{R}}(G)/4$.

PROOF OF PROPOSITION 8.6.2. We know from Corollary 7.4.10 that the spaces $\mathrm{H}^n_{\mathrm{cb}}(G, E)$ and $\mathrm{H}^n_{\mathrm{cb}}(H, E)$ can be realized isometrically on the complexes

$$\begin{array}{ccccccccc}
0 & \longrightarrow & C_{\mathrm{b}}(G, E)^G & \longrightarrow & C_{\mathrm{b}}(G^2, E)^G & \longrightarrow & C_{\mathrm{b}}(G^3, E)^G & \longrightarrow & \cdots \\
 & & \big\downarrow & & \big\downarrow & & \big\downarrow & & \\
0 & \longrightarrow & C_{\mathrm{b}}(G, E)^H & \longrightarrow & C_{\mathrm{b}}(G^2, E)^H & \longrightarrow & C_{\mathrm{b}}(G^3, E)^H & \longrightarrow & \cdots
\end{array}$$

Moreover, the Proposition 8.4.2 implies that the restriction

$$\mathrm{H}^n_{\mathrm{cb}}(G, E) \longrightarrow \mathrm{H}^n_{\mathrm{cb}}(H, E)$$

coincides with the map induced by the vertical inclusions in the diagram above. We define for all $n \geq 0$ a map

$$\mathrm{trans}^n: C_{\mathrm{b}}(G^{n+1}, E)^H \longrightarrow C_{\mathrm{b}}(G^{n+1}, E)^G$$

as follows. Let μ be the left invariant probability measure on G/H and set for every $f \in \mathrm{C_b}(G^{n+1}, E)^H$

$$\mathrm{trans}^n(f)(y) = \int_{G/H}^{(B)} (\lambda_\pi(x) f)(y)\, d\mu(x). \qquad (y \in G^{n+1})$$

The map $\mathrm{trans}^n(f)$ is G-invariant because one can commute π with the Bochner integral, so that we have only to check that $\mathrm{trans}^n(f)$ is continuous ; this verification is demoted to the independent Lemma 8.6.5 below. Thus the transfer trans^n gives indeed a norm one inverse to the above inclusions. Since it is obtained by *diagonal* integration, $\mathrm{trans}^\bullet$ is compatible with the homogeneous coboundary, and therefore induces a map

$$\mathrm{trans} : \mathrm{H}^\bullet_{\mathrm{cb}}(H, E) \longrightarrow \mathrm{H}^\bullet_{\mathrm{cb}}(G, E)$$

such that $\mathrm{trans} \circ \mathrm{res} = Id$. Therefore the norm non-increasing map res is injective. Since each $\mathrm{trans}^\bullet$ is of norm one, the transfer does not increase the semi-norm either, so that the restriction is isometric. □

REMARK 8.6.4. In the above proof, it is as though we were integrating $\lambda_\pi(x) f$ over $x \in G/H$, except that we did not bother to discuss such a thing as integration theory with values in spaces like $\mathrm{C_b}(G^{n+1}, E)$, and that anyways the map $x \mapsto \lambda_\pi(x) f$ lacks regularity when x ranges in G. Therefore we integrated pointwise $(\lambda_\pi(x) f)(y)$, taking advantage of the Bochner integral, and then used Lemma 8.6.5 in order to conclude that the resulting function is still in $\mathrm{C_b}(G^{n+1}, E)$. This justifies a posteriori the insight that the integration over G/H yields a projection of $\mathrm{C_b}(G^{n+1}, E)^H$ onto $\mathrm{C_b}(G^{n+1}, E)^G$.

The remaining verification is similar to that of Lemma 4.5.5 :

LEMMA 8.6.5.— *The map*

$$\mathrm{trans}^n(f) : G^{n+1} \longrightarrow E$$

of the above proof is continuous.

PROOF. We keep the notations of the proof of Proposition 8.6.2.

Fix $y \in G^{n+1}$ and $\varepsilon > 0$. Since μ is a finite Radon measure, there is a compact subset $K \subset G/H$ such that

$$\mu\big((G/H) \setminus K\big) \cdot \|f\|_\infty \leq \varepsilon/4.$$

The continuity of the map

$$G/H \times G^{n+1} \longrightarrow E$$
$$(x', y') \longmapsto \pi(x') f(x'^{-1} y')$$

implies in particular that for every $x \in G/H$ there is an open neighbourhood U_x of x in G/H and a neighbourhood V_x of y in G^{n+1} such that for all (x', y') in $U_x \times V_x$ one has

$$\big\|\pi(x) f(x^{-1} y) - \pi(x') f(x'^{-1} y')\big\|_E \leq \varepsilon/4. \qquad (*)$$

The family $(U_x)_{x \in G/H}$ is an open cover of K so that by compactness there is a finite subset $F \subset G/H$ with $K \subset \bigcup_{x \in F} U_x$. We pick for every $x \in K$ an element

$\bar{x} \in F$ with $x \in U_{\bar{x}}$ and define the neighbourhood $V = \bigcap_{\bar{x}\in F} V_{\bar{x}}$ of y. We claim now that for every $y' \in V$ the estimate

$$\|\mathrm{trans}^n(f)(y) - \mathrm{trans}^n(f)(y')\|_E \leq \varepsilon$$

holds, finishing the proof :

notice first that this term is bounded by

$$\varepsilon/2 + \int_K \left\| \pi(x)f(x^{-1}y) - \pi(x)f(x^{-1}y') \right\|_E d\mu(x)$$

because of the choice of K. The integrand in the right summand is bounded by

$$\left\| \pi(\bar{x})f(\bar{x}^{-1}y) - \pi(x)f(x^{-1}y) \right\|_E + \left\| \pi(\bar{x})f(\bar{x}^{-1}y) - \pi(x)f(x^{-1}y') \right\|_E,$$

wherein each term is bounded by $\varepsilon/4$ due to (*). Thus we have obtained the bound $\varepsilon/2 + \mu(K)\varepsilon/2 \leq \varepsilon$, as claimed. □

The next statement is much more powerful then Proposition 8.6.2, as shows the Example 8.6.7 following it.

PROPOSITION 8.6.6.— *Let H be a closed subgroup of the locally compact second countable group G and let (π, E) be a coefficient G-module. If H is co-Følner in G, then the restriction*

$$\mathrm{res} : \mathrm{H}^n_{\mathrm{cb}}(G, E) \longrightarrow \mathrm{H}^n_{\mathrm{cb}}(H, E)$$

is isometrically injective for all $n \geq 0$.

EXAMPLE 8.6.7. Suppose Γ is a countable group and φ is an automorphism of Γ. Considering the corresponding semi-direct product extension

$$1 \longrightarrow \Gamma \longrightarrow \Gamma \rtimes_\varphi \mathbf{Z} \longrightarrow \mathbf{Z} \longrightarrow 0,$$

we see that Proposition 8.6.6 implies the injectivity of the restriction map

$$\mathrm{res} : \mathrm{H}^\bullet_{\mathrm{b}}(\Gamma \rtimes_\varphi \mathbf{Z}) \longrightarrow \mathrm{H}^\bullet_{\mathrm{b}}(\Gamma).$$

For usual cohomology, this injectivity fails already in the simplest case, namely when Γ is the trivial group. It will turn out later one that the image of the restriction map can be exactly determined, see Example 8.8.7 below.

REMARKS 8.6.8.

(i) The statement of Proposition 8.6.6 does not hold for general Banach G-modules. For instance, applied to the above extension $\Gamma \rtimes_\varphi \mathbf{Z}$ with Γ trivial, it would contradict the fact that $\mathbf{Z}$ may have non-trivial $\mathrm{H}^1_{\mathrm{b}}(\mathbf{Z}, E)$ for certain E, as pointed out in point (ii) of Remarks 7.5.13.

(ii) Another approach to the injectivity of the restriction will be possible after the introduction of the *induction*, see Corollary 10.1.7 below.

PROOF OF PROPOSITION 8.6.6. The proof is similar to the proof of Proposition 8.6.2, except for some technicalities due to the fact that we replace integrals by means for the definition of the transfer. This is taken care of by the following claim :

Let $\mathfrak{m}$ be an invariant mean on $L^\infty(G/H)$. Then for every coefficient G-module F there is an *adjointly natural* G-equivariant mean

$$\mathfrak{m}_F : L^\infty_{\mathrm{w*}}(G/H, F) \longrightarrow F.$$

By adjointly natural, we mean that any adjoint G-morphism $\alpha : F \to F'$ of coefficient G-modules induces a commutative diagram

$$\begin{array}{ccc} L^\infty_{\mathrm{w*}}(G/H, F) & \xrightarrow{\mathfrak{m}_F} & F \\ \Big\downarrow{\scriptstyle \alpha_*} & & \Big\downarrow{\scriptstyle \alpha} \\ L^\infty_{\mathrm{w*}}(G/H, F') & \xrightarrow{\mathfrak{m}_{F'}} & F' \end{array}$$

($\mathfrak{m}_F$ itself is not adjoint in general).

Indeed, if for $f \in L^\infty_{\mathrm{w*}}(G/H, F)$ and u in the chosen predual $F^\flat$ of F we define $f_u \in L^\infty(G/H)$ almost everywhere by $f_u(\cdot) = \langle f(\cdot)|u\rangle$, we obtain the desired $\mathfrak{m}_F$ by $\langle \mathfrak{m}_F(f)|u\rangle = \mathfrak{m}(f_u)$; as $F^\flat$ is separable, it is enough to consider countably many elements u, settling the "almost everywhere" problem. If now $\alpha : F \to F'$ is as above, with predual $\alpha^\flat : F'^\flat \to F^\flat$, we check for $v \in F'^\flat$ the relation

$$\langle \alpha\mathfrak{m}_F(f)|v\rangle = \langle \mathfrak{m}_F(f)|\alpha^\flat v\rangle = \mathfrak{m}(f_{\alpha^\flat v}) = \mathfrak{m}(\alpha_* f_v) = \langle \mathfrak{m}_{F'}(\alpha_* f)|v\rangle,$$

where the third equality follows from $\langle f(\cdot)|\alpha^\flat v\rangle = \langle (\alpha_* f)(\cdot)|v\rangle$. This proves the claim.

We appeal now to Corollary 7.5.9 and Proposition 8.4.2 in order to deduce that the restriction $\mathrm{H}^n_{\mathrm{cb}}(G, E) \to \mathrm{H}^n_{\mathrm{cb}}(H, E)$ is realized, together with its operator semi-norm, by the inclusion $\iota^\bullet$ of complexes

$$\begin{array}{ccccccccc} 0 & \longrightarrow & L^\infty_{\mathrm{w*}}(G, E)^G & \longrightarrow & L^\infty_{\mathrm{w*}}(G^2, E)^G & \longrightarrow & L^\infty_{\mathrm{w*}}(G^3, E)^G & \longrightarrow & \cdots \\ & & \Big\downarrow{\scriptstyle \iota^0} & & \Big\downarrow{\scriptstyle \iota^1} & & \Big\downarrow{\scriptstyle \iota^2} & & \\ 0 & \longrightarrow & L^\infty_{\mathrm{w*}}(G, E)^H & \longrightarrow & L^\infty_{\mathrm{w*}}(G^2, E)^H & \longrightarrow & L^\infty_{\mathrm{w*}}(G^3, E)^H & \longrightarrow & \cdots \end{array}$$

We set $F^n = L^\infty_{\mathrm{w*}}(G^{n+1}, E)$ and consider the corresponding maps $\mathfrak{m}_{F^n}$. We define for every $f \in (F^n)^H$ the element $\tau^n f$ of $L^\infty_{\mathrm{w*}}(G/H, F^n)$ by $\tau^n f(gH) = \lambda_\pi(g)f$. One checks that the norm one map

$$\tau^n : (F^n)^H \longrightarrow L^\infty_{\mathrm{w*}}(G/H, F^n)$$

ranges actually in $L^\infty_{\mathrm{w*}}(G/H, F^n)^G$; moreover one has $\tau^n\iota^n = \epsilon$. Composing τ^n with $\mathfrak{m}_{F^n}$, we see that we have obtained a norm one left inverse $\mathrm{trans}^n_{\mathfrak{m}} = \mathfrak{m}_{F^n}\tau^n$ to the inclusion ι^n realizing the restriction since

$$\mathrm{trans}^n_{\mathfrak{m}}\iota^n = \mathfrak{m}_{F^n}\tau^n\iota^n = \mathfrak{m}_{F^n}\varepsilon = Id.$$

On the other hand, we have $\tau^n d^n = (d^n)_* \tau^{n-1}$ so that the naturality claim above ensures that $\mathfrak{m}_{F^\bullet}\tau^\bullet$ is a morphism of complexes because the differentials $d^\bullet$ are adjoint maps. Therefore it induces a left inverse of semi-norm at most one

$$\mathrm{trans}_{\mathfrak{m}} : \mathrm{H}^\bullet_{\mathrm{cb}}(H,E) \longrightarrow \mathrm{H}^\bullet_{\mathrm{cb}}(G,E)$$

to the restriction, finishing the proof. □

SCHOLIUM 8.6.9. We have decorated the transfer $\mathrm{trans}_{\mathfrak{m}}$ arising in the proof of Proposition 8.6.6 with the invariant mean $\mathfrak{m}$ for the following reason. Whilst in the setting of Proposition 8.6.2 the normalized invariant measure used in the construction of the transfer map is unique, the invariant mean $\mathfrak{m}$ is far from being unique in general : for instance, considering the Example 8.6.7, there are uncountably many invariant means on $\mathbf{Z}$. We do not see a priori why $\mathrm{trans}_{\mathfrak{m}}$ should not depend on the choice of such a mean $\mathfrak{m}$.

The general question of the uniqueness of invariant means has given rise to an extensive literature upon which we shall not touch. Let us only comment on the special case of those subgroups $H < G$ which satisfy the assumptions of both Propositions 8.6.2 and 8.6.6 : in that situation, it has been observed by G.A. Margulis [**102**] and D. Sullivan [**134**] that the uniqueness question is related to the question of the weak containment in $L^2_0(G/H)$ of the trivial representation, where $L^2_0(G/H) = (\mathbf{C}\mathbf{1}_{G/H})^\perp$ is the orthogonal to the constants (see e.g. Lubotzky's book [**97**], Theorem 3.4.5). Using this method, it has been shown e.g. that the normalized measure on G/H is the unique invariant mean when H is a lattice in a real semi-simple Lie group with finite centre and without compact factors [**11**, Theorem B]. In particular, the transfer of Proposition 8.6.2 must in this case coincide with the transfer $\mathrm{trans}_{\mathfrak{m}}$ of Proposition 8.6.6.

8.7. The conjugation action on cohomology. We focus now for a moment on contravariance for one single group.

Let thus H be a topological group and (π, E) be a Banach H-module ; we know from the preceding Section 8.4 that any automorphism ψ of H (i.e. topological group automorphisms) induces an isomorphism

$$\mathrm{H}^\bullet_{\mathrm{cb}}(H,E) \longrightarrow \mathrm{H}^\bullet_{\mathrm{cb}}(H,\psi^* E).$$

Therefore, denoting by $\mathrm{Aut}(H)$ the group of automorphisms of H, the contravariance of the functor $\mathrm{H}^n_{\mathrm{cb}}(-,E)$ provides us for all n with an isometric groupoid bundle $\mathcal{B}$ of semi-normed spaces

$$\mathcal{B} = \Big\{ \mathrm{H}^n_{\mathrm{cb}}(G,\psi^* E) \ : \ \psi \in \mathrm{Aut}(H) \Big\}$$

over $\mathrm{Aut}(H)$. In order to get this information back into a groupoid-free setting, we want to keep the same Banach module E instead of going to isomorphic copies $\psi^* E$. To achieve this, we shall change the viewpoint and suppose that $H \lhd G$ is actually a normal subgroup of some topological group G acting on H by conjugation, and that (π, E) is a Banach G-module.

For every $g \in G$, denote by $\psi_g \in \mathrm{Aut}(H)$ the automorphism $\psi_g(x) = g^{-1}xg$, so that $\psi_{(\cdot)}$ is a representation of the opposite group G^{op} in $\mathrm{Aut}(H)$. Observing now that

$$\pi(g): \psi_g^* E \longrightarrow E$$

is a G- and hence also an H-isomorphism, we conclude that the assignment

$$g \longmapsto \mathrm{H}_{\mathrm{cb}}^n(\psi_g, \pi(g))$$

defines an isometric representation of G on $\mathrm{H}_{\mathrm{cb}}^n(H, E)$.

DEFINITION 8.7.1. This isometric G-representation on the semi-normed spaces $\mathrm{H}_{\mathrm{cb}}^n(H, E)$ shall be called the *conjugation action on cohomology*, or simply the *action on cohomology* when this causes no confusion.

For reasonable classes of groups, the Section 8.8 will give alternative approaches to this action, more suited to certain applications ; for the time being, we prefer to keep our complete degree of generality.

The following fact concerning the case $G = H$ is the exact analogue of its pendant in usual cohomology :

LEMMA 8.7.2.— *Let H be a topological group and let (π, E) be a Banach H-module. Then the above defined representation of H on $\mathrm{H}_{\mathrm{cb}}^n(H, E)$ is trivial for all $n \geq 0$.*

PROOF. Let $h \in H$. By definition, the map $\mathrm{H}_{\mathrm{cb}}^n(\psi_h, \pi(h))$ comes from the morphism $R^\bullet$ defined on the canonical augmented resolution $\mathrm{C}_{\mathrm{b}}^\bullet(H, E)$ as follows : for $n \geq 0$, f in $\mathrm{C}_{\mathrm{b}}^n(H, E)$ and $x_0, \ldots, x_n \in H$ one has

$$R^n f(x_0) \ldots (x_n) = \pi(h) f(h^{-1} x_0 h) \ldots (h^{-1} x_n h).$$

On the other hand, denoting by ϱ^n the right diagonal translation on $\mathrm{C}_{\mathrm{b}}^n(H, E)$, we obtain a G-morphism ϱ^n of the canonical augmented resolution. Since $\varrho^n f = R^n f$ for all H-invariant elements f, both morphisms of complexes coincide on the invariants $\mathrm{C}_{\mathrm{b}}^\bullet(H, E)^H$ and hence induce the same maps in cohomology. Therefore, it is enough to show that $\varrho^\bullet$ is H-homotopic to the identity. The argument is very standard, but we spell it out for completeness :

For every $n \geq 1$ and $0 \leq i \leq n-1$ define the map

$$\sigma_i^n : \mathrm{C}_{\mathrm{b}}^n(H, E) \longrightarrow \mathrm{C}_{\mathrm{b}}^{n-1}(H, E)$$

by

$$\sigma_i^n f(x_0) \ldots (x_{n-1}) = f(x_0 h) \ldots (x_i h)(x_i) \ldots (x_{n-1}).$$

In order to verify that $\sigma_i^n f$ satisfies all the continuity conditions to be indeed in $\mathrm{C}_{\mathrm{b}}^{n-1}(H, E)$, the only non-trivial point is that the map

$$\sigma_i^n f(x_0) \ldots (x_{i-1}) : H \longrightarrow \mathrm{C}_{\mathrm{b}}^{n-i-2}(H, E)$$

is continuous on H for every fixed $x_0, \ldots, x_{i-1} \in H$. Thus we pick $x_i \in H$ and let $k \to e \in H$. The expression

$$\left\| \sigma_i^n f(x_0) \ldots (x_{i-1})(kx_i) - \sigma_i^n f(x_0) \ldots (x_{i-1})(x_i) \right\|_\infty =$$

$$= \sup_{x_{i+1}, \ldots, x_{n-1} \in H} \Big\| f(x_0 h) \ldots (x_{i-1} h)(kx_i h)(kx_i)(x_{i+1}) \ldots (x_{n-1})$$

$$- f(x_0 h) \ldots (x_{i-1} h)(x_i h)(x_i)(x_{i+1}) \ldots (x_{n-1}) \Big\|_E$$

is bounded above by

$$\left\| f(x_0 h) \ldots (x_{i-1} h)(kx_i h) - f(x_0 h) \ldots (x_{i-1} h)(x_i h) \right\|_\infty +$$

$$+ \left\| f(x_0 h) \ldots (x_{i-1} h)(x_i h)(kx_i) - f(x_0 h) \ldots (x_{i-1} h)(x_i h)(x_i) \right\|_\infty.$$

Here the first summand converges to zero because $f(x_0 h) \ldots (x_{i-1 h})$ is in $\mathrm{C}_\mathrm{b}^{n-i}(H, E)$, whilst the second converges to zero because on the other hand $f(x_0 h) \ldots (x_{i-1 h})(x_i h)$ is in $\mathrm{C}_\mathrm{b}^{n-i-1}(H, E)$. This settles the continuity question.

We claim now that $\sigma^n = \sum_{i=0}^{n-1} (-1)^i \sigma_i^n$ yields a H-homotopy equivalence from $\varrho^\bullet$ to Id. To check this, we define for every $-1 \leq j \leq n$ the map

$$\varrho_j^n : \mathrm{C}_\mathrm{b}^n(H, E) \longrightarrow \mathrm{C}_\mathrm{b}^n(H, E)$$

by $\varrho_{-1}^n f = f$ and otherwise

$$\varrho_j^n f(x_0) \ldots (x_{n-1}) = f(x_0 h) \ldots (x_j h)(x_{j+1}) \ldots (x_{n-1})$$

(thus $\varrho_n^n = \varrho^n$). The above definitions imply readily the following relations :

$$\sigma_i^n d_j^n = \begin{cases} d_j^{n-1} \sigma_{j-1}^{n-1} & \text{for } j+1 \leq i, \\ \varrho_{i-1}^{n-1} & \text{for } j = i, \\ \varrho_i^{n-1} & \text{for } j = i+1, \\ d_{j-1}^{n-1} \sigma_j^{n-1} & \text{for } j \geq i+2. \end{cases}$$

Replacing these in

$$\sigma^n d^n = \sum_{i=0}^{n-1} \sum_{j=0}^{n} (-1)^{i+j} \sigma_i^n d_j^n$$

we find

$$\sigma^n d^n = -d^{n-1} \sigma^{n-1} + \sum_{i=0}^{n-1} (\varrho_{i-1}^{n-1} - \varrho_i^{n-1}) = -d^{n-1} \sigma^{n-1} + Id - \varrho^n,$$

so that $\sigma^\bullet$ is indeed a H-homotopy equivalence from $\varrho^\bullet$ to Id. □

Because of the functoriality of $\mathrm{H}_\mathrm{cb}^\bullet(-, E)$, the above Lemma 8.7.2 has the following immediate consequence :

COROLLARY 8.7.3.— *Let $H \lhd G$ be a normal subgroup of a topological group G and let (π, E) be a Banach G-module. Then the isometric G-representation on $\mathrm{H}_\mathrm{cb}^\bullet(H, E)$ descends to a G/H-action.* □

An important consequence of Lemma 8.7.2 is the following

COROLLARY 8.7.4.— *Let $H \lhd G$ be a normal subgroup of a topological group G and let (π, E) be a Banach G-module. Then the restriction ranges in the G/H-invariants :*

$$\mathrm{res}: \mathrm{H}^\bullet_{\mathrm{cb}}(G,E) \longrightarrow \mathrm{H}^\bullet_{\mathrm{cb}}(H,E)^{G/H}.$$

We shall see later on (Corollary 8.8.6) that in certain cases the right hand side above actually coincides with the image of the restriction map.

PROOF OF COROLLARY 8.7.4. We pick $g \in G$ and denote by $R^\bullet_H$ the morphism of the canonical augmented resolution for H which defines the isometry $\mathrm{H}^n_{\mathrm{cb}}\big(\psi_g, \pi(g)\big)$ of $\mathrm{H}^\bullet_{\mathrm{cb}}(H,E)$, namely

$$R^n_H f(x_0)\ldots(x_n) = \pi(g) f(g^{-1}x_0 g)\ldots(g^{-1}x_n g).$$

Likewise, $R^\bullet_G$ is to denote the morphism of the resolution for G which induces $\mathrm{H}^n_{\mathrm{cb}}\big(\psi_g, \pi(g)\big)$ on $\mathrm{H}^\bullet_{\mathrm{cb}}(G,E)$. Then we have a commutative diagram

$$\begin{array}{ccc} \mathrm{C}^n_{\mathrm{b}}(G,E)^G & \xrightarrow{R^n_G} & \mathrm{C}^n_{\mathrm{b}}(G,E)^G \\ \downarrow{\scriptstyle \iota_*} & & \downarrow{\scriptstyle \iota_*} \\ \mathrm{C}^n_{\mathrm{b}}(H,E)^H & \xrightarrow{R^n_H} & \mathrm{C}^n_{\mathrm{b}}(H,E)^H \end{array}$$

where $\iota : H \to G$ is the inclusion. This induces the commutative diagram

$$\begin{array}{ccc} \mathrm{H}^n_{\mathrm{cb}}(G,E) & = & \mathrm{H}^n_{\mathrm{cb}}(G,E) \\ \downarrow{\scriptstyle \mathrm{res}} & & \downarrow{\scriptstyle \mathrm{res}} \\ \mathrm{H}^n_{\mathrm{cb}}(H,E) & \xrightarrow{\mathrm{H}^n_{\mathrm{cb}}(\psi_g, \pi(g))} & \mathrm{H}^n_{\mathrm{cb}}(H,E) \end{array}$$

where the upper map is the identity by Lemma 8.7.2. This proves the corollary. □

The above arguments extend readily to the following situation which brings us back to the discussion of Section 8.3.

NOTATION 8.7.5. Let $H \lhd G$ be as before a normal subgroup of the topological group G and E a Banach G-module. We denote by $K_G(H)$ the kernel of the representation of G

$$G \longrightarrow \mathrm{Out}(H) = \mathrm{Aut}(H)/\mathrm{Int}(H)$$

in the group $\mathrm{Out}(H)$ of outer automorphisms of H.

Observe that one can write $K_G(H)$ as a not necessarily direct product

$$K_G(H) = Z_G(H) \cdot H = H \cdot Z_G(H),$$

and that there is a quotient map

$$K_G(H) \twoheadrightarrow Z_G(H)/Z(H), \quad zh \mapsto zZ(H) \qquad (z \in Z_G(H), h \in H).$$

Now we identify the restriction to $K_G(H)$ of the G-action on $\mathrm{H}^\bullet_{\mathrm{cb}}(H,E)$.

COROLLARY 8.7.6.— *Let $H \lhd G$ be a normal subgroup of the topological group G and (π, E) a Banach G-module. Then the $K_G(H)$-action on $\mathrm{H}^\bullet_{\mathrm{cb}}(H,E)$ descends to a $Z_G(H)/Z(H)$-action.*

Moreover, the coefficient representation of $Z_G(H)$ on $\mathrm{H}^\bullet_{\mathrm{cb}}(H,E)$ (see Section 8.3) descends to the same $Z_G(H)/Z(H)$-action.

PROOF. Let $z \in Z_G(H)$ and $h \in H$. In analogy with proofs above, denote by $R^n(zh)$ the morphism realizing $\mathrm{H}^n_{\mathrm{cb}}(\psi_{zh}, \pi(zh))$ on $\mathrm{C}^n_{\mathrm{b}}(H,E)$, namely

$$(R^n(zh)f)(x_0)\ldots(x_n) = \pi(zh)f(h^{-1}z^{-1}x_0zh)\ldots(h^{-1}z^{-1}x_nhz).$$

We have then $R^n(zh) = \pi(z)_* R^n(h)$. Supposing now that f is an H-invariant cocycle, we have seen in the proof of Lemma 8.7.2 that $R^n(h)f$ is H-equivariantly homotopic to f ; more precisely, we have shown the existence of $\sigma^\bullet$ such that for this f one has $R^n(h)f = \varrho^n(h)f = f - \sigma^{n+1}d^{n+1}f - d^n\sigma^n f = f - d^n\sigma^n f$. Therefore the class of $R^n(zh)$ coincides with the class of $\pi(z)_* f$. Since this holds for all z, h the statements follow. □

EXAMPLE 8.7.7. Let $G = G_1 \times G_2$ and let (π, E) be a Banach G-module. Then the G_1-action on $\mathrm{H}^\bullet_{\mathrm{cb}}(G_2,E)$ coincides with the coefficient representation of G_1 on $\mathrm{H}^\bullet_{\mathrm{cb}}(G_2,E)$ induced by the family $\pi|_{G_1}$ of G_2-morphisms of E.

8.8. A closer look at the action on cohomology. We present now a few complements concerning the action on cohomology ; the first is about its compatibility with the functorial approach.

The action on cohomology defined in the foregoing section was introduced as a special case of co- and contravariance ; therefore it comes as no surprise that we can use the previously established functoriality and deduce :

PROPOSITION 8.8.1.— *Let $H \lhd G$ be a normal subgroup of the topological group G and let (π, E) be a Banach G-module. Suppose that there is a G-resolution $(\mathfrak{a}, E^\bullet)$ of E which is a strong H-resolution by H-relatively injective Banach G-modules. Then the G-action on $(\mathfrak{a}, E^\bullet)$ induces on $\mathrm{H}^\bullet_{\mathrm{cb}}(H,E)$ the conjugation action up to the canonical isomorphisms of Theorem 7.2.1.*

So this proposition tells us in particular that the G-action on the non-augmented complex of H-invariants

$$0 \longrightarrow (E^0)^H \longrightarrow (E^1)^H \longrightarrow (E^2)^H \longrightarrow (E^3)^H \longrightarrow \cdots$$

induces the conjugation action ; notice that the above complex is G-invariant since H is normal in G. Incidentally, we obtain trivially the statement of Lemma 8.7.2.

It is however not clear whether a resolution as in Proposition 8.8.1 always exists in this generality, even though there is always a strong H-resolution of E by relatively injective Banach H-modules and also a strong G-resolution of E by relatively injective Banach G-modules.

Therefore, we shall give below two families of instances to which Proposition 8.8.1 applies ; these instances cover indeed the very general setting of locally compact groups.

Let us first take care of the

PROOF OF PROPOSITION 8.8.1. Let $g \in G$ and denote as before by ψ_g the automorphism $h \mapsto g^{-1}hg$ of H. According to Proposition 8.4.2, the maps

$$\mathrm{H}^n_{\mathrm{cb}}(\psi, E) : \mathrm{H}^n_{\mathrm{cb}}(H, E) \longrightarrow \mathrm{H}^n_{\mathrm{cb}}(H, \psi_g^* E)$$

are induced, modulo the canonical topological isomorphisms, by the morphism of complexes (going downwards)

$$\begin{array}{ccccccccccc} 0 & \longrightarrow & E & \xrightarrow{\mathfrak{a}} & E^0 & \longrightarrow & E^1 & \longrightarrow & E^2 & \longrightarrow & \cdots \\ & & \| & & \| & & \| & & \| & & \\ 0 & \longrightarrow & \psi_g^* E & \xrightarrow{\mathfrak{a}} & \psi_g^* E^0 & \longrightarrow & \psi_g^* E^1 & \longrightarrow & \psi_g^* E^2 & \longrightarrow & \cdots \end{array}$$

which is equivariant with respect to ψ_g. Recall now that the conjugation action of g on $\mathrm{H}^n_{\mathrm{cb}}(H, E)$ as defined in Section 8.7 consists in composing this map $\mathrm{H}^n_{\mathrm{cb}}(\psi, E)$ with the map

$$\mathrm{H}^n_{\mathrm{cb}}(H, \pi(g)) : \mathrm{H}^n_{\mathrm{cb}}(H, \psi_g^* E) \longrightarrow \mathrm{H}^n_{\mathrm{cb}}(H, E)$$

induced by the isomorphism $\pi(g) : \psi_g^* E \to E$. However, according to Proposition 8.1.1, this map $\mathrm{H}^n_{\mathrm{cb}}(H, \pi(g))$ is also induced by the following H-morphism of complexes

$$\begin{array}{ccccccccccc} 0 & \longrightarrow & \psi_g^* E & \xrightarrow{\mathfrak{a}} & \psi_g^* E^0 & \longrightarrow & \psi_g^* E^1 & \longrightarrow & \psi_g^* E^2 & \longrightarrow & \cdots \\ & & \downarrow{\scriptstyle \pi(g)} & & \downarrow{\scriptstyle \pi_0(g)} & & \downarrow{\scriptstyle \pi_1(g)} & & \downarrow{\scriptstyle \pi_2(g)} & & \\ 0 & \longrightarrow & E & \xrightarrow{\mathfrak{a}} & E^0 & \longrightarrow & E^1 & \longrightarrow & E^2 & \longrightarrow & \cdots \end{array}$$

to within the canonical topological isomorphisms ; here π_j is the G-representation on E^j and we observe that $\pi_j(g) : \psi_g^* E^j \to E^j$ is indeed a G-and H-isomorphism. These isomorphism intertwine the differentials because $(\mathfrak{a}, E^\bullet)$ is assumed to be a G-resolution.

Summing up, we conclude that the compound morphism of complexes $\pi_\bullet(g) : E^\bullet \to E^\bullet$ induces the conjugation action attached to $g \in G$ to within the canonical isomorphisms of Theorem 7.2.1. □

Here are two examples to which Proposition 8.8.1 applies.

COROLLARY 8.8.2.— *Let G be a locally compact group, $H \lhd G$ a closed normal subgroup and (π, E) be a Banach G-module. Then the regular G-action λ_π on the complex*

$$0 \longrightarrow \mathrm{C_b}(G, E)^H \longrightarrow \mathrm{C_b}(G^2, E)^H \longrightarrow \mathrm{C_b}(G^3, E)^H \longrightarrow \cdots$$

induces the conjugation action on $\mathrm{H}^\bullet_{\mathrm{cb}}(H,E)$ *to within to canonical isomorphisms. If more generally* $K < G$ *is a compact subgroup, then the same holds for regular* G*-action* λ_π *on*

$$0 \longrightarrow \mathrm{C_b}\big(G/K,E\big)^H \longrightarrow \mathrm{C_b}\big((G/K)^2,E\big)^H \longrightarrow \mathrm{C_b}\big((G/K)^3,E\big)^H \longrightarrow \cdots$$

and the corresponding statements also hold for the sub-complexes of alternating cochains.

Proof. The complex

$$0 \longrightarrow E \stackrel{\epsilon}{\longrightarrow} \mathrm{C_b}\big(G/K,E\big) \longrightarrow \mathrm{C_b}\big((G/K)^2,E\big) \longrightarrow \cdots$$

(with K trivial for the first statement) is a G-resolution ; therefore, the Corollary 7.4.10 yields all the assumptions that one needs to apply Proposition 8.8.1. □

Corollary 8.8.3.— *Let* G *be a locally compact second countable group,* S *an amenable regular* G*-space and* E *a coefficient* G*-module. Let* $H \lhd G$ *be a closed normal subgroup. Then the regular* G*-action* λ_π *on the complex*

$$0 \longrightarrow L^\infty_{\mathrm{w*}}(S,E)^H \longrightarrow L^\infty_{\mathrm{w*}}(S^2,E)^H \longrightarrow L^\infty_{\mathrm{w*}}(S^3,E)^H \longrightarrow \cdots$$

induces the conjugation action on $\mathrm{H}^\bullet_{\mathrm{cb}}(H,E)$ *to within to canonical isomorphisms. The corresponding statements hold for the sub-resolution and subcomplex of alternating cochains.*

Proof. As in the previous case, the complex

$$0 \longrightarrow E \stackrel{\epsilon}{\longrightarrow} L^\infty_{\mathrm{w*}}(S,E) \longrightarrow L^\infty_{\mathrm{w*}}(S^2,E) \longrightarrow L^\infty_{\mathrm{w*}}(S^3,E) \longrightarrow \cdots$$

is a G-resolution. By Lemma 5.4.3, the H-action on S is amenable. Thus the Theorem 7.5.3 puts us in position to apply Proposition 8.8.1. □

We have mentioned before that one weakness of the theory of [continuous] bounded cohomology is that the functors $\mathrm{H}^\bullet_{\mathrm{cb}}$ may take us away from the category used to define them, namely Banach modules. However, this is not always the case, as there are situations in which one can show that the [continuous] bounded cohomology is Hausdorff ; such results will be established in Section 11.4.

More precisely, let G be a topological group and $H \lhd G$ a normal subgroup ; let E be a Banach G-module. If for some $n \geq 0$ the space $\mathrm{H}^n_{\mathrm{cb}}(H,E)$ is Hausdorff, then the conjugation action turns it into a Banach G-module, and actually into a Banach G/H-module by Corollary 8.7.3.

Remark 8.8.4. Suppose moreover that G is locally compact second countable, H closed in G, and that E is a coefficient G-module. Then, given any amenable regular G-space S, the complex

$$0 \longrightarrow L^\infty_{\mathrm{w*}}(S,E)^H \longrightarrow L^\infty_{\mathrm{w*}}(S^2,E)^H \longrightarrow L^\infty_{\mathrm{w*}}(S^3,E)^H \longrightarrow \cdots \qquad (*)$$

endows the space $\mathrm{H}^n_{\mathrm{cb}}(H,E)$ with a canonical (given S and $E^\flat$) coefficient G-module structure. Indeed, we have supposed $\mathrm{H}^n_{\mathrm{cb}}(H,E)$ Hausdorff, which implies that the image of $L^\infty_{\mathrm{w*}}(S^n,E)^H$ in $L^\infty_{\mathrm{w*}}(S^{n+1},E)$ under the homogeneous differential d is closed. However, the map d is adjoint ; therefore, the closed range theorem [**126**, I 4.14] implies that this image is weak-* closed, so that the quotient defining the cohomology of (*) in degree n inherits the weak-* topology from $L^{\infty}_{w*}(S^{n+1}, E)$.

This coefficient module structure depends of course on the predual of E. We insist however that the coefficient module structure depends a priori also on the amenable space S. Indeed, the equivariant means studied in Section 5 are not adjoint in general.

Finally, we come back to the question of the image of the restriction map. We begin with the elementary case of finite index subgroups, where the Proposition 8.8.1 allows us to benefit some more of the idea of transfer :

COROLLARY 8.8.5.— *Keep the notation of Proposition 8.8.1 and suppose H of finite index in G. Then for every $n \geq 0$ the restriction*

$$\mathrm{res}: \ \mathrm{H}^n_{\mathrm{cb}}(G, E) \longrightarrow \mathrm{H}^n_{\mathrm{cb}}(H, E)^{G/H}$$

is an isometric isomorphism on the G/H-invariants.

PROOF. The restriction is induced by the inclusion $\iota^\bullet$ of complexes

$$\begin{array}{ccccccccc} 0 \longrightarrow & (E^0)^G & \longrightarrow & (E^1)^G & \longrightarrow & (E^2)^G & \longrightarrow & (E^3)^G & \longrightarrow \cdots \\ & \downarrow{\scriptstyle \iota^0} & & \downarrow{\scriptstyle \iota^1} & & \downarrow{\scriptstyle \iota^2} & & \downarrow{\scriptstyle \iota^3} & \\ 0 \longrightarrow & (E^0)^H & \longrightarrow & (E^1)^H & \longrightarrow & (E^2)^H & \longrightarrow & (E^3)^H & \longrightarrow \cdots \end{array}$$

Define a transfer map $\mathrm{trans}^n : (E^n)^H \to (E^n)^G$ by

$$\mathrm{trans}^n(v) = \frac{1}{[G:H]} \sum_{gH \in G/H} \pi_n(g)v.$$

Now trans^n is a left inverse of norm at most one to the norm one map ι^n, so it is sufficient to show that for every $v \in (E^n)^H$ representing a G/H-invariant cohomology class, the element $\iota^n\mathrm{trans}^n(v)$ is cohomologous to v. But, according to Proposition 8.8.1, the G/H-invariance means that for every $hG \in G/H$ there is an element $v_g \in (E^{n-1})^H$ with $v - \pi_n(g)v = du_{gH}$. The linearity of d implies now $v - \iota^n\mathrm{trans}^n(v) = du$ for

$$u = \frac{1}{[G:H]} \sum_{gH \in G/H} u_{gH}.$$

□

When dealing with coefficient modules, we can implement a similar idea for the much more general setting of amenable quotients. However, since we are not taking finite averages any more, we will have to suppose that the relevant cohomology space is Hausdorff ; we shall see later on (Section 11.4) that in most cases this assumption is satisfied in degrees up to two.

COROLLARY 8.8.6.— *Let $H \lhd G$ be a closed normal subgroup of the locally compact second countable group G and let (π, E) be a coefficient G-module. Assume that $\mathrm{H}^n_{\mathrm{cb}}(H,E)$ is Hausdorff. If the quotient G/H is amenable, then the restriction yields an isometric isomorphism*

$$\mathrm{H}^n_{\mathrm{cb}}(G,E) \cong \mathrm{H}^n_{\mathrm{cb}}(H,E)^{G/H}.$$

PROOF. The proof consists in using the explicit realization of the G-action on $\mathrm{H}^n_{\mathrm{cb}}(H,E)$ provided by Corollary 8.8.3 in order to benefit further from the transfer construction used in the proof of Proposition 8.6.6. Therefore, we borrow the whole notation from this latter proof. In particular, we recall that the restriction is realized by

$$\begin{array}{ccccccccc}
0 & \longrightarrow & L^\infty_{\mathrm{w*}}(G,E)^G & \longrightarrow & L^\infty_{\mathrm{w*}}(G^2,E)^G & \longrightarrow & L^\infty_{\mathrm{w*}}(G^3,E)^G & \longrightarrow & \cdots \\
 & & \downarrow{\scriptstyle \iota^0} & & \downarrow{\scriptstyle \iota^1} & & \downarrow{\scriptstyle \iota^2} & & \\
0 & \longrightarrow & L^\infty_{\mathrm{w*}}(G,E)^H & \longrightarrow & L^\infty_{\mathrm{w*}}(G^2,E)^H & \longrightarrow & L^\infty_{\mathrm{w*}}(G^3,E)^H & \longrightarrow & \cdots
\end{array}$$

and that we constructed for every coefficient G-module F a G-equivariant mean

$$\mathfrak{m}_F :\ L^\infty_{\mathrm{w*}}(G/H,F) \longrightarrow F$$

which is *adjointly natural* in F. Further, we had set $F^n = L^\infty_{\mathrm{w*}}(G^{n+1},E)$ and defined

$$\tau^n :\ (F^n)^H \longrightarrow L^\infty_{\mathrm{w*}}(G/H,F^n)^G, \qquad \tau^n f(gH) = \lambda_\pi(g)f$$

so that $\mathrm{trans}^n_{\mathfrak{m}} = \mathfrak{m}_{F^n}\tau^n$ gives a norm one left inverse to ι^n which commutes to the differentials $d^\bullet$.

Turning now to Corollary 8.8.6, we have only to show that the restriction is surjective onto the G/H-invariants. Thus we want to show that every map $f \in L^\infty_{\mathrm{w*}}(G^{n+1},E)^H$ representing a G/H-invariant class in $\mathrm{H}^n_{\mathrm{cb}}(H,E)^{G/H}$ is in the space

$$L^\infty_{\mathrm{w*}}(G^{n+1},E)^G + d\Big(L^\infty_{\mathrm{w*}}(G^n,E)^H\Big).$$

To this end, we shall show that $\mathrm{trans}^n_{\mathfrak{m}} f - f$ is in $d\big(L^\infty_{\mathrm{w*}}(G^n,E)^H\big)$. Since $\mathfrak{m}_{F^n}$ is a mean, we have $f = \mathfrak{m}_{F^n}(f\mathbf{1}_{G/H})$. Therefore we have

$$\mathrm{trans}^n_{\mathfrak{m}} f - f = \mathfrak{m}_{F^n}\big(\tau^n f - f\mathbf{1}_{G/H}\big).$$

Now Corollary 8.8.3 tells us that the G-action on $\mathrm{H}^n_{\mathrm{cb}}(H,E)$ is also induced by the G-representation λ_π on F^n. Therefore, the element

$$\lambda_\pi(g)f - f = \big(\tau^n f - f\mathbf{1}_{G/H}\big)(gH)$$

is in $d\big(L^\infty_{\mathrm{w*}}(G^n,E)^H\big)$ for every $g \in G$. Since the latter space is closed in norm by assumption and since the differential is adjoint, the closed range theorem implies that $d\big(L^\infty_{\mathrm{w*}}(G^n,E)^H\big)$ is weak-* closed in the space of cocycles and hence in F^n. Thus, by the duality principle, the inclusion map

$$d\Big(L^\infty_{\mathrm{w*}}(G^n,E)^H\Big) \longrightarrow F^n$$

is adjoint, and therefore the adjoint naturality of $\mathfrak{m}_{F^n}$ established in the proof of Proposition 8.6.6 implies that $\mathfrak{m}_{F^n}\big(\tau^n f - f\mathbf{1}_{G/H}\big)$ is indeed in $d\big(L^\infty_{\mathrm{w*}}(G^n, E)^H\big)$, as claimed. □

EXAMPLE 8.8.7. In the Example 8.6.7, we have considered the case of an automorphism φ of a countable group Γ. As a consequence of the above Corollary 8.8.6, the corresponding semi-direct product extension

$$1 \longrightarrow \Gamma \longrightarrow \Gamma \rtimes_\varphi \mathbf{Z} \longrightarrow \mathbf{Z} \longrightarrow 0$$

leads to the isometric isomorphism

$$\mathrm{H}^n_{\mathrm{b}}(\Gamma \rtimes_\varphi \mathbf{Z}) \cong \mathrm{H}^n_{\mathrm{cb}}(\Gamma)^{<\varphi>}$$

as soon as $\mathrm{H}^n_{\mathrm{cb}}(\Gamma)$ is Hausdorff.

9. Continuous cohomology and the comparison map

One of the important features of the theory of continuous bounded cohomology is that the bi-functors $\mathrm{H}^\bullet_{\mathrm{cb}}$ come with a natural transformation

$$\Psi^\bullet : \mathrm{H}^\bullet_{\mathrm{cb}} \longrightarrow \mathrm{H}^\bullet_{\mathrm{c}}$$

to the continuous cohomology $\mathrm{H}^\bullet_{\mathrm{c}}$ or to the Eilenberg-MacLane cohomology if the group under consideration is discrete. As a matter of fact, most applications of bounded continuous cohomology are intimately connected with the properties of these *comparison maps* :

On one hand, the kernel of Ψ^2 has very concrete interpretations in terms of *quasimorphisms* and more generally of *rough actions*, as we shall see in Section 13.3.

On the other hand, since the maps $\Psi^\bullet$ link [continuous] bounded cohomology to Eilenberg-MacLane [resp. continuous] cohomology, they really present the former as valuable additional and refined information about the brave old cohomology. We mention two instances of this.

(1) Suppose a very classical cohomology class $\omega \in \mathrm{H}^n_{\mathrm{c}}(G)$ turns out to be in the image $\Psi^n\mathrm{H}^n_{\mathrm{cb}}(G)$ (see the examples in Section 9.3 below). Then ω inherits a canonical quotient semi-norm from the semi-norm on $\mathrm{H}^n_{\mathrm{cb}}(G)$, thus providing G with a new numerical invariant, the *Gromov semi-norm*.

In the case where $G = \pi_1(M)$ is the fundamental group of a manifold M, M. Gromov has shown in [**69**] the relevance of this invariant to geometric characteristics of M such as the minimal volume.

(2) It happens in many situations that $\omega \in \mathrm{H}^n_{\mathrm{c}}(G)$ is actually given in a natural way by a particular bounded cocycle. There is then a distinguished pre-image ω_{b} in $\mathrm{H}^n_{\mathrm{cb}}(G)$ for ω under Ψ^n. Since this map is in general not injective, the class ω_{b} may provide more information than the sole class ω. This happens for the Euler class ω associated to an action on the circle : while ω gives only

the obstruction to lift the action to the line, É. Ghys has shown in [**63**] that the natural "bounded" Euler class ω_b classifies reasonably the topology of the action in the sense that ω_b is a complete invariant of topological semi-conjugacy, see Section 13.5 below.

9.1. Continuous cohomology. There are several comprehensive sources where the continuous cohomology of *locally compact* topological groups is presented ; we refer to A. Guichardet's book [**74**] and to the exposition [**83**] by G. Hochschild and G.D Mostow.

Even though our main interest is in this setting, we shall define the comparison map for general topological groups, as this is the generality in which we have defined continuous bounded cohomology. For such groups, most of the techniques developed in [**74**] and [**83**] fail, and actually the corresponding continuous cohomology appears very scarcely in the literature. We shall briefly recall below the definitions given by G.D. Mostow in [**116**, §2] for general topological groups, but first we give a crude definition :

DEFINITION 9.1.1. Let G be a topological group and (π, E) a Banach G-module. The *continuous cohomology* of G with coefficients in E, denoted by $\mathrm{H}^\bullet_\mathrm{c}(G,E)$, is by definition the cohomology of the complex

$$0 \longrightarrow \mathrm{C}(G,E)^G \stackrel{d}{\longrightarrow} \mathrm{C}(G^2,E)^G \stackrel{d}{\longrightarrow} \mathrm{C}(G^3,E)^G \stackrel{d}{\longrightarrow} \cdots$$

of invariant continuous functions with the homogeneous differential of Notation 7.1.1. Invariance is understood with respect to the left regular representation λ_π.

REMARK 9.1.2. It follows immediately from the definition of $\mathrm{H}^\bullet_\mathrm{c}$ that continuous cohomology generalizes the classical Eilenberg-MacLane cohomology of groups. That is, if the topological group G is discrete, then one has $\mathrm{H}^\bullet_\mathrm{c}(G,-) = \mathrm{H}^\bullet(\underline{G},-)$, where $\underline{G}$ denotes the group underlying G and $\mathrm{H}^\bullet$ is the classical Eilenberg-MacLane cohomology.

Applying post- and pre-composition to the cochains, we see that $\mathrm{H}^\bullet_\mathrm{c}(G,E)$ is covariant in E and contravariant in G.

PROPOSITION 9.1.3.— *The inclusion map* $\mathcal{C}E \to E$ *induces an identification* $\mathrm{H}^\bullet_\mathrm{c}(G,\mathcal{C}E) \cong \mathrm{H}^\bullet_\mathrm{c}(G,E)$.

PROOF. It is enough to check that for all $n \geq 0$ any element of $\mathrm{C}(G^{n+1},E)^G$ ranges in $\mathcal{C}E$. Let $x \in G^{n+1}$ and let $(g_\alpha)_{\alpha\in A}$ be a net converging to $e \in G$. Then

$$\|\pi(g_\alpha)f(x) - f(x)\|_E = \|\pi(g_\alpha)f(x) - \pi(g_\alpha)f(g_\alpha^{-1}x)\|_E = \|f(x) - f(g_\alpha^{-1}x)\|_E$$

tends to zero. □

Now we turn back to Mostow's exposition : in [**116**], a *topological module* E over a topological ring R is a module for which all structure maps are continuous. In particular, a Banach space is a topological module over $\mathbf{C}$. If G is a topological group acting by topological automorphisms on the topological module E over R, then E is called a *topological G-R module* if the action map $G \times E \to E$ is continuous. More generally, it is a *continuous G-R module* if all orbital maps are continuous.

In view of Lemma 1.1.1, these two notions coincide for a Banach G-module and amount to the concept of *continuous* Banach G-module.

G.D. Mostow defines $\mathrm{H}_{\mathrm{c}}^{\bullet}(G,-)$ exactly as in Definition 9.1.1, but only for continuous G-R modules ; in view of Proposition 9.1.3 above, this is irrelevant to us. The modules $\mathrm{C}(G^{n+1},E)$, for E continuous, are given a structure of continuous G-R module by means of the topology of pointwise convergence.

Now, one finds in [**116**] the basics of a functorial approach to continuous cohomology, relying on a weakened notion of injectivity and on the notion of R-split sequences. Further developments of such a theory were restricted to the locally compact case, and the paper of Hochschild-Mostow [**83**] amends Mostow's original [**116**] in this way, as explained in the introduction of [**83**].

All this methodology is a bit simpler than our Section 7.2 for the reasons exposed in the head of Section 7.

The only important upshot for us in the next Section 9.2 is that the continuous subcomplex of our strong augmented resolutions are R-split in Mostow's sense.

We notice in passing that the continuous cohomology also admits inhomogeneous complexes :

LEMMA 9.1.4.— *Let G be a topological group and (π, E) a Banach G-module. Then the inhomogeneous complex*

$$0 \longrightarrow CE \xrightarrow{\delta_\pi} \mathrm{C}(G,CE) \xrightarrow{\delta_\pi} \mathrm{C}(G^2,CE) \xrightarrow{\delta_\pi} \mathrm{C}(G^3,CE) \longrightarrow \cdots$$

realizes $\mathrm{H}_{\mathrm{c}}^{\bullet}(G,E)$.

PROOF. The very formulae given in the proof of Proposition 7.4.12 in the locally compact case for the isomorphisms $U^\bullet$ and $V^\bullet$ apply also in the present case, since we have seen in the proof of Proposition 9.1.3 that the elements of $\mathrm{C}(G^{n+1},E)^G$ range in CE. □

In the particular case of Lie groups, there are many known characterizations of the continuous bounded cohomology, using notably invariant differential forms or Lie algebra relative cohomology in the general case. The latter identification is the object of the van Est theorem, and can be viewed as a particular case of the functorial approach of G. Hochschild and G.D. Mostow ([**83**], or see the book of A. Borel and N. Wallach [**19**]).

We quote for instance the following general version of van Est's theorem :

THEOREM 9.1.5.— *Let G be a real analytic group, $K < G$ a maximal compact subgroup and V a locally convex integrable differentiable G-module. There is a canonical isomorphism*

$$\mathrm{H}_{\mathrm{b}}^{\bullet}(G,V) \cong \mathrm{H}^{\bullet}(\mathfrak{g},\mathfrak{k},V),$$

where the right hand side is the relative Lie algebra cohomology. These spaces are further canonically isomorphic to the cohomology of the complex

$$0 \longrightarrow \Omega^0(G/K,V)^G \longrightarrow \Omega^1(G/K,V)^G \longrightarrow \Omega^2(G/K,V)^G \longrightarrow \cdots$$

of invariant V-valued differential forms on G/K.

The ***integrability*** of V is a technical condition that ensures that the appropriate version of a V-valued Poincaré lemma holds. We do not explicit this here since the above theorem is quoted as an illustration.

On the proof. This is Theorem 6.1 in [**83**]. Alternatively, see [**19**, IX.5.6]. □

We emphasize that no analogous result is known for continuous bounded cohomology, and there is strong evidence indeed that any similar statement be false.

9.2. The comparison map. We first give a concrete definition of the comparison maps for the sake of definiteness, and then put it into its natural functorial context.

Let G be a topological group and E a Banach G-module. Recall that $\mathrm{H}^\bullet_{\mathrm{cb}}(G,E)$ is defined by the canonical complex

$$0 \longrightarrow \mathrm{C}^0_{\mathrm{b}}(G,E)^G \xrightarrow{\partial^1} \mathrm{C}^1_{\mathrm{b}}(G,E)^G \xrightarrow{\partial^2} \mathrm{C}^2_{\mathrm{b}}(G,E)^G \longrightarrow \cdots$$

and that there is a canonical isometric embedding of the latter into the homogeneous complex

$$0 \longrightarrow \mathrm{C}_{\mathrm{b}}(G,E)^G \xrightarrow{d} \mathrm{C}_{\mathrm{b}}(G^2,E)^G \xrightarrow{d} \mathrm{C}_{\mathrm{b}}(G^3,E)^G \xrightarrow{d} \cdots$$

(see Remark 6.1.2). On the other hand, the latter is a subcomplex of the complex

$$0 \longrightarrow \mathrm{C}(G,E)^G \xrightarrow{d} \mathrm{C}(G^2,E)^G \xrightarrow{d} \mathrm{C}(G^3,E)^G \xrightarrow{d} \cdots$$

defining $\mathrm{H}^\bullet_{\mathrm{c}}(G,E)$. We have therefore a canonical collection of linear maps

$$\Psi^n : \mathrm{H}^n_{\mathrm{cb}}(G,E) \longrightarrow \mathrm{H}^n_{\mathrm{c}}(G,E).$$

Definition 9.2.1. We call $\Psi^\bullet$ the *comparison maps*. If necessary, we write $\Psi^\bullet_G$.

We shall occasionly use the following shorthand :

Definition 9.2.2. The kernel of the map Ψ^n is called the *exact part* of $\mathrm{H}^n_{\mathrm{cb}}(G,E)$ and is denoted by $\mathrm{EH}^n_{\mathrm{cb}}(G,E)$. One defines accordingly $\mathrm{EH}^n_{\mathrm{b}}$.

A first observation is that the comparison maps are compatible with co- and contravariance in asmuch as they intertwine the corresponding transformations of the bifunctors $\mathrm{H}^\bullet_{\mathrm{cb}}$ and $\mathrm{H}^\bullet_{\mathrm{c}}$; this follows from the very definitions since in both cases functoriality comes from post- and pre-composition.

However, the comparison maps are natural in a stronger sense :

PROPOSITION 9.2.3.— *Let G be a topological group, E a Banach G-module and $(\mathfrak{a}, E^\bullet)$ a strong resolution of E by relatively injective Banach G-modules. Suppose there is a weaker locally convex topology $\mathcal{T}$ on E and $\mathcal{T}^n$ on each of the E^n for which they are all continuous G-$\mathbf{C}$ modules and such that the augmented resolution is still a $\mathbf{C}$-split complex for these topologies. Let $(\mathfrak{b}, F^\bullet)$ be a $\mathbf{C}$-split resolution of E, in its topology $\mathcal{T}$, by continuous G-$\mathbf{C}$ modules that are injective in Mostow's sense.*

Then there is a sequence of G-equivariant linear maps continuous for $\mathcal{T}, \mathcal{T}^n$

$$\begin{array}{ccccccccc}
0 & \longrightarrow & E & \stackrel{\mathfrak{a}}{\longrightarrow} & E^0 & \longrightarrow & E^1 & \longrightarrow & E^2 & \longrightarrow & \cdots \\
 & & \| & & \downarrow & & \downarrow & & \downarrow & & \\
0 & \longrightarrow & E & \stackrel{\mathfrak{b}}{\longrightarrow} & F^0 & \longrightarrow & F^1 & \longrightarrow & F^2 & \longrightarrow & \cdots
\end{array}$$

commuting with the differentials. Any such sequence induces $\Psi^\bullet$ to within canonical topological isomorphisms.

PROOF. This follows from the conjunction of our Theorem 7.2.1 with G.D. Mostow's functoriality [**116**]. □

REMARK 9.2.4. The statement is a bit lengthy, but keep in mind that the auxiliary topologies may very well just be the initial Banach topology if e.g. one uses the continuous sub-resolution.

EXAMPLE 9.2.5. Let G be a locally compact group and E a Banach G-module. Then $\Psi^\bullet : \mathrm{H}^\bullet_{\mathrm{cb}}(G, E) \to \mathrm{H}^\bullet_{\mathrm{c}}(G, E)$ is induced by the inclusion of complexes

$$\begin{array}{ccccccccccc}
0 & \longrightarrow & E & \longrightarrow & \mathrm{C_b}(G, E) & \longrightarrow & \mathrm{C_b}(G^2, E) & \longrightarrow & \mathrm{C_b}(G^3, E) & \longrightarrow & \cdots \\
 & & \| & & \downarrow & & \downarrow & & \downarrow & & \\
0 & \longrightarrow & E & \longrightarrow & \mathrm{C}(G, E) & \longrightarrow & \mathrm{C}(G^2, E) & \longrightarrow & \mathrm{C}(G^3, E) & \longrightarrow & \cdots
\end{array}$$

Indeed, there is no loss of generality in supposing E continuous, and at the level of invariants the inclusions above induce the same maps than the inclusions

$$\mathcal{C}\mathrm{C_b}(G^{n+1}, E) \longrightarrow \mathrm{C}(G^{n+1}, E).$$

We may thus apply the Proposition 9.2.3 with the original Banach topology on the continuous sub-resolution since by Corollary 7.4.7 the first line in the above diagram is a strong augmented resolution of E by relatively injective Banach G-modules.

REMARK 9.2.6. It follows from this example that for a locally compact group G the inclusion of complexes

$$\begin{array}{ccccccccccc}
0 & \longrightarrow & \mathcal{C}E & \stackrel{\delta_\pi}{\longrightarrow} & \mathrm{C_b}(G, \mathcal{C}E) & \stackrel{\delta_\pi}{\longrightarrow} & \mathrm{C_b}(G^2, \mathcal{C}E) & \stackrel{\delta_\pi}{\longrightarrow} & \mathrm{C_b}(G^3, \mathcal{C}E) & \stackrel{\delta_\pi}{\longrightarrow} & \cdots \\
 & & \| & & \downarrow & & \downarrow & & \downarrow & & \\
0 & \longrightarrow & \mathcal{C}E & \stackrel{\delta_\pi}{\longrightarrow} & \mathrm{C}(G, \mathcal{C}E) & \stackrel{\delta_\pi}{\longrightarrow} & \mathrm{C}(G^2, \mathcal{C}E) & \stackrel{\delta_\pi}{\longrightarrow} & \mathrm{C}(G^3, \mathcal{C}E) & \stackrel{\delta_\pi}{\longrightarrow} & \cdots
\end{array}$$

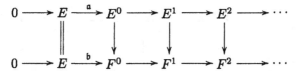

induces also $\Psi^\bullet$. Indeed, the isomorphisms between these complexes and the homogeneous complexes in the above example are given exactly by the same formulae (Proposition 7.4.12 and Lemma 9.1.4).

EXAMPLE 9.2.7. Let G be a locally compact second countable group and E a separable coefficient G-module. We recall from Lemma 3.3.3 that in this situation the weak-* Borel structure and the norm Borel structure on E coincide. Thus we may write $L^\infty(-,E)$ for $L^\infty_{w*}(-,E)$, but more importantly, fixing $1 < p < \infty$, there are inclusions

$$\begin{array}{ccccccccccc} 0 & \longrightarrow & E & \longrightarrow & L^\infty(G,E) & \longrightarrow & L^\infty(G^2,E) & \longrightarrow & L^\infty(G^3,E) & \longrightarrow & \cdots \\ & & \| & & \downarrow & & \downarrow & & \downarrow & & \\ 0 & \longrightarrow & E & \longrightarrow & L^p_{\mathrm{loc}}(G,E) & \longrightarrow & L^p_{\mathrm{loc}}(G^2,E) & \longrightarrow & L^p_{\mathrm{loc}}(G^3,E) & \longrightarrow & \cdots \end{array}$$

These inclusions induce the comparison maps $\Psi^\bullet$, and the same holds for alternating cochains. Indeed, we may apply Proposition 9.2.3 by taking the weak-* topologies as auxiliary topologies : the differentials are adjoint, the contracting homotopy of Lemma 7.5.5 is adjoint because it comes from integration, and the inclusions

$$L^\infty(G^{n+1},E) \hookrightarrow L^p_{\mathrm{loc}}(G^{n+1},E)$$

are adjoint to

$$L^q_{00}(G^{n+1},E^\flat) \hookrightarrow L^1(G^{n+1},E^\flat),$$

where $1/p+1/q=1$ and L^q_{00} denotes the Fréchet space of compactly supported q-integrable function classes. The remaining assumptions on the first line are satisfied by Proposition 7.5.1, while the injectivity, in Mostow's sense, of the second line is due to Ph. Blanc : Théorème 3.4 in [**13**].

9.3. Examples and remarks. Given an ordinary cohomology class ω, we have insisted in the head of Section 9 upon the importance for ω of being in the image of $\Psi^\bullet$. This happens in several interesting cases.

EXAMPLE 9.3.1. If G is a connected semi-simple real rank one Lie group, then $\Psi^n : \mathrm{H}^n_{\mathrm{cb}}(G) \to \mathrm{H}^n_{\mathrm{c}}(G)$ is surjective for any $n \geq 2$ (see M. Gromov [**69**]). This is due to the fact that the geodesic simplices in the corresponding symmetric space have uniformly bounded volume.

M. Gromov proposed a more general statement :

EXAMPLE 9.3.2. If Γ is a non-elementary Gromov-hyperbolic group, then

$$\Psi^n : \mathrm{H}^n_{\mathrm{b}}(\Gamma,E) \to \mathrm{H}^n(\Gamma,E)$$

is surjective for every $n \geq 2$ and every Banach Γ-module E (I. Mineyev [**111**]). I. Mineyev's argument uses a homological bi-combing in order to capture the geometric intuition behind the previous example. It turns out that this surjectivity characterizes non-elementary Gromov-hyperbolic groups [**112**].

EXAMPLE 9.3.3. Let G be any connected simple real Lie group. A. Guichardet and D. Wigner show in [**75**] that the space $\mathrm{H}_{\mathrm{c}}^2(G)$ is either zero or one dimensional. The latter is the case exactly when the associated symmetric space is of Hermitian type. Then Guichardet and Wigner proceed to give an explicit cocycle generating $\mathrm{H}_{\mathrm{c}}^2(G)$ (see also [**51**]). This cocycle arises as the obstruction to extend to G a complex character defined on a maximal compact subgroup, and therefore it is bounded. Thus the comparison map $\Psi^2 : \mathrm{H}_{\mathrm{cb}}^2(G) \to \mathrm{H}_{\mathrm{c}}^2(G)$ is surjective. On the other hand, one can show that this comparison map is injective : this has been observed by M. Burger and the author (Lemma 6.1 in [**35**]) and for the convenience of the reader we shall give the proof below (Lemma 13.3.4).

Thus Ψ^2 is an isomorphism $\mathrm{H}_{\mathrm{cb}}^2(G) \cong \mathrm{H}_{\mathrm{c}}^2(G)$ and both spaces have dimension one or zero according to wether the associated symmetric space is of Hermitian type ; notice that the case of connected almost simple real Lie groups follows at once.

REMARK 9.3.4. For the semi-simple case, it is well known that $\mathrm{H}_{\mathrm{c}}^2(G)$ is the sum of the $\mathrm{H}_{\mathrm{c}}^2$ spaces of its almost simple factors. One way of seeing this is to use van Est's theorem (Theorem 9.1.5 above), because in relative Lie algebra cohomology with finite dimensional coefficients there is a Künneth formula, see Section II.9 in [**19**]. As a result, Example 9.3.3 implies that the dimension of $\mathrm{H}_{\mathrm{c}}^2(G)$ is the number of almost simple factors of Hermitian type occuring in G.

We however insist that there is no general Künneth formula for less trivial coefficients, even in degree one. As for continuous bounded cohomology, we shall establish a strong analogue of a degree two Künneth formula in Section 12.

EXAMPLE 9.3.5. Taking over the preceding Example 9.3.3 to the non Archimedean world, we get the following. Let $G = \mathbf{G}(k)$, where $\mathbf{G}$ is a connected, simply connected, almost simple k-isotropic group and k a non Archimedean local field. Then $\mathrm{H}_{\mathrm{cb}}^2(G) = 0$.

Indeed the comparison map is again injective as we shall prove in Lemma 13.3.4 below and the space $\mathrm{H}_{\mathrm{c}}^2(G)$ vanishes (see Theorem 4.12 of Chapter X in the book [**19**] of Borel-Wallch).

EXAMPLE 9.3.6. As mentioned in Examples 8.6.3, the volume form of the symmetric space corresponding to $G = \mathrm{SL}_n(\mathbf{R})$ yields for every $n \geq 2$ a non-trivial cohomology class in $\mathrm{H}_{\mathrm{c}}^d(G)$ for $d = n(n+1)/2 - 1$. Savage's result [**127**] shows that this class is in the image of Ψ^d. Observe that the case $n = 2$ falls into Example 9.3.1.

In Example 9.3.1, the particular case of the three dimensional hyperbolic space is of particular interest : it is known that the corresponding volume form cohomology class can be expressed in terms of the imaginary part of the dilogarithm ; the cocycle equation corresponds then to the Abel-Spence five terms functional equation[2]. For background on these connections, see for instance [**14, 50, 61, 95**].

[2]For a cohomological interpretation of the real part, we shall publish some remarks with M.Burger in [**38**].

Now, instead of considering this instance as the three dimensional case of the family of rank one examples, we recall that the connected component of the corresponding isometry group is isomorphic to $\mathrm{SL}_2(\mathbf{C})$. Thus there is another possible generalization if we increase the rank :

EXAMPLE 9.3.7. A. Goncharov's study of functional equations for the trilogarithm [**67**] shows that the Borel class generating $\mathrm{H}^5_{\mathrm{c}}(\mathrm{SL}_3(\mathbf{C}))$ can be represented by a bounded cocycle and hence Ψ^5 is surjective.

This last example is crying out for further generalization. Recall that the continuous cohomology ring $\mathrm{H}^\bullet_{\mathrm{c}}(\mathrm{SL}_n(\mathbf{C}))$ is an exterior algebra

$$\mathrm{H}^\bullet_{\mathrm{c}}(\mathrm{SL}_n(\mathbf{C})) = \bigwedge \langle x_{n,2j+1} \,|\, 1 \leq j \leq n-1 \rangle$$

over the *Borel classes* $x_{n,2j+1}$ of degree $2j+1$, see [**15**]. A. Goncharov gives also formulae expressing cocycles for theses classes in terms of higher polylogarithms ; we could however not check wether theses cocycles are bounded. We propose the

CONJECTURE 9.3.8.— *The comparison map*

$$\Psi^\bullet : \mathrm{H}^\bullet_{\mathrm{cb}}(\mathrm{SL}_n(\mathbf{C})) \to \mathrm{H}^\bullet_{\mathrm{c}}(\mathrm{SL}_n(\mathbf{C}))$$

is surjective in every degree and for all $n \geq 2$.

More generally, one could formulate Conjecture 9.3.8 for every semi-simple connected real Lie group G with finite centre. J.L. Dupont proposed in [**49**] a conjecture that turns out to be even stronger ; in order to state it, we have to recall that the integration over geodesic simplices of invariant differential forms on the associated symmetric space X yields the *Dupont isomorphism*

$$\mathcal{D}: \ \Omega^\bullet(X)^G \longrightarrow \mathrm{H}^\bullet_{\mathrm{c}}(G),$$

where $\Omega^\bullet(X)^G$ is the graded space of G-invariant differential forms.

CONJECTURE 9.3.9 (J.L. Dupont).— *The cocycles for* $\mathrm{H}^\bullet_{\mathrm{c}}(G)$ *obtained by the integration* $\mathcal{D}$ *are bounded.*

We should also give a non-trivial example where the comparison map is far from being surjective. We borrow the following observation from our joint article [**37**] with M. Burger :

EXAMPLE 9.3.10. Let $\Gamma < G = \mathrm{SL}_2(\mathbf{R}) \times \mathrm{SL}_2(\mathbf{R})$ be a co-compact torsion free irreducible lattice. Then, as we shall prove below (Corollary 14.3.3), the space $\mathrm{H}^2_{\mathrm{b}}(\Gamma)$ has dimension two. On the other hand, the computation of ℓ^2 Betti numbers shows that

$$\dim \mathrm{H}^2(\Gamma) = c\mathrm{Vol}(\Gamma\backslash G) - 2,$$

wherein c is an absolute constant. In particular, this dimension goes to infinity when we let Γ range over the finite index subgroups of a given such lattice.

On a more anecdotal level, we mention that it is important to require that the Lie groups under consideration have finite centre :

EXAMPLE 9.3.11. Let $G = \mathrm{SL}_2(\mathbf{R})$ and denote by $\pi : \widetilde{G} \to G$ its universal covering ; $\widetilde{G}$ is a non-linear Lie group with same Lie algebra $\mathfrak{sl}_2(\mathbf{R})$. We claim :

(i) The comparison map $\mathrm{H}^2_{\mathrm{cb}}(\widetilde{G}) \to \mathrm{H}^2_{\mathrm{c}}(\widetilde{G})$ is not injective.

(ii) The comparison map $\mathrm{H}^3_{\mathrm{cb}}(\widetilde{G}) \to \mathrm{H}^3_{\mathrm{c}}(\widetilde{G})$ is not surjective.

For the first claim, we consider the commutative diagram

$$\begin{array}{ccc} \mathrm{H}^2_{\mathrm{cb}}(G) & \xrightarrow{\Psi^2_G} & \mathrm{H}^2_{\mathrm{c}}(G) \\ {\scriptstyle\mathrm{inf}}\downarrow & & \downarrow{\scriptstyle\mathrm{inf}} \\ \mathrm{H}^2_{\mathrm{cb}}(\widetilde{G}) & \xrightarrow[\Psi^2_{\widetilde{G}}]{} & \mathrm{H}^2_{\mathrm{c}}(\widetilde{G}) \end{array}$$

Since the inflation is induced by the map π which has amenable kernel $\mathbf{Z}$, we know from Corollary 8.5.2 that the left arrow is injective. However, the inflation on the right is a zero map, since $\mathrm{H}^2_{\mathrm{c}}(G)$ is generated precisely by the class corresponding to the extension

$$1 \longrightarrow \mathbf{Z} \longrightarrow \widetilde{G} \longrightarrow G \longrightarrow 1.$$

Therefore, the map $\Psi^{\bullet}_{\widetilde{G}}$ cannot be injective.

As for (ii), van Est's theorem yields that $\mathrm{H}^3_{\mathrm{c}}(\widetilde{G})$ has dimension one (see also [**50**] for more on this). On the other hand, Corollary 8.5.2 identifies $\mathrm{H}^3_{\mathrm{cb}}(\widetilde{G})$ with $\mathrm{H}^3_{\mathrm{cb}}(G)$, and we have shown with M. Burger [**38**] that the latter vanishes.

CHAPTER IV

Cohomological techniques

The functoriality painfully collected through Sections 7 and 8 will now allow us to establish powerful cohomological techniques.

We will first see in Section 10 a couple of constructions which are very analogous to elementary procedures in Eilenberg-MacLane cohomology, except for some difficulties arising notably from the presence of a topology on the groups.

Then, in Section 11, we encounter a new phenomenon which pertains genuinely to the theory of [continuous] *bounded* cohomology : the existence of doubly ergodic amenable spaces. This phenomenon, which has been presented by M. Burger and the author in [**37**] as a generalization of Mautner's property, has remarkable consequences for bounded cohomology and its applications.

Finally, the Section 12 attacks the issue of Hochschild-Serre spectral sequences for group extensions. There, if the strategy is again inspired by usual cohomology, the outcome is very much influenced by the phenomenon of double ergodicity. In particular, we obtain an exact sequence up to degree three, with a refined term in degree two.

10. General techniques

10.1. Induction. The induction principle (more properly referred to as co-induction) is a generalization of the Eckmann–Shapiro lemma[1] in usual discrete group cohomology, which provides an isomorphism between the cohomology of a group and the cohomology of some ambient group, with coefficients in a suitable induction module. The original Eckmann–Shapiro lemma has been generalized : a version for the continuous cohomology of topological groups is given by Ph. Blanc [**13**, Théorème 8.7] or by A. Borel and N.R. Wallach [**19**, IX, Proposition 2.3].

We derive an analogous result in continuous bounded cohomology for coefficient modules over locally compact second countable groups. This setting has the advantage of allowing for flexibility in the definition of the induction map, benefiting from the variety of resolutions over amenable spaces provided by the

[1] This lemma has long been named after A. Shapiro. It appears however that this induction principle was introduced by B. Eckmann in [**52**].

Theorem 7.5.3 (of course, these various realizations of the induction map will turn out to coincide).

A very useful "deviant" variant of this induction will be introduced later in Section 10.2.

Let G be a locally compact second countable group and $H < G$ a closed subgroup. Let (π, E) be a coefficient H-module. Construct a coefficient G-module out of the Banach H-module E in the following way :

DEFINITION 10.1.1. The ***induction module*** $\mathrm{I}_H^G E$ is the space $\mathrm{I}_H^G E = L^\infty_{\mathrm{w}*}(G, E)^H$ endowed with the right translation G-representation ϱ.

REMARKS 10.1.2.

(i) As usual (Notation 2.4.7), the H-invariance considered in the definition of $\mathrm{I}_H^G E$ is with respect to λ_π.

(ii) We may consider $L^\infty_{\mathrm{w}*}(G, E)$ as a $H \times G$-module via $\lambda_\pi \times \varrho$, as in Scholium 2.4.3, except that here π is only defined on H. Now, applying the Lemma 1.2.9, we conclude that $(\varrho, \mathrm{I}_H^G E)$ is indeed a coefficient G-module.

(iii) Covariance : the definition $\mathrm{I}_H^G E = L^\infty_{\mathrm{w}*}(G, E)^H$ implies that for any adjoint H-morphism $\alpha : E \to F$ of coefficient H-modules the map $\alpha_* : L^\infty_{\mathrm{w}*}(G, E) \to L^\infty_{\mathrm{w}*}(G, F)$ defines an adjoint G-morphism $\alpha_* : \mathrm{I}_H^G E \to \mathrm{I}_H^G F$ depending functorially and additively on α.

(iv) Contravariance : let $\psi : G \to G'$ be a morphism to another locally compact second countable group G' and let $H' < G'$ be a closed subgroup containing $\psi(H)$. Let E be a coefficient H'-module. Then the map $\psi^* : \mathrm{I}_{H'}^{G'} E \to \mathrm{I}_H^G \psi^* E$ induced by pre-composition depends functorially on ψ and gives a G-morphism $\psi^* : \psi^* \mathrm{I}_{H'}^{G'} E \to \mathrm{I}_H^G \psi|_H^* E$ of coefficient G-modules. Moreover, if $\alpha : E \to F$ is an adjoint H'-morphism of coefficient H'-modules, then the diagram

$$\begin{array}{ccc} \mathrm{I}_{H'}^{G'} E & \xrightarrow{\alpha_*} & \mathrm{I}_{H'}^{G'} F \\ \downarrow{\scriptstyle \psi^*} & & \downarrow{\scriptstyle \psi^*} \\ \mathrm{I}_H^G \psi|_H^* E & \xrightarrow{\alpha_*} & \mathrm{I}_H^G \psi|_H^* F \end{array}$$

is commutative.

(v) Suppose that (π, E) is actually a coefficient G-module, not just an H-module. Then, by Corollary 2.4.5, there is an isometric isomorphism of coefficient G-modules

$$\alpha : \ (\varrho, \mathrm{I}_H^G E) \longrightarrow \Big(\lambda_\pi, L^\infty_{\mathrm{w}*}(G/H, E)\Big)$$

defined almost everywhere by $(\alpha v)(gH) = \pi(g) v(g^{-1})$ for $v \in \mathrm{I}_H^G E$.

Now the idea is to define the induction maps using the resolutions of Theorem 7.5.3. Thus let S be an amenable regular G-space, considered also as an amenable H-space (Lemma 5.4.3). If f is an element of $L^\infty_{\mathrm{w*}}(S^{n+1},E)^H$, we define $\mathbf{i}^n_S f : S^{n+1} \to \mathrm{I}^G_H$ almost everywhere by

$$\mathbf{i}^n_S f(s_0,\ldots,s_n)(x) = f(xs_0,\ldots,xs_n). \qquad (s_i \in S, x \in G)$$

We state now the analogue of the Eckmann–Shapiro lemma :

PROPOSITION 10.1.3.— *Let G be a locally compact second countable group, $H < G$ a closed subgroup, S an amenable regular G-space and (π, E) a coefficient H-module.*

(i) *The maps $\mathbf{i}^\bullet_S$ induce an isometric isomorphism*

$$\mathbf{i}_S : \mathrm{H}^\bullet_{\mathrm{cb}}(H,E) \longrightarrow \mathrm{H}^\bullet_{\mathrm{cb}}(G,\mathrm{I}^G_H E).$$

(ii) *Moreover, the isomorphism $\mathbf{i}_S$ does not depend on the choice of S. More precisely, let T be another amenable regular G-space and denote by "$\cong$" the canonical isometric isomorphisms obtained from Theorem 7.5.3. Then the following diagram is commutative :*

$$\begin{array}{ccccccc} & & \mathrm{H}^\bullet(L^\infty_{\mathrm{w*}}(S^\bullet,E)^H) & \xrightarrow{\mathbf{i}_S} & \mathrm{H}^\bullet(L^\infty_{\mathrm{w*}}(S^\bullet,\mathrm{I}^G_H E)^G) & & \\ & \nearrow\!\!\swarrow \cong & \updownarrow \cong & & \updownarrow \cong & \cong \nwarrow\!\!\searrow & \\ \mathrm{H}^\bullet_{\mathrm{cb}}(H,E) & & & & & & \mathrm{H}^\bullet_{\mathrm{cb}}(G,\mathrm{I}^G_H E) \\ & \searrow\!\!\nwarrow \cong & & & & \cong \swarrow\!\!\nearrow & \\ & & \mathrm{H}^\bullet(L^\infty_{\mathrm{w*}}(T^\bullet,E)^H) & \xrightarrow{\mathbf{i}_T} & \mathrm{H}^\bullet(L^\infty_{\mathrm{w*}}(T^\bullet,\mathrm{I}^G_H E)^G) & & \end{array}$$

DEFINITION 10.1.4. In view of the second point of the above proposition, we denote simply by $\mathbf{i}$ the obtained isometric isomorphism

$$\mathbf{i} : \mathrm{H}^\bullet_{\mathrm{cb}}(H,E) \longrightarrow \mathrm{H}^\bullet_{\mathrm{cb}}(G,\mathrm{I}^G_H E)$$

and call $\mathbf{i}$ as well as all maps $\mathbf{i}^n_S$ the *induction maps*.

The induction maps $\mathbf{i}^n_S$ are defined on the subcomplex of invariants, not on the corresponding augmented resolution. Therefore, the naturality of $\mathbf{i}$ is with respect to the maps of Remarks 10.1.2 (iii) and (iv) is not, at first sight, a formal consequence of its definition. However, rather then checking the naturality "by hand" (which is immediate for covariance but not quite so for contravariance), we shall see that the proof of the above proposition yields also the following supplementary points :

PROPOSITION 10.1.5.— *Keep the notation of Proposition 10.1.3 and Definition 10.1.4. The induction map $\mathbf{i}$ is natural in the following sense :*

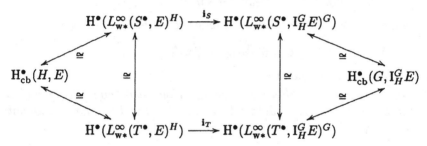

(iii) *If $\alpha : E \to F$ is an adjoint H-morphism of coefficient H-modules, then the diagram*

$$\begin{array}{ccc} \mathrm{H}^\bullet_{\mathrm{cb}}(H,E) & \xrightarrow{\ \mathrm{i}\ } & \mathrm{H}^\bullet_{\mathrm{cb}}(G,\mathrm{I}_H^G E) \\ {\scriptstyle \mathrm{H}^\bullet_{\mathrm{cb}}(H,\alpha)}\downarrow & & \downarrow{\scriptstyle \mathrm{H}^\bullet_{\mathrm{cb}}(G,\alpha_*)} \\ \mathrm{H}^\bullet_{\mathrm{cb}}(H,F) & \xrightarrow{\ \mathrm{i}\ } & \mathrm{H}^\bullet_{\mathrm{cb}}(G,\mathrm{I}_H^G F) \end{array}$$

commutes, wherein α_* is the G-morphism of Remarks 10.1.2 point (iii).

(iv) *Let $\psi : G \to G'$ be a morphism to another locally compact second countable group G', let $H' < G'$ be a closed subgroup containing $\psi(H)$ and let E be a coefficient H'-module. Then the diagram*

$$\begin{array}{ccccc} \mathrm{H}^\bullet_{\mathrm{cb}}(H',E) & \xrightarrow{\ \mathrm{i}\ } & \mathrm{H}^\bullet_{\mathrm{cb}}(G',\mathrm{I}_{H'}^{G'} E) & & \\ & & & \searrow{\scriptstyle \mathrm{H}^\bullet_{\mathrm{cb}}(\psi,E)} & \\ {\scriptstyle \mathrm{H}^\bullet_{\mathrm{cb}}(\psi|_H,E)}\downarrow & & & & \mathrm{H}^\bullet_{\mathrm{cb}}(G,\psi^*\mathrm{I}_{H'}^{G'} E) \\ & & & \swarrow{\scriptstyle \mathrm{H}^\bullet_{\mathrm{cb}}(G,\psi^*)} & \\ \mathrm{H}^\bullet_{\mathrm{cb}}(H,\psi|_H^* E) & \xrightarrow{\ \mathrm{i}\ } & \mathrm{H}^\bullet_{\mathrm{cb}}(G,\mathrm{I}_H^G \psi|_H^* E) & & \end{array}$$

commutes, wherein $\mathrm{H}^\bullet_{\mathrm{cb}}(G,\psi^*)$ is induced by the G-morphism

$$\psi^* : \psi^*\mathrm{I}_{H'}^{G'} E \to \mathrm{I}_H^G \psi|_H^* E$$

of Remarks 10.1.2 point (iv).

Moreover, in presence of both a morphism of topological groups and an adjoint morphism of coefficient H'-modules, the maps of (iii) and (iv) above commute with each other.

REMARK 10.1.6. The idea of the proof is to put ourselves in a position where the Fubini–Lebesgue theorem allows us to derive sort of a "Frobenius reciprocity" by intertwining two actions on a suitable $H \times G$-space. This strategy is virtually the same as Ph. Blanc's arguments on pages 160–163 of [**13**]. Surprisingly, he didn't interpret his construction in terms of $H \times G$-injectivity, thus obtaining the pendant of Proposition 10.1.3 point (i) without the naturality addenda of points (ii) to (iv).

PROOF OF PROPOSITIONS 10.1.3 AND 10.1.5. Let G, H, S and (π, E) be as in the statement of Proposition 10.1.3. We consider E also as a coefficient $H \times G$-module by extending π through the canonical map $H \times G \twoheadrightarrow H$ so that G act trivially. For all $n \geq 0$, we define two $H \times G$-actions ϑ_1, ϑ_2 on the measure space $S^{n+1} \times G$ almost everywhere by

$$\vartheta_1(h,g)(s,x) = (gs, hxg^{-1}), \qquad (h \in H, g, x \in G, s \in S^{n+1})$$
$$\vartheta_2(h,g)(s,x) = (hs, hxg^{-1}). \qquad \text{(idem)}$$

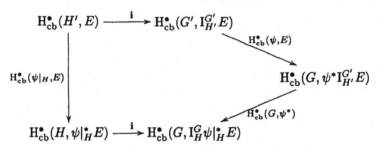

The measure class isomorphism

$$\begin{aligned}\Phi:\ S^{n+1}\times G &\longrightarrow S^{n+1}\times G\\ (s,x)&\longmapsto \Phi(s,x)=(xs,x)\end{aligned}$$

intertwines ϑ_1 with ϑ_2 :

$$\begin{aligned}\Phi\big(\vartheta_1(h,g)(s,x)\big) &= \Phi(gs,hxg^{-1}) = (hxs,hxg^{-1}) =\\ &= \vartheta_2(h,g)(xs,x) = \vartheta_2(h,g)\Phi(s,x).\end{aligned}$$

We claim that $S^{n+1}\times G$ is an amenable $H\times G$-space for both actions, and verify it using the amenability criterion of Theorem 5.3.9 for ϑ_2 ; because of the intertwining isomorphism Φ, the action ϑ_1 will then also be amenable.

On one hand, the equivalence relation induced by the $\vartheta_2(H\times G)$ action on $S^{n+1}\times G$ is amenable because it is isomorphic, as equivalence relation, to the relation induced on S^{n+1} by H. The latter is amenable because of the amenability of the G-action on S^{n+1}. On the other hand, the stabilizer of $(s,x)\in S^{n+1}\times G$ for ϑ_2 is

$$L=\big\{(h,x^{-1}hx) : h\in \mathrm{Stab}_H(s)\big\}.$$

Since L is for all $x\in G$ topologically isomorphic to $\mathrm{Stab}_H(s)$, which is amenable for almost every $s\in S^{n+1}$ independently of x, we deduce that L is amenable for almost every $(s,x)\in S^{n+1}\times G$. Thus the action ϑ_2 is amenable, whence the claim.

We set now $F^n=L^\infty_{\mathrm{w*}}(S^{n+1}\times G,E)$ and endow it with the regular $H\times G$-actions ω_1,ω_2 determined by the $H\times G$-action π on E and the actions ϑ_1 and ϑ_2, respectively. In both cases, the above amenability claim allows us to apply Theorem 5.7.1 and deduce that F^n is $H\times G$-relatively injective. Moreover, Φ induces by pre-composition an isometric adjoint $H\times G$-isomorphism

$$\Phi^n:\ (\omega_2,F^n)\longrightarrow(\omega_1,F^n).$$

On the other hand, we have the canonical isometric identification

$$F^n\cong L^\infty_{\mathrm{w*}}\Big(S^{n+1},L^\infty_{\mathrm{w*}}(G,E)\Big)$$

of Corollary 2.3.3. Endowing $L^\infty_{\mathrm{w*}}(G,E)$ with the coefficient $H\times G$-module structure η given by

$$(\eta(h,g)f)(x)=\pi(h)f(h^{-1}xg),\qquad\qquad (f\in L^\infty_{\mathrm{w*}}(G,E))$$

we see that the above identification is $H\times G$-equivariant when F^n is endowed with ω_1,ω_2 and the right hand side with λ_η and $\lambda_\eta|_H$, respectively. Moreover, the homogeneous differentials $d: F^n\to F^{n+1}$ defined via this identification are both ω_1-to-ω_1 and ω_2-to-ω_2 equivariant. Last, we notice that the coefficient inclusion

$$\epsilon:\ L^\infty_{\mathrm{w*}}(G,E)\longrightarrow L^\infty_{\mathrm{w*}}\Big(S,L^\infty_{\mathrm{w*}}(G,E)\Big)\cong F^0$$

is at the same time η-to-ω_1 and η-to-ω_2 equivariant. Indeed, for $f \in L^\infty_{\mathrm{w}*}(G,E)$, $(h,g) \in H \times G$ and $i = 1,2$ we have for respectively $z = g$ or $z = h$

$$(\omega_i(h,g)\epsilon f)(s,x) = \pi(h)\epsilon f(z^{-1}s, h^{-1}xg) = \\ = \pi(h)f(h^{-1}xg) = (\eta(h,g)f)(x),$$

whether z is g or h. Summing up, we are in position to apply the Theorem 7.5.3 and deduce that the augmented resolution

$$0 \longrightarrow L^\infty_{\mathrm{w}*}(G,E) \xrightarrow{\ \epsilon\ } F^0 \longrightarrow F^1 \longrightarrow F^2 \longrightarrow \cdots$$

realizes canonically isometrically $\mathrm{H}^\bullet_{\mathrm{cb}}\big(H \times G, L^\infty_{\mathrm{w}*}(G,E)\big)$ for both ω_1 and ω_2. It is not that much the latter cohomology space that matters to us, but rather the fact that the above resolution comes with all the functoriality and naturality worked out through Section 8. Indeed, as a result, if we can identify canonically isometrically the H-invariants of this resolution with a resolution as in Theorem 7.5.3 for the G-module $\mathrm{I}^G_H E$ and the G-invariants with a resolution as in Theorem 7.5.3 for the H-module H, then we will have at the same time the induction statement (i) of Proposition 10.1.3 *and* all the statements (ii) to (iv) of Propositions 10.1.3 and 10.1.5.

Thus we proceed do to these identifications. Considering the $\omega_1|_H$-invariants of F^n, we find

$$L^\infty_{\mathrm{w}*}\Big(S^{n+1}, L^\infty_{\mathrm{w}*}(G,E)\Big)^{\omega_1|_H(H)} = L^\infty_{\mathrm{w}*}\Big(S^{n+1}, L^\infty_{\mathrm{w}*}(G,E)^H\Big) = \\ = L^\infty_{\mathrm{w}*}(S^{n+1}, \mathrm{I}^G_H E),$$

and the remaining $\omega_1|_G$-action coincides with the G-action λ_ϱ on the right hand side. This gives a resolution of the type of Theorem 7.5.3 for the G-module $\mathrm{I}^G_H E$.

Considering now the the $\omega_2|_G$-invariants of F^n, we find

$$L^\infty_{\mathrm{w}*}\Big(S^{n+1}, L^\infty_{\mathrm{w}*}(G,E)\Big)^{\omega_2|_G(G)} = L^\infty_{\mathrm{w}*}(S^{n+1}, E\mathbf{1}_G) \cong L^\infty_{\mathrm{w}*}(S^{n+1}, E),$$

and the remaining $\omega_2|_H$-action is the H-action λ_π. This in turn yields a resolution as in Theorem 7.5.3 for the H-module E. We observe in conclusion that the morphism of complexes $\Phi^\bullet$ induced by Φ coincides with $\mathbf{i}^\bullet_S$ on the $H \times G$-invariants. □

Let us come back to the point (v) of Remarks 10.1.2. Namely, we suppose that H is a closed subgroup of the locally compact second countable group G and that (π, E) is a coefficient G-module, so that we have the isomorphism

$$\alpha:\ (\varrho, \mathrm{I}^G_H E) \longrightarrow \Big(\lambda_\pi, L^\infty_{\mathrm{w}*}(G/H, E)\Big).$$

defined almost everywhere by $(\alpha v)(gH) = \pi(g)v(g^{-1})$. Further, denote by $\epsilon: E \to L^\infty_{\mathrm{w}*}(G/H,E)$ the coefficient inclusion.

If f is in $L^\infty_{\mathrm{w}*}(G^{n+1},E)^H$, the map $\alpha_*\mathbf{i}^n_G f$ is defined almost everywhere by

$$(\alpha_*\mathbf{i}^n_G f)(x_0,\ldots,x_n)(x) = \pi(x)f(x^{-1}x_0,\ldots,x^{-1}x_n).$$

In particular, if f is G-invariant for λ_π, the map $(\alpha_* \mathrm{i}_G^n f)(x_0, \ldots, x_n)$ is constant on G/H. Hence we have a commutative diagram

$$\begin{array}{ccc} \mathrm{H}^\bullet_{\mathrm{cb}}(G,E) & \xrightarrow{\mathrm{H}^\bullet_{\mathrm{cb}}(G,\epsilon)} & \mathrm{H}^\bullet_{\mathrm{cb}}\Big(G, L^\infty_{\mathrm{w*}}(G/H,E)\Big) \\ {\scriptstyle \mathrm{res}}\big\downarrow & & \big\uparrow{\scriptstyle \mathrm{H}^\bullet_{\mathrm{cb}}(G,\alpha)} \\ \mathrm{H}^\bullet_{\mathrm{cb}}(H,E) & \xrightarrow{\quad \mathrm{i} \quad} & \mathrm{H}^\bullet_{\mathrm{cb}}(G, \mathrm{I}_H^G E) \end{array}$$

Recall now that if there is a G-invariant mean on G/H, then ϵ admits a (norm one) left inverse G-morphism, and hence (Definition 4.2.7) the exact sequence

$$0 \longrightarrow E \xrightarrow{\epsilon} L^\infty_{\mathrm{w*}}(G/H,E) \longrightarrow L^\infty_{\mathrm{w*}}(G/H,E)/\epsilon(E) \longrightarrow 0 \qquad (*)$$

is G-split. Therefore, by Proposition 8.2.9, the long exact sequence associated to $(*)$ degenerates to

$$0 \longrightarrow \mathrm{H}^\bullet_{\mathrm{cb}}(G,E) \xrightarrow{\mathrm{H}^\bullet_{\mathrm{cb}}(G,\epsilon)} \mathrm{H}^\bullet_{\mathrm{cb}}\Big(G, L^\infty_{\mathrm{w*}}(G/H,E)\Big) \longrightarrow$$
$$\longrightarrow \mathrm{H}^\bullet_{\mathrm{cb}}\Big(G, L^\infty_{\mathrm{w*}}(G/H,E)/\epsilon(E)\Big) \longrightarrow 0,$$

so that in particular the upper arrow $\mathrm{H}^\bullet_{\mathrm{cb}}(G,\epsilon)$ of the above commutative diagram is injective. As a result, the restriction is injective, as had already been shown in Proposition 8.6.6. The discussion above, however, depends only on the map $\mathrm{H}^\bullet_{\mathrm{cb}}(G,\epsilon)$ being injective. Therefore, even if there is no invariant mean on G/H, we consider the long exact sequence attached to $(*)$ by Proposition 8.2.1 point (ii) :

$$\cdots \longrightarrow \mathrm{H}^{n-1}_{\mathrm{cb}}\Big(G, L^\infty_{\mathrm{w*}}(G/H,E)/\epsilon(E)\Big) \longrightarrow \mathrm{H}^n_{\mathrm{cb}}(G,E) \longrightarrow$$
$$\longrightarrow \mathrm{H}^n_{\mathrm{cb}}\Big(G, L^\infty_{\mathrm{w*}}(G/H,E)\Big) \longrightarrow \cdots$$

This together with the commutative rectangle above yields :

COROLLARY 10.1.7.— *Let H be a closed subgroup of the locally compact second countable group G and let (π, E) be a coefficient G-module. Then the restriction fits into an infinite exact sequence*

$$\cdots \longrightarrow \mathrm{H}^{n-1}_{\mathrm{cb}}(G,E_0) \longrightarrow \mathrm{H}^n_{\mathrm{cb}}(G,E) \xrightarrow{\mathrm{res}} \mathrm{H}^n_{\mathrm{cb}}(H,E) \longrightarrow \mathrm{H}^n_{\mathrm{cb}}(G,E_0) \longrightarrow \cdots$$

wherein $E_0 = L^\infty_{\mathrm{w*}}(G/H,E)/\epsilon(E)$. □

In particular, we see that for every $n \geq 0$ the following assertions are equivalent :

(i) The restriction $\mathrm{H}^n_{\mathrm{cb}}(G,E) \to \mathrm{H}^n_{\mathrm{cb}}(H,E)$ is injective.

(ii) The transgression $\mathrm{H}^{n-1}_{\mathrm{cb}}(G,E_0) \to \mathrm{H}^n_{\mathrm{cb}}(G,E)$ vanishes.

(In case $n = 0$, the convention $\mathrm{H}^{-1}_{\mathrm{cb}}(G,E_0) = 0$ is understood.)

We want now to collect a useful formula for further reference :

EXAMPLE 10.1.8. Let G be a locally compact second countable group, $\Gamma < G$ a lattice and (π, E) a coefficient Γ-module. Let $\sigma : G \to \Gamma$ be a left Γ-equivariant map. If we combine the induction formula with the morphism of complexes given in Example 7.5.14, we obtain that the isometric induction isomorphism

$$\mathbf{i} : \ \mathrm{H}^\bullet_{\mathrm{b}}(\Gamma, E) \longrightarrow \mathrm{H}^\bullet_{\mathrm{cb}}(G, \mathrm{I}_\Gamma^G E)$$

of Proposition 10.1.3 and Definition 10.1.4 is realized by the morphism of complexes consisting of

$$\mathbf{i}\sigma^* : \ \ell^\infty(\Gamma^{n+1}, E) \longrightarrow L^\infty_{\mathrm{w*}}(G^{n+1}, \mathrm{I}_\Gamma^G E) \qquad (n \geq 0)$$

and the explicit formula for $\omega \in \ell^\infty(\Gamma^{n+1}, E)$ is

$$\mathbf{i}\sigma^*\omega(g_0, \ldots, g_n)(g) = \omega\big(\sigma(gg_0), \ldots, \sigma(gg_n)\big).$$

10.2. L^p induction. The induction principle as introduced in Section 10.1 has two flaws. First, the induction module is in general difficult to understand. If one starts with a rather nice Banach H-module E, say, a unitary representation of H, then the induction module $\mathrm{I}_H^G E$ will usually be intractable by means of operator theory, on top of not being a continuous module. Even if E is the trivial module $\mathbf{C}$, the induction module $\mathrm{I}_H^G \mathbf{C}$ is already the unfriendly space $L^\infty(G/H)$.

The second drawback concerns the interplay between induction and the comparison map. For suitable H-modules E, the induction modules appearing in usual (continuous) cohomology can have e.g. the form $L^p_{\mathrm{loc}}(G, E)^H$ for $p < \infty$ (see [**13**, Section 8]), but certainly cannot be $\mathrm{I}_H^G E$ since one is inducing unbounded cochains. Even though there is for E separable a natural inclusion map $\mathrm{I}_H^G E \to L^p_{\mathrm{loc}}(G, E)^H$ making the diagram

$$\begin{array}{ccc} \mathrm{H}^\bullet_{\mathrm{cb}}(H, E) & \longrightarrow & \mathrm{H}^\bullet_{\mathrm{cb}}(G, \mathrm{I}_H^G E) \\ \big\downarrow & & \big\downarrow \\ \mathrm{H}^\bullet_{\mathrm{c}}(H, E) & \longrightarrow & \mathrm{H}^\bullet_{\mathrm{c}}\big(G, L^p_{\mathrm{loc}}(G, E)^H\big) \end{array}$$

commutative, it would be preferable to have the same G-modules as coefficients on the right hand side, or at least more related modules.

We shall show in this section that it is possible to some extent to remedy these flaws in certain situations by continuing the induction towards L^p induction modules.

Let G be a locally compact second countable group, $H < G$ a closed subgroup such that $H\backslash G$ has finite invariant measure and (π, E) either a coefficient H-module or a separable Banach H-module. We extend the definition of the action λ_π to any measurable function class $f : G \to E$ by the usual formula $\lambda_\pi(h)f(x) = \pi(h)f(h^{-1}x)$. Since π is isometric, any $\lambda_\pi(H)$-invariant function class f yields a well defined $\|f\|_E : H\backslash G \to \mathbf{R}$. Moreover, the latter is measurable because of the separability of the predual of E (same argument as for Lemma 2.2.3) or of E itself in the second case considered. Using these two facts, we define for every $1 \leq p < \infty$ the space $L^{[p]}(G, E)^H$ to be the space of those

H-invariant element for which $\|f\|_E$ is in $L^p(H\backslash G)$. The notation $L^{[p]}$ is to distinguish this space from the space of H-invariants in $L^p(G,E)$.

DEFINITION 10.2.1. The L^p *induction module* is the space

$$L^p\mathrm{I}_H^G E = L^{[p]}(G,E)^H$$

endowed with the right G-translation ϱ and with the norm $\|f\|_{[p]} = \|(\|f\|_E)\|_p$, where $\|\cdot\|_p$ is the usual p-norm for the normalized measure on $H\backslash G$.

This definition is justified by the following proposition.

PROPOSITION 10.2.2.— *Let G be a locally compact second countable group, $H < G$ a closed subgroup such that $H\backslash G$ has finite invariant measure and (π, E) a separable coefficient H-module. Let $1 < p < \infty$.*

The norm and representation of Definition 10.2.1 turn $L^p\mathrm{I}_H^G E$ into a separable coefficient G-module. Moreover, if E is reflexive, then so is $L^p\mathrm{I}_H^G E$.

PROOF. Fix a separable predual $E^\flat$ of E and recall that by Proposition 3.3.2, the Banach G-module E is continuous. At the level of Banach spaces, we have an isometric isomorphism $L^p\mathrm{I}_H^G E \cong L^p(H\backslash G, E)$. Since E is a separable dual, it has the Radon-Nikodým property (see [**44**, VII 7]). Therefore, there is a canonical isometric isomorphism

$$L^p(H\backslash G, E) = L^q(H\backslash G, E^\flat)^\sharp \qquad (1/p + 1/q = 1)$$

(Theorem 1 in [**44**, IV 1]) which induces an isomorphism $L^p\mathrm{I}_H^G E \cong (L^q\mathrm{I}_H^G E^\flat)^\sharp$. Moreover, $L^p\mathrm{I}_H^G E$ is separable because G is second countable and E separable. In order to show that it is a coefficient module, it remains only to observe that the action on $L^q\mathrm{I}_H^G E^\flat$ is continuous, which follows from $q < \infty$ since $E^\flat$ is continuous. □

REMARKS 10.2.3.

(i) Recall that by Proposition 3.3.2 the separable coefficient module $L^p\mathrm{I}_H^G E$ of the above proposition is continuous.

(ii) This proof shows of course also that $L^2\mathrm{I}_H^G E$ is a Hilbert space if E is a Hilbert space ; in this case, $L^2\mathrm{I}_H^G E$ is nothing else than Mackey's unitarily induced representation.

(iii) If (π, E) is actually a coefficient G-module, then the argument of point (v) in Remarks 10.1.2 shows that there is an isomorphic isomorphism of coefficient G-modules

$$(\varrho, L^p\mathrm{I}_H^G) \cong \Big(\lambda_\pi, L^p(G/H, E)\Big).$$

(In the general case, there is anyway an isometric isomorphism of Banach spaces between the above.)

(iv) The covariance and contravariance of $L^p\mathrm{I}_H^G E$ goes exactly as in Remarks 10.1.2 above.

Since E is separable, weak-* measurability coincides with normic measurability (Lemma 3.3.3). Since moreover G/H has finite invariant measure, we have the inclusion

$$L^\infty_{\mathrm{w}*}(G,E)^H \subset L^{[p]}(G,E)^H$$

and hence a natural map

$$\mathrm{H}^\bullet_{\mathrm{cb}}(G, \mathrm{I}^G_H E) \longrightarrow \mathrm{H}^\bullet_{\mathrm{cb}}(G, L^p\mathrm{I}^G_H E).$$

DEFINITION 10.2.4. We denote by $L^p\mathbf{i}$ the composition

$$L^p\mathbf{i}:\ \mathrm{H}^\bullet_{\mathrm{cb}}(H,E) \longrightarrow \mathrm{H}^\bullet_{\mathrm{cb}}(G, L^p\mathrm{I}^G_H E)$$

of the above map with the induction $\mathbf{i}$ and call $L^p\mathbf{i}$ the *L^p induction map.*

We remark that since $\mathbf{i}$ is natural in the group and the coefficients (points (iii) and (iv) of Proposition 10.1.5), and since one has also naturality for the map induced by the inclusion $L^\infty \subset L^{[p]}$, the L^p induction map $L^p\mathbf{i}$ is also natural. In particular, the second part of Proposition 10.1.3 implies that $L^p\mathbf{i}$ can be realized on any amenable regular G-space S by the cochain map

$$L^p\mathbf{i}^n_S:\ L^\infty_{\mathrm{w}*}(S^{n+1},E)^H \longrightarrow L^\infty_{\mathrm{w}*}(S^{n+1}, L^p\mathrm{I}^G_H E)^G$$

defined almost everywhere by

$$L^p\mathbf{i}^n_S f(s_0,\dots,s_n)(x) = f(xs_0,\dots,xs_n).$$

REMARKS 10.2.5.

(i) Since we have taken the *normalized* G-invariant measure on $H\backslash G$, the map $L^p\mathbf{i}: \mathrm{H}^n_{\mathrm{cb}}(H,E) \to \mathrm{H}^n_{\mathrm{cb}}(G, L^p\mathrm{I}^G_H E)$ does not increase the semi-norm ; in particular, it is continuous.

(ii) The formula given in Example 10.1.8 has the following application. Let G be a locally compact second countable group, $\Gamma < G$ a lattice and (π, E) a separable coefficient Γ-module. Let $\sigma: G \to \Gamma$ be a left Γ-equivariant map. Then the L^p induction from $\mathrm{H}^\bullet_{\mathrm{b}}(\Gamma,E)$ to $\mathrm{H}^\bullet_{\mathrm{cb}}(G, L^p\mathrm{I}^G_\Gamma E)$ is realized on the complexes

$$\ell^\infty(\Gamma^{n+1},E) \longrightarrow L^\infty_{\mathrm{w}*}(G^{n+1}, L^p\mathrm{I}^G_\Gamma E) \qquad (n \geq 0)$$

by

$$\mathbf{i}\sigma^*\omega(g_0,\dots,g_n)(g) = \omega\big(\sigma(gg_0),\dots,\sigma(gg_n)\big)$$

for $\omega \in \ell^\infty(\Gamma^{n+1},E)$.

The price to pay for having replaced the L^∞ space $\mathrm{I}^G_H E$ by nicer spaces like e.g. $L^2\mathrm{I}^G_H E$ is that we do not have a general Eckmann–Shapiro isomorphism. However, there is still the following injectivity statement, which turns out to be most valuable in applications :

PROPOSITION 10.2.6.— *Let G be a locally compact second countable group, $H < G$ a closed subgroup such that $H\backslash G$ has finite invariant measure, E a separable coefficient H-module.*

If there exists some amenable regular G-space S so that the diagonal H-action on S^n is ergodic, then the L^p induction

$$L^p\mathbf{i}: \mathrm{H}^n_{\mathrm{cb}}(H,E) \longrightarrow \mathrm{H}^n_{\mathrm{cb}}(G, L^p\mathrm{I}_H^G E)$$

is injective for all $1 < p < \infty$.

The whole point of this proposition is that we assume ergodicity on S^n only, whilst the cochains relevant to $\mathrm{H}^n_{\mathrm{cb}}(H,E)$ are defined on S^{n+1}.

We shall see later that the assumption on the existence of such a space S is not merely whishful thinking, but is always satisfied up to finite index for $n = 2$ if G is compactly generated (Theorem 11.1.3). But having such an assumption up to finite index is still sufficient : indeed we generalize the Proposition 10.2.6 to the more general form of Proposition 10.2.7 below. As a result, we shall actually obtain later on the injectivity of the L^p induction in degree two for compactly generated groups (Corollary 11.1.5 below).

PROPOSITION 10.2.7.— *Let G be a locally compact second countable group, $H < G$ a closed subgroup such that $H\backslash G$ has finite invariant measure, E a separable coefficient H-module.*

If for some finite index open subgroup $G' < G$ there exists an amenable regular G'-space S so that the diagonal $H \cap G'$-action on S^n is ergodic, then the L^p induction

$$L^p\mathbf{i}: \mathrm{H}^n_{\mathrm{cb}}(H,E) \longrightarrow \mathrm{H}^n_{\mathrm{cb}}(G, L^p\mathrm{I}_H^G E)$$

is injective for all $1 < p < \infty$.

PROOF OF PROPOSITION 10.2.6. Let S be an amenable regular G-space as in the statement. An element of the kernel of $L^p\mathbf{i}$ is then represented by an element f of $L^\infty_{\mathrm{w*}}(S^{n+1}, E)^H$ such that $L^p\mathbf{i}^n_S f = db$ for some b in $L^\infty_{\mathrm{w*}}(S^n, L^p\mathrm{I}_H^G E)^G$. In order to "disinduce" b, we proceed now to give a more general Frobenius reciprocity. Let us use $\widetilde{L}(-,-)$ to denote the vector space of *all* weak-* measurable function classes. As in the proof of Proposition 10.1.3, we denote by Φ the isomorphism of $S^n \times G$ given by $\Phi(s,x) = (xs,x)$ and consider the corresponding automorphism Φ^{n-1} of $\widetilde{L}(S^n \times G, E)$. Taking the $H \times G$-invariants exactly as in the proof of Proposition 10.1.3, we find by Fubini–Lebesgue that this automorphism descends to an isomorphism

$$\widetilde{\mathbf{i}}^{n-1}_S : \widetilde{L}(S^n,E)^H \longrightarrow \widetilde{L}\Big(S^n, \widetilde{L}(G,E)^H\Big)^G$$

commuting with differentials. The right hand side of the above contains

$$L^\infty_{\mathrm{w*}}(S^n, L^p\mathrm{I}_H^G E)^G$$

and moreover the restriction of $\widetilde{\mathbf{i}}^{n-1}_S$ to $L^\infty_{\mathrm{w*}}(S^n,E)^H$ is but $L^p\mathbf{i}^{n-1}_S$.

Now we can write b as $b = \widetilde{\mathbf{i}}^{n-1}_S b'$, where b' is in $\widetilde{L}(S^n,E)^H$. We have $db' = f$, so that it remains only to show that b' is essentially bounded. Since the map

$\|b'\|_E : S^n \to \mathbf{R}$ is measurable and H-invariant, the assumption on S implies that the norm of b' is essentially constant, hence bounded. □

PROOF OF PROPOSITION 10.2.7. Let $G' < G$ and S be as in the statement and set $H' = H \cap G'$. Since G' is open in G, there is a restriction morphism

$$r : L^2\mathrm{I}_H^G F \longrightarrow L^2\mathrm{I}_{H'}^{G'} F$$

making the following diagram commutative

$$\begin{array}{ccc} \mathrm{H}^2_{\mathrm{cb}}(H,F) & \xrightarrow{\ \mathbf{i}\ } & \mathrm{H}^2_{\mathrm{cb}}(G, L^2\mathrm{I}_H^G F) \\ \Big\downarrow{\scriptstyle \mathrm{res}} & & \Big\downarrow{\scriptstyle r_*\circ\mathrm{res}} \\ \mathrm{H}^2_{\mathrm{cb}}(H',F) & \xrightarrow{\ \mathbf{i}\ } & \mathrm{H}^2_{\mathrm{cb}}(G', L^2\mathrm{I}_{H'}^{G'} F) \end{array}$$

Since H' is of finite index in H, the left restriction arrow is injective by Proposition 8.6.6. Therefore, it is enough to show the injectivity of the lower induction map. This is exactly the content of Proposition 10.2.6. □

The definition of the L^p induction map $L^p\mathbf{i}$ requires only H to admit a finite invariant measure on G/H, while in usual continuous cohomology there is a priori no induction map

$$\mathrm{H}^\bullet_{\mathrm{c}}(H,E) \longrightarrow \mathrm{H}_{\mathrm{c}}(G, L^p\mathrm{I}_H^G E)$$

unless H is co-compact in G. However, essentially the same formalism as used above yields Ph. Blanc's version of the Eckmann–Shapiro lemma, namely an isomorphism

$$\mathrm{H}^\bullet_{\mathrm{c}}(H,E) \longrightarrow \mathrm{H}_{\mathrm{c}}\Big(G, L^p_{\mathrm{loc}}(G,E)^H\Big)$$

where the Fréchet space $L^p_{\mathrm{loc}}(G,E)^H$ of locally p-summable function classes is endowed with the G-action ϱ. This is Théorème 8.7 in [**13**], with other conventions for the actions. Now the inclusion map $\iota : L^p\mathrm{I}_H^G E \to L^p_{\mathrm{loc}}(G,E)^H$ makes the following diagram commute

$$\begin{array}{ccccc} \mathrm{H}^\bullet_{\mathrm{cb}}(H,E) & & \longrightarrow & & \mathrm{H}^\bullet_{\mathrm{c}}(H,E) \\ \Big\downarrow{\scriptstyle L^p\mathbf{i}} & & & & \Big\downarrow{\scriptstyle \mathbf{i}} \\ \mathrm{H}^\bullet_{\mathrm{cb}}(G, L^p\mathrm{I}_H^G E) & \longrightarrow & \mathrm{H}^\bullet_{\mathrm{c}}(G, L^p\mathrm{I}_H^G E) & \xrightarrow{\mathrm{H}^\bullet_{\mathrm{c}}(G,\iota)} & \mathrm{H}^\bullet_{\mathrm{c}}\big(G, L^p_{\mathrm{loc}}(G,E)^H\big) \end{array}$$

so that in case H is co-compact in G we have the simpler diagram

$$\begin{array}{ccc} \mathrm{H}^\bullet_{\mathrm{cb}}(H,E) & \longrightarrow & \mathrm{H}^\bullet_{\mathrm{c}}(H,E) \\ \Big\downarrow{\scriptstyle L^p\mathbf{i}} & & \Big\downarrow{\scriptstyle \cong} \\ \mathrm{H}^\bullet_{\mathrm{cb}}(G, L^p\mathrm{I}_H^G E) & \longrightarrow & \mathrm{H}^\bullet_{\mathrm{c}}(G, L^p\mathrm{I}_H^G E) \end{array}$$

10.3. Dimension shifting. The acyclicity of relatively injective Banach G-modules established in Proposition 7.4.1 has a useful immediate consequence. Suppose namely that a Banach G-module fits into a short exact sequence

$$0 \longrightarrow E \longrightarrow F \longrightarrow D \longrightarrow 0$$

with F relatively injective. Suppose moreover that this sequence satisfies one of the sufficient conditions given in Section 8.2 for having a long exact sequence in cohomology. Then the vanishing of $\mathrm{H}^n_{\mathrm{cb}}(G, F)$ makes this long exact sequence degenerate in such a way that the transgression maps give isomorphisms

$$\mathrm{H}^n_{\mathrm{cb}}(G, D) \cong \mathrm{H}^{n+1}_{\mathrm{cb}}(G, E)$$

for all $n \geq 1$. Repeating the procedure will further increase the degree.

This motivates the following definition :

DEFINITION 10.3.1. Let G be a topological group and E a Banach G-module. The *shifting module* $\mathfrak{S}E$ is the quotient Banach G-module

$$\mathfrak{S}E = \mathcal{C}\mathrm{C}_{\mathrm{b}}(G, E)/\epsilon(\mathcal{C}E).$$

More generally, we define for $k \geq 1$ the kth shifting module $\mathfrak{S}^k E$ inductively by $\mathfrak{S}^1 E = \mathfrak{S}E$ and then $\mathfrak{S}^k E = \mathfrak{S}\mathfrak{S}^{k-1} E$.

The map ϵ has closed image, so that the quotient defining $\mathfrak{S}E$ is indeed a Banach G-module. Observe moreover that $\mathfrak{S}E$ is a *continuous* Banach G-module since it is a quotient of the continuous Banach G-module $\mathcal{C}\mathrm{C}_{\mathrm{b}}(G, E)$. Likewise, all $\mathfrak{S}^k E$ are continuous. The shifting modules have the expected covariance in E and contravariance in G.

Since evaluation (e.g. at $e \in G$) yields a continuous linear section to the coefficient inclusion ϵ (see Lemma 1.3.1), we see that the short exact sequence of continuous Banach G-modules

$$0 \longrightarrow \mathcal{C}E \longrightarrow \mathcal{C}\mathrm{C}_{\mathrm{b}}(G, E) \longrightarrow \mathfrak{S}E \longrightarrow 0 \qquad (*)$$

is weakly admissible (Definition 4.2.5). Thus the Proposition 8.2.7 yields a long exact sequence corresponding to $(*)$. By Proposition 4.4.1, the Banach G-module $\mathrm{C}_{\mathrm{b}}(G, E)$, hence also $\mathcal{C}\mathrm{C}_{\mathrm{b}}(G, E)$, are relatively injective. Therefore, the continuous bounded cohomology of G with coefficients in $\mathcal{C}\mathrm{C}_{\mathrm{b}}(G, E)$ vanish in degree $n \geq 1$ by Proposition 7.4.1, so that the long exact sequence degenerates to

$$0 \longrightarrow \mathrm{H}^n_{\mathrm{cb}}(G, \mathfrak{S}E) \longrightarrow \mathrm{H}^{n+1}_{\mathrm{cb}}(G, E) \longrightarrow 0$$

for all $n \geq 1$ (the map in the middle is the transgression). Repeating this k times, we obtain the following statement :

PROPOSITION 10.3.2.— *Let G be a topological group and E a Banach G-module. The transgression maps induce a continuous isomorphism*

$$\mathrm{H}^n_{\mathrm{cb}}(G, \mathfrak{S}^k E) \longrightarrow \mathrm{H}^{n+k}_{\mathrm{cb}}(G, E)$$

for all $n, k \geq 1$. Moreover, this isomorphism is natural in E and G. □

(The naturality in E is contained in Proposition 8.2.7, while the naturality in G follows from Proposition 8.4.2 and the commutativity of the diagram preceding it).

EXAMPLE 10.3.3. Let Γ be a non-elementary Gromov-hyperbolic group. Then the space

$$\mathrm{H}^1_{\mathrm{b}}(\Gamma, \ell^\infty(\Gamma)/\mathbf{C})$$

is infinite dimensional. Indeed, the coefficient Γ-module $\ell^\infty(\Gamma)/\mathbf{C}$ is $\mathfrak{S}\mathbf{C}$. Thus, by Proposition 10.3.2, this $\mathrm{H}^1_{\mathrm{b}}$ space is isomorphic to $\mathrm{H}^2_{\mathrm{b}}(\Gamma)$, which is infinite-dimensional by the result [**53**] of Epstein-Fujiwara.

This example stands in contrast to the fact that $\mathrm{H}^1_{\mathrm{cb}}$ (hence $\mathrm{H}^1_{\mathrm{b}}$) always vanishes for reflexive coefficients (Proposition 6.2.1).

As explained in the beginning of this section, the shifting procedure depends essentially just on $\mathrm{C_b}(G, E)$ being relatively injective and on the fundamental long exact sequence. Therefore, we can produce many more interesting shifting modules in the case of coefficient modules :

DEFINITION 10.3.4. Let G be a locally compact second countable group and E a coefficient G-module. For every amenable regular G-space S, we define the *shifting module over* S, denoted by $\mathfrak{S}_S E$, as the quotient Banach G-module

$$\mathfrak{S}_S E = L^\infty_{\mathrm{w*}}(S, E)/\epsilon(E).$$

More generally, we define for $k \geq 1$ the kth shifting module over S inductively by $\mathfrak{S}^1_S E = \mathfrak{S}_S E$ and then $\mathfrak{S}^k_S E = \mathfrak{S}_S \mathfrak{S}^{k-1}_S E$.

Since $\epsilon(E) = E\mathbf{1}_S$ is weak-* closed in $L^\infty_{\mathrm{w*}}(S, E)$, we see that $\mathfrak{S}_S E$ and hence also all $\mathfrak{S}^k_S E$ are coefficient modules. In particular, the short exact sequence

$$0 \longrightarrow E \longrightarrow L^\infty_{\mathrm{w*}}(S, E) \longrightarrow \mathfrak{S}_S \longrightarrow 0$$

satisfies the hypothesis (ii) of Proposition 8.2.1, so that we have again a long exact sequence in cohomology and deduce exactly as before the

PROPOSITION 10.3.5.— *Let G be a locally compact second countable group and E a coefficient G-module. For every amenable regular G-space S, the transgression maps induce a continuous isomorphism*

$$\mathrm{H}^n_{\mathrm{cb}}(G, \mathfrak{S}^k_S E) \longrightarrow \mathrm{H}^{n+k}_{\mathrm{cb}}(G, E)$$

for all $n, k \geq 1$. Moreover, this isomorphism is natural in E and G. □

The shifting modules over amenable spaces behave particularly nicely with respect to iteration ; that is, we can identify $\mathfrak{S}^k_S$ very concretely :

LEMMA 10.3.6.— *Let G be a locally compact second countable group, E a coefficient G-module and S an amenable regular G-space. Let $k \geq 1$.*

The coefficient G-module $\mathfrak{S}^k_S$ identifies canonically isometrically and equivariantly to the contragredient of

$$\underbrace{L^1_0(\mu)\widehat{\otimes}\cdots\widehat{\otimes}L^1_0(\mu)}_{k\ times}\widehat{\otimes}E^\flat,$$

where L^1_0 denotes the space of function classes of integral zero for a measure μ as in Definition 2.1.1.

PROOF. Recall that the duality between $L^\infty_{w*}(S,E)$ and $L^1(\mu,E^\flat) \cong L^1(\mu)\widehat{\otimes}E^\flat$ is obtained by integrating a duality on $E \times E^\flat$ with respect to μ (see Proposition 2.3.1). This duality identifies the subspace $\epsilon(E) = E\mathbf{1}_S$ as the annihilator of $L^1_0(\mu)\widehat{\otimes}E^\flat$. Thus, by the duality principle (see [**126**, I, Theorem 4.9]), the coefficient G-module $\mathfrak{S}_S E$ is contragredient to $L^1_0(\mu)\widehat{\otimes}E^\flat$, and the claim follows by induction on k. □

We come back to Example 10.3.3 with these more general shifting modules :

EXAMPLE 10.3.7. Let Γ be a non-elementary Gromov-hyperbolic group and endow its boundary at infinity $\partial_\infty\Gamma$ with the class of any quasi-invariant measure. Then the space

$$\mathrm{H}^1_\mathrm{b}(\Gamma, L^\infty(\partial_\infty\Gamma)/\mathbf{C})$$

is infinite dimensional. Indeed, (point (viii) of Example 5.4.1), the Γ-action on $\partial_\infty\Gamma$ is amenable. Thus we apply Proposition 10.3.5 to get

$$\mathrm{H}^1_\mathrm{b}(\Gamma, L^\infty(\partial_\infty\Gamma)/\mathbf{C}) \cong \mathrm{H}^2_\mathrm{b}(\Gamma)$$

and conclude as in Example 10.3.3 with the result [**53**] of Epstein-Fujiwara.

For the same reason, we have that

$$\mathrm{H}^1_\mathrm{b}(\Gamma, L^\infty(S)/\mathbf{C})$$

is infinite dimensional for any amenable Γ-space S.

11. DOUBLE ERGODICITY

Our characterization of amenable actions in terms of relative injectivity (Theorem 5.7.1) shows that such actions play a central rôle from a cohomological viewpoint. This is reflected in Theorem 7.5.3, which provides us with a variety of resolutions realizing isometrically the continuous bounded cohomology.

It is therefore natural to seek the simplest possible amenable actions of a given group. This quest turns out fruitful in asmuch as one can find "small" amenable regular spaces related to *all* compactly generated locally compact second countable groups (Theorem 11.1.3). Already in the case of "abstract" (non topological) groups, this result yields a very useful tool for the study of bounded cohomology.

11.1. Definition and elementary properties. Let G be a locally compact second countable group and S an amenable regular G-space. In simple words, amenability is a condition that forces S to be in a sense "large" with respect to the group action. We want to investigate to what extend we could get the space S as "small" as possible, yet keeping the G-action amenable. When can an amenable action be "more than ergodic" ?

One way in this direction would be to require ergodicity of the diagonal G-action on $S \times S$. The following definition introduces an even stronger requirement.

DEFINITION 11.1.1. Let $\mathfrak{X}$ be any class of coefficient Banach modules, G a topological group and S a regular G-space. We say that the G-action on S is *doubly $\mathfrak{X}$-ergodic* if for every coefficient G-module E in $\mathfrak{X}$, any weak-* measurable G-equivariant function class

$$f : S \times S \longrightarrow E$$

is essentially constant ($S \times S$ is endowed with the diagonal action).

We say synonymously that S is a doubly $\mathfrak{X}$-ergodic G-space and write simply "doubly E-ergodic" if $\mathfrak{X}$ is reduced to a single coefficient module E.

The reader will have noticed that in this definition, the class $\mathfrak{X}$ does not a priori depend on the group under consideration. This allows for the following convenient notations :

DEFINITION 11.1.2. We write $\mathfrak{X}^{\text{Hilb}}$ for the class of all unitary coefficient modules over locally compact groups. Likewise, $\mathfrak{X}^{\text{refl}}$ is the class of all reflexive coefficient modules, $\mathfrak{X}^{\text{sep}}$ the class of all separable coefficient modules and $\mathfrak{X}^{\text{cont}}$ the class of all continuous coefficient modules, in each case over locally compact groups.

We observe that one has the following inclusions of classes :

$$\mathfrak{X}^{\text{Hilb}} \subset \mathfrak{X}^{\text{refl}} \subset \mathfrak{X}^{\text{sep}} \subset \mathfrak{X}^{\text{cont}}.$$

The only non-trivial inclusion here is the last one, which follows from Proposition 3.3.2. We see e.g. that a coefficient G-module in $\mathfrak{X}^{\text{Hilb}}$ is but a continuous unitary representation on a separable Hilbert space.

The main result of this Section 11 is :

THEOREM 11.1.3.— *Let G be a compactly generated locally compact group. There exists a canonical topologically characteristic finite index open subgroup $G^* \lhd G$ and a regular G^*-space S such that*

(i) *The G^*-action on S is amenable.*

(ii) *The G^*-action on S is doubly $\mathfrak{X}^{\text{sep}}$-ergodic.*

(Proof in Section 11.3.)

The construction of S uses notably the solution to Hilbert's fifth problem in order to take advantage, on one hand, of Furstenberg's boundary, and on the other hand of Poisson boundaries for random walks on totally disconnected groups. It will turn out (Remark 11.3.2) that if G is either connected or totally diconnected, then $G^* = G$. Moreover,

(i) If G is already a connected semi-simple group, then $G^* = G$ and S is Furstenberg's boundary.

(ii) If G is discrete, then $G^* = G$ and S is the Poisson boundary of a random walks on G.

REMARK 11.1.4. The group G in the statement of Theorem 11.1.3 above is not supposed second countable ; we recall that we have extended in the Definition 5.3.10 the notion of an amenable action to certain non second countable groups. This definition is to be understood here. As we shall see, second countability will be needed below in order to apply L^p induction, but Theorem 11.1.3 will be reduced to that case.

Only the most basic facts will be treated for groups that are not necessarily compactly generated nor locally compact.

The Theorem 11.1.3 will turn out to be most useful for a number of applications. As a first instance, we observe right away that it gives a sound justification to the concept of L^p induction introduced in Section 10.2 :

COROLLARY 11.1.5.— *Let G be a compactly generated locally compact second countable group, $H < G$ a closed subgroup such that $H\backslash G$ has finite invariant measure and E a separable coefficient H-module. Then the L^p induction*

$$L^p\mathbf{i}: \ \mathrm{H}^2_{\mathrm{cb}}(H,E) \longrightarrow \mathrm{H}^2_{\mathrm{cb}}(G, L^p\mathrm{I}_H^G E)$$

is injective for all $1 < p < \infty$.

(The proof is given after the Lemma 11.1.11.)

We turn now to the first elementary properties of the concepts introduced above. It is apparent that double $\mathbf{C}$-ergodicity of a topological group G on S is nothing but ergodicity on $S \times S$. More generally,

LEMMA 11.1.6.— *If $E \neq 0$ is a coefficient G-module with trivial G-action and S a regular G-space, then the diagonal G-action is ergodic on $S \times S$ if and only if S is doubly E-ergodic.*

PROOF. The "if" part is obvious. Conversely, suppose G ergodic on $S \times S$ and let $f : S \times S \rightarrow E$ be an invariant weak-* measurable function class. By the definition of coefficient modules, there is a countable dense subset C in the predual $E^\flat$. By weak-* measurability, one has for every $c \in C$ a measurable scalar function class f_c on $S \times S$ defined by $f_c(\cdot) = f(\cdot)(c)$. Now f_c is invariant, so by ergodicity there is a null-set $N_c \subset S \times S$ away from which f_c is constant. Hence f is constant outside $\cup_{c \in C} N_c$, which is a null-set since C is countable. □

For a general coefficient G-module E it is however possible that G acts ergodically on $S \times S$ but that the G-action on S is not doubly E-ergodic. An example of this situation will be given in Remarks 11.4.3.

In connection with Lemma 11.1.6, we make the following remark :

LEMMA 11.1.7.— *Let S be a regular G-space such that the diagonal G-action on $S \times S$ is ergodic. Then either S is reduced to a single atom or S is purely non-atomic.*

PROOF. Suppose for a contradiction that S has an atom $s \in S$ and is not reduced to $\{s\}$. Then there is a measurable non-null subset $E \subset S$ not containing s. Denote by $\Delta(S) \subset S \times S$ the diagonal. Since (s, s) is in $\Delta(S)$, the latter is not a null-set. On the other hand, $s \notin E$ implies $E \times \{s\} \subset S \times S \setminus \Delta(S)$, which hence is not null neither. Thus there are two non-null disjoint G-invariant subsets in $S \times S$, contradicting the assumption. □

Here are two elementary closure properties for double ergodicity :

LEMMA 11.1.8.— *Let $\mathfrak{X}$ be a class of coefficient modules, G_1, G_2 topological groups and S_1, S_2 doubly $\mathfrak{X}$-ergodic G_1- respectively G_2-spaces. Set $G = G_1 \times G_2$ and suppose $\mathfrak{X}$ closed under taking pull-backs through $G_i \hookrightarrow G$. Then $S = S_1 \times S_2$ is a doubly $\mathfrak{X}$-ergodic G-space.*

PROOF. Let $E \in \mathfrak{X}$ be a coefficient G-module and let $f : S \times S \to E$ represent a weak-* measurable G-equivariant function class. By Fubini-Lebesgue, f yields for almost all $x \in S_1 \times S_1$ a weak-* measurable G_2-equivariant map $S_2 \times S_2 \to E$, viewing E as a G_2-module. Hence this map has an essential value $F(x)$, which by Fubini-Lebesgue depends weak-* measurably on x. But $x \mapsto F(x)$ is G_1-equivariant by the G-equivariance of f, so F and hence f are essentially constant. □

LEMMA 11.1.9.— *Let $\mathfrak{X}$ be either $\mathfrak{X}^{\mathrm{sep}}$, $\mathfrak{X}^{\mathrm{cont}}$, $\mathfrak{X}^{\mathrm{refl}}$ or $\mathfrak{X}^{\mathrm{Hilb}}$. Let $H \lhd G$ be a closed normal subgroup of the locally compact group G and S a regular G/H-space.*

Then the G/H action on S is doubly $\mathfrak{X}$-ergodic if and only G action on S defined via $G \to G/H$ is also doubly $\mathfrak{X}$-ergodic.

PROOF. The "if" part is obvious. Conversely, suppose the G/H-action doubly ergodic. If now $E \in \mathfrak{X}$ is a coefficient G-module and $f : S \times S \to E$ is weak-* measurable and G-equivariant, then f ranges (outside some null-set) in E^H because the H-action on S is trivial. We observe that each of the classes considered for $\mathfrak{X}$ are closed under taking coefficient submodules. According to Lemma 1.2.9, the invariants E^H constitute a coefficient submodule of E, so that we have $E^H \in \mathfrak{X}$. Since f is now G/H-equivariant, we conclude that it is essentially constant. □

A less elementary but very important closure property is the following.

PROPOSITION 11.1.10.— *Suppose $\mathfrak{X}$ is either $\mathfrak{X}^{\mathrm{Hilb}}$, $\mathfrak{X}^{\mathrm{refl}}$ or $\mathfrak{X}^{\mathrm{sep}}$. Let G be a locally compact second countable group, S a doubly $\mathfrak{X}$-ergodic G-space and $H < G$ a closed subgroup. If $H \backslash G$ admits a finite invariant measure, then the H-action on S is also doubly $\mathfrak{X}$-ergodic.*

The proof relies on L^p induction, whence the need for second countability. We begin with a preliminary observation :

LEMMA 11.1.11.— *Keep the notation of Proposition 11.1.10. Then the H-action on $S \times S$ is ergodic.*

PROOF OF THE LEMMA. To test ergodicity, it is enough to consider a *bounded* H-invariant measurable function $b : S \times S \to \mathbf{C}$. Now the induced map $L^2\mathbf{i}_S^1 b$ introduced just below Definition 10.2.4 ranges in $L^2\mathrm{I}_H^G\mathbf{C} \cong L^2(H\backslash G)$ and the assumption on S implies that $\mathbf{i}b$, hence also b, is essentially constant. □

PROOF OF COROLLARY 11.1.5. Theorem 11.1.3 provides us with a group $G' = G^*$ and a space S as required in Proposition 10.2.7, since by the above lemma the $H \cap G'$-action on $S \times S$ is ergodic. □

PROOF OF PROPOSITION 11.1.10. We pick a weak-* measurable H-equivariant map $f : S \times S \to E$, with E in $\mathfrak{X}$. By the preceding lemma, H is ergodic on $S \times S$ and thus the measurable function $\|f\|_F$ is constant. Therefore f is bounded and we may consider the G-invariant element $L^2\mathrm{i}_S^1 f$ of $L^\infty_{\mathrm{w*}}(S^2, L^2\mathrm{I}_H^G E)$. The Proposition 10.2.2 and the first remark following it show that $L^2\mathrm{I}_H^G E$ is still in $\mathfrak{X}$. Therefore we may apply the assumption on S and conclude that $L^2\mathrm{i}_S^1 f$ is essentially constant. In consequence, its essential value in $L^2\mathrm{I}_H^G F$ is of the form $v\mathbf{1}_G$ for some $v \in F$ and hence f is essentially constant too. □

11.2. Classical examples. The most classical example of double ergodicity with coefficients is the double $\mathfrak{X}^{\mathrm{Hilb}}$-ergodicity of simple Lie groups on their Furstenberg boundary. This phenomenon is sometimes referred to as *Mautner's property* because the structure theory of such groups reduces it to the "generalized Mautner lemma", see Lemma 11.2.4 below. Indeed, as far as *ordinary* ergodicity (i.e. without coefficients) is concerned, it is an immediate consequence of the Bruhat decomposition that a simple Lie group is doubly ergodic on its Furstenberg boundary because there is an orbit of full measure. Now Mautner's lemma is just the argument needed to lift this ordinary ergodiciy to ergodicity with unitary coefficients.

It has been observed [**98**] that the same phenomenon takes place for groups of tree automorphisms. Moreover, the whole discussion can be carried out for continuous isometric representations rather than just unitary representations. One obtains thus the two following basic examples :

PROPOSITION 11.2.1.— *Let* $\mathbf{G}$ *be a connected simply connected almost simple isotropic algebraic group over* $\mathbf{R}$. *Let* $G = \mathbf{G}(\mathbf{R})$ *and let* Q *be a parabolic subgroup. Then the* G *action on* G/Q *is doubly* $\mathfrak{X}^{\mathrm{cont}}$*-ergodic.*

PROPOSITION 11.2.2.— *Let* $\mathcal{T}$ *be a locally finite regular or bi-regular tree,* G *its group of automorphisms and* P *the stabilizer of a point at infinity. Then the* G*-action on* G/P *is doubly* $\mathfrak{X}^{\mathrm{cont}}$*-ergodic.*

We observe right away that Proposition 11.2.1 generalizes to the following setting :

COROLLARY 11.2.3.— *Let* G *be a connected semi-simple adjoint real Lie group. Then the Furstenberg boundary* $B(G)$ *of* G *is a doubly* $\mathfrak{X}^{\mathrm{cont}}$*-ergodic* G*-space.*

PROOF OF THE COROLLARY. Let $G = G_1 \times \cdots \times G_n$ be the decomposition of G into its almost simple factors G_j, $1 \le j \le n$. The adjoint group G_j is of the form $G_j = \widetilde{G_j}/Z(\widetilde{G_j})$, with $\widetilde{G_j}$ simply connected in the algebraic sense. Now the doubly $\mathfrak{X}^{\mathrm{cont}}$-ergodic $\widetilde{G_j}$-space of the form $\widetilde{G_j}/P_j$ of Proposition 11.2.1 is actually a G_j-space, because $Z(\widetilde{G_j}) < P_j$. Thus, by Lemma 11.1.9, the G_j-action on $\widetilde{G_j}/P_j$ is also doubly $\mathfrak{X}^{\mathrm{cont}}$-ergodic. Applying now Lemma 11.1.8, we deduce that the G-action on $\widetilde{G_1}/P_1 \times \cdots \times \widetilde{G_n}/P_n$ is doubly $\mathfrak{X}^{\mathrm{cont}}$-ergodic. On the other hand, this space is just Furstenberg's boundary $B(G)$ of G. □

Now we state the "generalized Mautner lemma" used in the proof of the two above propositions ; we quote faithfully Margulis' statement and proof (Lemma 3.2 in [**103**, chap. II]), replacing only unitarity by isometry :

LEMMA 11.2.4 (Mautner's lemma).— *Let H be a topological group and (π, E) a continuous Banach H-module. Let $x, y \in H$ be such that $\lim_{n\to\infty} x^n y x^{-n} = e$. Then any $v \in E$ fixed by x is fixed by y.*

PROOF OF THE LEMMA. Since v is fixed by x and π is isometric, we have for any $n \geq 0$

$$\|\pi(y)v - v\|_E = \|\pi(y)\pi(x^{-n})v - \pi(x^{-n})v\|_E = \|\pi(x^n y x^{-n})v - v\|_E.$$

Since (π, E) is continuous, $\pi(x^n y x^{-n})v$ converges to v and thus we conclude $\|\pi(y)v - v\|_E = 0$. □

ON THE PROOF OF PROPOSITIONS 11.2.1 AND 11.2.2. In the setting of Proposition 11.2.1, consider a weak-* measurable G-equivariant map $f : (G/Q)^2 \to E$ to a continuous Banach G-module (π, E). For every $a, b \in G$ the element $v = f(aQ, bQ)$ is fixed by the group

$$T = aQa^{-1} \cap bQb^{-1}.$$

Therefore, by Mautner's lemma in the form of Lemma 11.2.4, we deduce that w is fixed by all elements $y \in G$ that are contracted to e by inner automorphisms coming from T. Since T contains always a maximal split torus, the group generated by those $y \in G$ acts ergodically on $(G/Q)^2$. Thus, since f is equivariant, it must be essentially constant.

For Proposition 11.2.1, the argument in the unitary case is given in [**98**] ; the stabilizer of a pair of points at infinity plays the rôle of a torus. The case of continuous Banach G-modules follows exactly the same way since Lemma 11.2.4 above encompasses such modules. □

11.3. The group G^*. We present now the canonical construction of a subgroup G^* for any locally compact group G.

So let G be a locally compact group. The closure properties of the class of amenable locally compact groups (see [**122**]) imply that there is a unique maximal amenable closed normal subgroup, thus containing all amenable closed normal subgroups of G. We denote this subgroup by $A(G)$. Recall further that for a topological group G, the identity component is denoted by G^0. For $H \lhd G$, the group $K_G(H) = H.Z_G(H)$ has been defined in Notation 8.7.5.

DEFINITION 11.3.1. Let G be a locally compact group. We define

$$G^* = \pi^{-1}(K_L(L^0)),$$

where $L = G/A(G)$ and $\pi : G \to L$ is the quotient map.

REMARK 11.3.2. In other words, $G^* \lhd G$ is the kernel of the representation $G \to \mathrm{Out}(L^0)$ defined by

$$G \xrightarrow{\pi} L \longrightarrow \mathrm{Out}(L^0).$$

Therefore, if G is connected, we have $G^* = G$ because the map $L \to \mathrm{Out}(L^0)$ is trivial in view of $L^0 = L$. On the other extreme, if G is totally disconnected (for instance if G is discrete), we have again $G^* = G$: indeed, L is also totally disconnected (by Corollaire 3 in [**27**, III §4 N° 6]) ; therefore L^0 is trivial and $G^* = G$.

In the notations of the above definition, we have the

LEMMA 11.3.3.— *Let $M \lhd L$ be a closed normal subgroup of $L = G/A(G)$. Then $A(M)$ is trivial.*

PROOF. The map $\pi : G \to L$ yields a topological group extension

$$1 \longrightarrow A(G) \cap \pi^{-1}\big(A(M)\big) \longrightarrow \pi^{-1}\big(A(M)\big) \longrightarrow A(M) \longrightarrow 1.$$

The two extreme terms are amenable, hence $\pi^{-1}\big(A(M)\big)$ is amenable. We claim that this group is also normal in G, which finishes then the proof, for it forces $\pi^{-1}\big(A(M)\big) \subset A(G)$ whence $A(M) = 1$.

As for the claim, it is equivalent to say that $A(M)$ is normal in L. We have $A(M) \lhd M \lhd L$, so that any conjugate of $A(M)$ is normal in M and amenable, hence contained in $A(M)$. This proves the claim. □

The solution to Hilbert's fifth problem given by D. Montgomery and L. Zippin [**114**] allows us to analyze the quotient $G^*/A(G) = K_L(L^0)$. Then, using the finiteness of the group of automorphisms of semi-simple groups, we deduce that G^* is of finite index in G :

THEOREM 11.3.4.— *Let G be a locally compact group and define L, G^* as above.*

(i) *G^* is a topologically characteristic finite index open subgroup of the group G.*

(ii) *The group $G^*/A(G) = K_L(L^0)$ is the topological direct product $L^0.Z_L(L^0)$, where L^0 is a connected semi-simple adjoint real Lie group without compact factors.*

PROOF. Since L^0 is a connected locally compact group, there is by [**114**, Theorem 4.6] a compact normal subgroup $K \lhd L^0$ such that L^0/K is a connected real Lie group (this is the solution to Hilbert's fifth problem). Now $A(L^0) = 1$ (Lemma 11.3.3) implies $K = 1$, hence L^0 is a connected real Lie group. The triviality of $A(L^0)$ implies further that L^0 is semi-simple, adjoint and without compact factors. In particular, the group $\mathrm{Out}(L^0)$ is finite : indeed, this follows from the corresponding statement for the Lie algebra $\mathfrak{l} = \mathrm{Lie}(L^0)$, and it is well known that $\mathrm{Out}(\mathfrak{l})$ is isomorphic to the group of automorphism of a simple root system for $\mathfrak{l}$ (Theorem 1 in [**120**, chap. 4 §4]). Therefore G^* is open of finite index in G. Since L^0 has trivial centre, the product $L^0.Z_L(L^0)$ is direct.

It remains only to show that G^* is topologically characteristic, so let ψ be a topological automorphism of G. Since $\psi(A(G))$ is amenable, it is contained in $A(G)$ and hence ψ induces a topological automorphism $\overline{\psi}$ of L. Now $\overline{\psi}$ preserves L^0, hence also $Z_L(L^0)$, and therefore preserves $K_L(L^0)$. Thus ψ preserves G^*. □

We observe furthermore the

PROPOSITION 11.3.5.— *If G is σ-compact, e.g. if G is compactly generated, then the group L, and hence also $Z_L(L^0)$, is second countable.*

PROOF. By Satz 6 in [**90**], there is a compact normal subgroup $K \lhd G$ with G/K second countable. Since K is contained in $A(G)$, the group L is a quotient of G/K and hence it is also second countable. □

In view of the second statement of Theorem 11.3.4, the next important step is to produce an amenable doubly $\mathfrak{X}^{\mathrm{sep}}$-ergodic space for totally disconnected groups. To this end, we use a so-called Schreier graph to reduce the problem to the automorphism group of a locally finite regular tree. Here is the construction.

We adopt J.-P. Serre's conventions [**129**] for graphs, so that any graph $\mathfrak{g} = (V, E)$ consists of a vertex set V, a set E of oriented edges, a fixed point free involution $e \mapsto \overline{e}$ of E (the orientation reversal) and two boundary maps $o, t : E \to V$ satisfying $t(\overline{e}) = o(e)$.

Let G be a compactly generated totally disconnected locally compact group. By Corollaire 1 in [**27**, III §4 N° 6], we can find a compact open subgroup $U < G$. We fix now a compact generating set C of G ; upon replacing C by UCU, we may assume that $C = UCU$ holds. We proceed now to define the *Schreier graph* $\mathfrak{g}$ associated to this data. The set of vertices is $V = G/U$ and the set of edges is $E = E_+ \sqcup \overline{E_+}$, where

$$E_+ = \big\{(gU, gcU) : g \in G, c \in C\big\}.$$

It is enough to define the boundary maps on E_+, where we let $o(gU, gcU) = gU$ and $t(gU, gcU) = gcU$. Observing that by compactness of C the set C/U is finite, we denote by $r = |C/U|$ its cardinal. Thus the graph $\mathfrak{g}$ is regular of degree r ; moreover, it is connected because C generates G. Thus $\mathfrak{g}$ admits a r-regular tree $\mathcal{T}_r$ as universal covering, and we fix a simplicial universal covering projection $\mathcal{T}_r \to \mathfrak{g}$.

Now we let G act on the graph $\mathfrak{g}$ by left multiplication on V, which is compatible with the graph structure. The kernel of this action is $K = \bigcap_{g \in G} gUg^{-1}$, which is a compact normal subgroup. We write $G_1 = G/K$, so that we have an exact sequence

$$1 \longrightarrow \pi_1(\mathfrak{g}) \longrightarrow \widetilde{G} \longrightarrow G_1 \longrightarrow 1,$$

where $\widetilde{G}$ is co-compact in $\mathrm{Aut}(\mathcal{T}_r)$. Let $\partial_\infty \mathcal{T}_r$ be the boundary at infinity of the tree $\mathcal{T}_r$ with its $\mathrm{Aut}(\mathcal{T}_r)$-action. We endow $\partial_\infty \mathcal{T}_r$ with the class of the $\mathrm{Aut}(\mathcal{T}_r)$-quasi-invariant measures. We define now the probability G_1-space (B, ν) as the point realization of $L^\infty(\partial_\infty \mathcal{T}_r)^{\pi_1(\mathfrak{g})}$ (see the remarks preceding Lemma 5.4.4 and

Remark 5.7.2). Recall that B is a regular G_1-space given with a canonical C *-algebra isomorphism between $L^\infty(B)$ and the weak-* closed sub- C *-algebra $L^\infty(\partial_\infty \mathcal{T}_r)^{\pi_1(\mathfrak{g})}$ of $L^\infty(\partial_\infty \mathcal{T}_r)$, the isomorphism being induced by a measurable equivariant map $\partial_\infty \mathcal{T}_r \to B$ preserving the measure class. We consider B as a G-space via the canonical map $G \to G_1$.

PROPOSITION 11.3.6.— *Suppose that the compactly generated totally disconnected locally compact group G is second countable. Then*

(i) *The G-action on B is amenable.*

(ii) *The G-action on B is doubly $\mathfrak{X}^{\mathrm{sep}}$-ergodic.*

PROOF. Recall first that the $\mathrm{Aut}(\mathcal{T}_r)$-action on $\partial_\infty \mathcal{T}_r$ is amenable ; indeed, $\partial_\infty \mathcal{T}_r$ is of the form $\mathrm{Aut}(\mathcal{T}_r)/P$ where P is the stabilizer of a point at infinity, and this stabilizer is an amenable group, so we are in the situation (i) of the Examples 5.4.1. Thus the restricted $\widetilde{G}$-action is also amenable (Lemma 5.4.3). Now this implies that the $G_1 = \widetilde{G}/\pi_1(\mathfrak{g})$-action on (B, ν) is amenable (Lemma 5.4.4). Since G_1 is the quotient of G by the compact, hence amenable, kernel K, we conclude by point (vii) in Examples 5.4.1 that the G-action on B is amenable.

As for point (ii), it is enough by Lemma 11.1.9 to show that the G_1-action on B is doubly $\mathfrak{X}^{\mathrm{sep}}$-ergodic. If $f : B \times B \to F$ is a G_1-equivariant weak-* measurable map to a reflexive coefficient G_1-module F, we pull back through $\partial_\infty \mathcal{T}_r \to B$ and obtain a weak-* measurable $\widetilde{G}$-equivariant $f' : \partial_\infty \mathcal{T}_r \times \partial_\infty \mathcal{T}_r \to F$. Applying successively Proposition 11.1.10, which we may as G is second countable, and Proposition 11.2.2, we conclude that f' is essentially constant. Hence f is essentially constant, finishing the proof. □

If we collect the results established so far, we can complete the

PROOF OF THEOREM 11.1.3. Let G be a locally compact compactly generated group and recall the notations $L = G/A(G)$ and $G^* = \pi^{-1}\big(K_L(L^0)\big)$ of Theorem 11.3.4. Since G^* is closed of finite index in G, it is also compactly generated. Hence $K_L(L^0) = G^*/A(G)$ is compactly generated. By the second point of Theorem 11.3.4, $K_L(L^0) = L^0.Z_L(L^0)$ is a direct product, which implies that $Z_L(L^0) = K_L(L^0)/L^0$ is a totally disconnected compactly generated locally compact group. This group is moreover second countable by Proposition 11.3.5. Therefore there is an amenable doubly $\mathfrak{X}^{\mathrm{sep}}$-ergodic regular $Z_L(L^0)$-space B by Proposition 11.3.6.

On the other hand, we know that L^0 is a connected semi-simple adjoint real Lie group without compact factors. Thus Proposition 11.2.1 provides us with an amenable regular L^0-space $B(L^0)$ which is doubly $\mathfrak{X}^{\mathrm{cont}}$-ergodic, hence in particular also doubly $\mathfrak{X}^{\mathrm{sep}}$-ergodic. This space is of course nothing but the Furstenberg boundary of L^0.

Applying Lemma 11.1.8, we conclude that the second countable direct product $K_L(L^0) = L^0.Z_L(L^0)$ admits an amenable regular $K_L(L^0)$-space S which is doubly $\mathfrak{X}^{\mathrm{sep}}$-ergodic, namely $S = B \times B(L^0)$.

We view now S as a G^*-space via the canonical map $G^* \to G^*/A(G) = K_L(L^0)$ and conclude by Lemma 11.1.9 that S is a doubly $\mathfrak{X}^{\mathrm{sep}}$-ergodic G^*-space. It remains only to check that the G^*-action on S is amenable ; we have to use Definition 5.3.10 since G^* need not be second countable. Since G is compactly generated and hence σ-compact, the Satz 6 in [**90**] provides us with a compact normal subgroup $K \lhd G$ such that G/K is second countable. Since $K \subset A(G)$, it is also a (normal) subgroup of G^* and moreover its normality in G implies that the projection $G^* \to K_L(L^0)$ factors as

$$\begin{array}{ccccccccc} & & & & G^* & & & & \\ & & & & \downarrow & \searrow & & & \\ 1 & \longrightarrow & A(G)/K & \longrightarrow & G^*/K & \longrightarrow & K_L(L^0) & \longrightarrow & 1 \end{array}$$

where the row is an exact sequence of second countable groups. Since $A(G)/K$ is amenable, the amenability of the G^*/K-action on S is equivalent to the amenability of the $K_L(L^0)$-action, as pointed out in the point (vii) of Examples 5.4.1. □

11.4. Low degree, II. In low degree, double ergodicity allows us to gain a closer description of continuous bounded cohomology thanks to the explicit resolutions on amenable spaces of Theorem 7.5.3. There is of course nothing more to say about the degree zero, where $\mathrm{H}^0_{\mathrm{cb}}(G, E) = E^G$ holds for all groups G and all Banach G-modules E. But in degree one and two, we derive the following general statement from double ergodicity :

PROPOSITION 11.4.1.— *Let G be a compactly generated locally compact second countable group and $\alpha : E \to F$ an injective adjoint morphism of coefficient G-modules. Assume F separable. Then*

(i) $\mathrm{H}^1_{\mathrm{cb}}(G, E) = 0$.

(ii) *The induced map $\mathrm{H}^2_{\mathrm{cb}}(G, E) \to \mathrm{H}^2_{\mathrm{cb}}(G, F)$ is injective and both spaces are Banach spaces.*

In simpler terms, the main upshot of this proposition is the following special case :

COROLLARY 11.4.2.— *Let G be a compactly generated locally compact second countable group and E a separable coefficient G-module. Then $\mathrm{H}^1_{\mathrm{cb}}(G, E) = 0$ and $\mathrm{H}^2_{\mathrm{cb}}(G, E)$ is a Banach space.* □

PROOF OF PROPOSITION 11.4.1. Let $G^* \lhd G$ and S be as in Theorem 11.1.3. We shall first prove the proposition for G^* instead of G. By Theorem 7.5.3, the continuous bounded cohomology $\mathrm{H}^\bullet_{\mathrm{cb}}(G^*, E)$ and $\mathrm{H}^\bullet_{\mathrm{cb}}(G^*, F)$, together with the map induced by $\alpha : E \to F$, are realized on the complexes

$$\begin{array}{ccccccccc} 0 & \longrightarrow & L^\infty_{\mathrm{w*,alt}}(S, E)^{G^*} & \longrightarrow & L^\infty_{\mathrm{w*,alt}}(S^2, E)^{G^*} & \longrightarrow & L^\infty_{\mathrm{w*,alt}}(S^3, E)^{G^*} & \longrightarrow & \cdots \\ & & \downarrow{\scriptstyle \alpha^0_*} & & \downarrow{\scriptstyle \alpha^1_*} & & \downarrow{\scriptstyle \alpha^2_*} & & \\ 0 & \longrightarrow & L^\infty_{\mathrm{w*,alt}}(S, F)^{G^*} & \longrightarrow & L^\infty_{\mathrm{w*,alt}}(S^2, F)^{G^*} & \longrightarrow & L^\infty_{\mathrm{w*,alt}}(S^3, F)^{G^*} & \longrightarrow & \cdots \end{array}$$

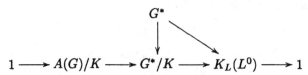

where $\alpha^\bullet_*$ denotes the post-composition by α, and thus is injective in all degrees. Since the G^*-action on S is doubly $\mathfrak{X}^{\mathrm{sep}}$-ergodic, the space $L^\infty_{\mathrm{w*}}(S^2, F)^{G^*}$ contains only essentially constant functions. Thus, the only *alternating* element of $L^\infty_{\mathrm{w*}}(S^2, F)^{G^*}$ is the zero constant. This implies at once that $\mathrm{H}^1_{\mathrm{cb}}(G^*, F)$ vanishes and that $\mathrm{H}^2_{\mathrm{cb}}(G^*, F)$ identifies with the Banach subspace of G^*-invariant cocycles in $L^\infty_{\mathrm{w*}}(S^3, F)$. The injectivity of $\alpha^\bullet_*$ now implies the remaining statements for the group G^*.

Now we come back to G, and recall from Proposition 8.6.6 that the restriction $\mathrm{H}^\bullet_{\mathrm{cb}}(G, E) \to \mathrm{H}^\bullet_{\mathrm{cb}}(G^*, E)$ is injective since G^* is of finite index in G. Therefore, we deduce $\mathrm{H}^1_{\mathrm{cb}}(G, E) = 0$. Since the restriction is moreover continuous (in fact isometric by Proposition 8.6.6), it sends the closure of zero in $\mathrm{H}^2_{\mathrm{cb}}(G, E)$ to zero in $\mathrm{H}^2_{\mathrm{cb}}(G^*, E)$, so that $\mathrm{H}^2_{\mathrm{cb}}(G, E)$ is Hausdorff and hence a Banach space. The same argument holds for $\mathrm{H}^2_{\mathrm{cb}}(G, F)$. We have now a commutative diagram

$$\begin{array}{ccc}
\mathrm{H}^2_{\mathrm{cb}}(G, E) & \xhookrightarrow{\ \mathrm{res}\ } & \mathrm{H}^2_{\mathrm{cb}}(G^*, E) \\
{\scriptstyle \mathrm{H}^2_{\mathrm{cb}}(G,\alpha)} \big\downarrow & & \big\downarrow {\scriptstyle \mathrm{H}^2_{\mathrm{cb}}(G^*,\alpha)} \\
\mathrm{H}^2_{\mathrm{cb}}(G, F) & \xhookrightarrow{\ \mathrm{res}\ } & \mathrm{H}^2_{\mathrm{cb}}(G^*, F)
\end{array}$$

and therefore the map $\mathrm{H}^2_{\mathrm{cb}}(G, \alpha)$ is injective. □

REMARKS 11.4.3. We insist once again on the importance of the class $\mathfrak{X}$ in the notion of double $\mathfrak{X}$-ergodicity. In contrast to Corollary 11.4.2, we recall from Example 10.3.7 that the space $\mathrm{H}^1_{\mathrm{b}}(\Gamma, \mathfrak{S}_S\mathbf{C})$ is infinite dimensional for every non-elementary Gromov-hyperbolic group Γ and any amenable Γ-space S. As is well known, such a group Γ is finitely generated, hence this group would fit the assumption of the corollary.

In particular, we deduce that whenever S is an amenable Γ-space, then no amenable Γ-space T is doubly $\mathfrak{S}_S\mathbf{C}$-ergodic.

This example also allows us to show the necessity of restricting the class of coefficient modules for the second statement of Corollary 11.4.2. Indeed, if Γ is the fundamental group of a surface of genus at least two, then T. Soma has shown in [**132**] and [**133**] that $\mathrm{H}^3_{\mathrm{b}}(\Gamma)$ is not Hausdorff — actually, the closure of zero is infinite dimensional. Thus the space $\mathrm{H}^2_{\mathrm{b}}(\Gamma, \mathfrak{S}_S\mathbf{C})$ in non Hausdorff for any amenable Γ-space S.

There is another issue in low degree upon which double ergodicity sheds new light, namely the action on cohomology and its relation with the coefficient representation. We recall from Sections 8.3 and 8.7 that if $H \lhd G$ is a closed normal subgroup of the topological group G and (π, E) is a G-modules, then we can one one hand consider the G-action on $\mathrm{H}^\bullet_{\mathrm{cb}}(H, E)$ (actually a G/H-action), and on the other hand there is on $\mathrm{H}^\bullet_{\mathrm{cb}}(H, E)$ the coefficient representation of $Z_G(H)$ defined by π. The purpose of Corollary 8.7.6 was to show that these actions coincide when suitably restricted.

This situation becomes of interest for instance in connection with the restriction map. We know that the restriction $\mathrm{H}^\bullet_{\mathrm{cb}}(G, E) \to \mathrm{H}^\bullet_{\mathrm{cb}}(H, E)$ ranges in the

G-invariants $\mathrm{H}^\bullet_{\mathrm{cb}}(H,E)^G$ (Corollary 8.7.4). These invariants, by the above remarks, are in particular invariant under the coefficient representation of $Z_G(H)$.

However, there is no reason a priori that such cohomology classes should actually range in the coefficient H-module $E^{Z_G(H)}$ of $Z_G(H)$-invariants. The following theorem shows that this happens however in degree two.

THEOREM 11.4.4.— *Let G be a locally compact second countable group, $H \lhd G$ a compactly generated closed subgroup and (π, E) a separable coefficient G-module.*

Then the inclusion $E^{Z_G(H)} \to E$ induces a canonical isometric identification $\mathrm{H}^2_{\mathrm{cb}}(H, E^{Z_G(H)})^G \cong \mathrm{H}^2_{\mathrm{cb}}(H,E)^G$.

Thanks to Theorem 11.1.3, the main part of the proof will be the following :

PROPOSITION 11.4.5.— *Let G be a locally compact second countable group, $H \lhd G$ a closed subgroup and (π, E) a coefficient G-module.*

If H admits an amenable doubly E-ergodic regular space S, then the inclusion $E^{Z_G(H)} \to E$ induces a canonical isometric identification $\mathrm{H}^2_{\mathrm{cb}}(H, E^{Z_G(H)}) \cong \mathrm{H}^2_{\mathrm{cb}}(H,E)^{K_G(H)}$.

PROOF OF PROPOSITION 11.4.5. In view of Corollary 8.7.6, it only remains to show that any class $\omega \in \mathrm{H}^2_{\mathrm{cb}}(H,E)$ invariant under the coefficient $Z_G(H)$-action is represented by a cocycle ranging in $E^{Z_G(H)}$. We realize $\mathrm{H}^\bullet_{\mathrm{cb}}(H,E)$ on the complex $L^\infty_{\mathrm{w*}}(S^\bullet, E)^H$ in accordance with Theorem 7.5.3. Thus ω can be represented by a cocycle $f \in L^\infty_{\mathrm{w*}}(S^3,E)^H$, and for each $z \in Z_G(H)$ there is $b_z \in L^\infty_{\mathrm{w*}}(S^2,E)^H$ with $\pi(z) \circ f = f + db_z$. One can take f and b_z alternating, but since b_z is essentially constant, this forces $b_z = 0$ and hence f ranges in $E^{Z_G(H)}$. □

A diagram chase will now allow us to obtain the

PROOF OF THEOREM 11.4.4. We let $H^* < H$ be as in Theorem 11.1.3. We have then the following natural diagram, where α, β, η are the maps induced by the corresponding inclusions of coefficients (observe that $Z_G(H^*) \supset Z_G(H)$ implies $E^{Z_G(H^*)} \subset E^{Z_G(H)}$). The theorem is about α.

$$\begin{array}{ccccc}
\mathrm{H}^2_{\mathrm{cb}}(H, E^{Z_G(H)})^G & & \xrightarrow{\quad\alpha\quad} & & \mathrm{H}^2_{\mathrm{cb}}(H,E)^G \\
\cong \big\downarrow \mathrm{res} & & & & \mathrm{res} \big\downarrow \cong \\
\mathrm{H}^2_{\mathrm{cb}}(H^*, E^{Z_G(H)})^G & \xleftarrow{\ \beta\ } & \mathrm{H}^2_{\mathrm{cb}}(H^*, E^{Z_G(H^*)})^G & \xrightarrow[\cong]{\ \eta\ } & \mathrm{H}^2_{\mathrm{cb}}(H^*, E)^G
\end{array}$$

The map η is an isomorphism by Proposition 11.4.5, and the two restrictions by Proposition 8.6.6). Since all maps above are obtained either by covariance or contravariance, all possible commutation relations hold. Thus it is enough to show that β is bijective. But

$$\mathrm{res} = \beta \circ \eta^{-1} \circ \mathrm{res} \circ \alpha$$

implies surjectivity, while

$$\eta = \mathrm{res} \circ \alpha \circ \mathrm{res}^{-1} \circ \beta$$

entails injectivity. □

The Theorem 11.4.4 has the following particular case :

COROLLARY 11.4.6.— *Let G be a locally compact second countable group, $H \lhd G$ a compactly generated closed subgroup and (π, E) a separable coefficient G-module.*

If $K_G(H) = G$ (e.g. if H is a direct factor of G), then the inclusion $E^{Z_G(H)} \to E$ induces a canonical isometric identification $\mathrm{H}^2_{\mathrm{cb}}(H, E^{Z_G(H)}) \cong \mathrm{H}^2_{\mathrm{cb}}(H, E)^G$.

PROOF. If $K_G(H) = G$, the Corollary 8.7.6 shows that the G-action coincides with the $Z_G(H)$-action by coefficient representation, which is trivial on $E^{Z_G(H)}$, whence $\mathrm{H}^2_{\mathrm{cb}}(H, E^{Z_G(H)}) = \mathrm{H}^2_{\mathrm{cb}}(H, E^{Z_G(H)})^G$, so that one can conclude with Theorem 11.4.4. □

12. Hochschild-Serre spectral sequence

The purpose of this section is to derive an analogue of the Hochschild-Serre spectral sequence in [continuous] bounded cohomology. That is, given a topological group extension

$$1 \longrightarrow N \longrightarrow G \longrightarrow Q \longrightarrow 1,$$

we establish relations between the continuous bounded cohomology of G with coefficients and the corresponding bounded cohomology of N and Q. This allows us to derive exact sequences paralleling the Lyndon-Hochschild-Serre five term exact sequence. As a matter of illustration, we mention briefly discrete groups and an application to topological fibrations in Section 12.4.

REMARK 12.0.1. In this section, we will consider locally compact second countable groups. Since such groups are in particular σ-compact, it is a well known consequence of Baire's category theorem that a short exact sequence of morphisms of topological groups $1 \to N \xrightarrow{i} G \to Q \to 1$ is topologically isomorphic to

$$1 \longrightarrow i(N) \longrightarrow G \longrightarrow G/i(N) \longrightarrow 1$$

(see e.g. the Corollary 3.11 in [**46**, III]). Therefore, we shall always consider the case where N is a closed normal subgroup of G and $Q = G/N$ the quotient.

Actually, a first exact sequence in low degree (Proposition 12.3.1) is the exact analogue of the usual Lyndon-Hochschild-Serre sequence. But a second sequence, valid under very general assumptions, takes advantage of the vanishing in degree one to yields information up to degree three :

THEOREM 12.0.2.— *Let G be a locally compact second countable group, $N \lhd G$ a closed subgroup and $Q = G/N$ the quotient. Let (π, F) be a coefficient G-module.*

If $\mathrm{H}^1_{\mathrm{cb}}(N,F) = 0$, then inflation and restriction fit into a natural sequence

$$0 \longrightarrow \mathrm{H}^2_{\mathrm{cb}}(Q, F^N) \xrightarrow{\mathrm{inf}} \mathrm{H}^2_{\mathrm{cb}}(G,F) \xrightarrow{\mathrm{res}} \mathrm{H}^2_{\mathrm{cb}}(N,F)^Q \longrightarrow$$
$$\longrightarrow \mathrm{H}^3_{\mathrm{cb}}(Q, F^N) \xrightarrow{\mathrm{inf}} \mathrm{H}^3_{\mathrm{cb}}(G,F).$$

(Throughout this section, we shall avoid the letter E for coefficient modules, in order not to prejudice legibility in presence of spectral sequences $\mathrm{E}^{\bullet,\bullet}_{\bullet}$).

Bringing in the consequences of double ergodicity established in Section 11.4, we deduce a sequence which is more precise in the $\mathrm{H}^2_{\mathrm{cb}}(N,-)$ term, and incidentally is freed from the assumption on $\mathrm{H}^1_{\mathrm{cb}}(N,F)$:

THEOREM 12.0.3.— *Let G be a locally compact second countable group, $N \lhd G$ a compactly generated closed subgroup and $Q = G/N$ the quotient. Let (π, F) be a separable coefficient G-module.*

Then we have the exact sequence

$$0 \longrightarrow \mathrm{H}^2_{\mathrm{cb}}(Q, F^N) \xrightarrow{\mathrm{inf}} \mathrm{H}^2_{\mathrm{cb}}(G,F) \xrightarrow{\mathrm{res}} \mathrm{H}^2_{\mathrm{cb}}(N,F^{Z_G(N)})^Q \longrightarrow$$
$$\longrightarrow \mathrm{H}^3_{\mathrm{cb}}(Q, F^N) \xrightarrow{\mathrm{inf}} \mathrm{H}^3_{\mathrm{cb}}(G,F).$$

As a consequence we have the following

COROLLARY 12.0.4.— *Let $G_1, \ldots, G_n$ be compactly generated locally compact second countable groups and let $G = \prod_{j=1}^n G_j$, $G'_j = \prod_{i\neq j} G_i$. Let (π, E) be a separable coefficient G-module. Then we have a canonical topological isomorphism*

$$\mathrm{H}^2_{\mathrm{cb}}(G,F) \cong \bigoplus_{j=1}^{n} \mathrm{H}^2_{\mathrm{cb}}(G_j, F^{G'_j})$$

realized by $\oplus_{j=1}^n \mathrm{res}$ with inverse $\sum_{j=1}^n \mathrm{inf}$. The result still holds true if one of the groups G_j is not compactly generated.

(Proofs are given in Section 12.3.)

This result is stronger then a Künneth type formula, since the coefficient G-module F is not assumed to be a tensor product.

EXAMPLE 12.0.5. In the setting of Corollary 12.0.4, denote by F_0 the G-invariant subspace $F_0 = \sum_{j=1}^n F^{G'_j}$ of F. We shall see below (Lemma 12.3.5) that F_0 is closed and even weak-* closed, so that it is a coefficient submodule of F. Since $F_0^{G'_j} = F^{G'_j}$, we apply the Corollary 12.0.4 successively to F and to F_0 and deduce an isomorphism

$$\mathrm{H}^2_{\mathrm{cb}}(G,F) \cong \mathrm{H}^2_{\mathrm{cb}}(G,F_0).$$

In particular, if for all j the space $F^{G'_j}$ vanishes, we have $\mathrm{H}^2_{\mathrm{cb}}(G,F) = 0$.

REMARK 12.0.6. It is somewhat surprising that the statements of Corollary 12.0.4 and of Example 12.0.5 allow for *one* non compactly generated group, whilst the rest of the statement is completely symmetric in the groups G_j. The reason is that their proof involves an inductive application of Theorem 12.0.3 ; the latter requires the kernel N of the extension to be compactly generated in order to ensure the existence of a doubly ergodic amenable space. This is however not needed for the quotient Q of the extension.

The liberty to admit one non compactly generated group will turn out useful in connection with adélic groups (Section 14.5).

The technology of spectral sequences cannot be used as smoothly as in usual cohomology of discrete groups. At an intuitive level, one can say that spectral sequences involve taking cohomology, then the cohomology of the cohomology, and so on. Since bounded cohomology is not necessarily separated, we might get stuck already at the first iteration. This difficulty is actually familiar in the theory of continuous cohomology, where it is usually bypassed simply by assuming the relevant spaces to be Hausdorff. This ploy is not quite satisfactory here, as bounded cohomology is typically non Hausdorff in degree three (see [**132**] and [**133**]). However, it turns out that this phenomenon, even though it impedes a uniform interpretation of the second tableau in the spectral sequence under consideration, does not obstruct to the proof of the exact sequences.

REMARK 12.0.7. A bibliographical remark is in order, as there are similar attempts for discrete groups. In the end of his thesis [**23**], A. Bouarich proposes a Hochschild-Serre spectral sequence for the bounded cohomology of discrete groups. Unfortunately, his discussion seems to be erroneous. In particular, the phenomenon mentioned below in Remark 12.2.4 is not taken into account. However, without the use of spectral sequences, he derives in the first part of his work a four-term exact sequence for the cohomology of discrete groups with *trivial* coefficients, as published in [**22**]. In [**118**], G.A. Noskov gives a spectral sequence for the bounded cohomology of discrete groups (following closely [**19**, IX §4] ; then abandoning the spectral sequence, he derives by hand the discrete analogue of our first exact sequence (Proposition 12.3.1). The article [**121**] of D. Paul presents a spectral sequence for the discrete case with Q *finite* ; this paper appears to contain many false statements about unfortunate definitions.

12.1. General setup. Throughout this Section 12, G is a locally compact second countable group, $N \lhd G$ a closed normal subgroup and Q the quotient G/N.

We use standard conventions for the spectral sequence associated to a filtered cochain complex, and more specifically for the two filtrations, horizontal and vertical, of the total complex associated to a first quadrant double complex. We take the notations from the book of I. Gelfand and Yu. Manin [**60**, III.7], which coincide with the notations of [**21**, III §14]. An alternative source (with dual notations) is [**137**, chap. 5]. The reader should be warned against the numerous misprints of [**60**].

Unless otherwise stated, the spectral sequence arguments, direct sums etc. are taking place in the linear category, other issues being postponed until the interpretation of the results.

Let (π, F) be a coefficient G-module, S an amenable regular G-space and T an amenable regular Q-space. We consider without further notational ado F as a coefficient N-module, F^N as a coefficient Q-module (Lemma 1.2.9), S as an amenable N-space (Lemma 5.4.3) and T as a regular G-space. We define a first quadrant double complex $(L^{\bullet,\bullet}, {}^I\!d, {}^{II}\!d)$ as follows. For all $p, q \geq 0$ set

$$L^{p,q} = L^\infty_{\mathrm{w*}}(S^{p+1} \times T^{q+1}, F)^G$$

and $L^{p,q} = 0$ if p or q is negative. Let now

$$d:\ L^\infty_{\mathrm{w*}}\Big(S^{p+1}, L^\infty_{\mathrm{w*}}(T^{q+1}, F)\Big) \longrightarrow L^\infty_{\mathrm{w*}}\Big(S^{p+2}, L^\infty_{\mathrm{w*}}(T^{q+1}, F)\Big)$$

be the standard homogeneous differential. We define the horizontal differential ${}^I\!d$ of bi-degree $(1,0)$ by restricting d to the G-invariants via the canonical identification arising from in Corollary 2.3.3. Similarly, switching the factors, we let

$$d':\ L^\infty_{\mathrm{w*}}\Big(T^{q+1}, L^\infty_{\mathrm{w*}}(S^{p+1}, F)\Big) \longrightarrow L^\infty_{\mathrm{w*}}\Big(T^{q+2}, L^\infty_{\mathrm{w*}}(S^{p+1}, F)\Big)$$

be the standard homogeneous differential and define the vertical differential ${}^{II}\!d$ of bi-degree $(0,1)$ by restricting $(-1)^{(p+1)}d'$ to the G-invariants via the corresponding identification. The sign convention ensures ${}^I\!d\,{}^{II}\!d + {}^{II}\!d\,{}^I\!d = 0$ and hence the total differential $D = {}^I\!d + {}^{II}\!d$ turns indeed the total graded space

$$TL^n = \bigoplus_{p+q=n} L^{p,q}$$

into a cochain complex. The horizontal and vertical filtrations of $TL^\bullet$ are respectively

$${}^I\!F^m TL^n = \bigoplus_{\substack{p+q=n \\ p \geq m}} L^{p,q}, \qquad {}^{II}\!F^m TL^n = \bigoplus_{\substack{p+q=n \\ q \geq m}} L^{p,q}.$$

We get thus two spectral sequences ${}^I\mathrm{E}_\bullet^{\bullet,\bullet}$ and ${}^{II}\mathrm{E}_\bullet^{\bullet,\bullet}$ starting respectively with

$${}^I\mathrm{E}_1^{p,q} = \mathrm{H}^{p,q}(L^{p,\bullet}, {}^{II}\!d), \qquad {}^{II}\mathrm{E}_1^{p,q} = \mathrm{H}^{q,p}(L^{\bullet,p}, {}^I\!d)$$

and converging both (in the category of linear spaces) to the cohomology of the total complex $(TL^\bullet, D)$. Recall that for both spectral sequences the differentials are of the form

$$\mathrm{E}_r^{p-r,q+r-1} \longrightarrow \mathrm{E}_r^{p,q} \longrightarrow \mathrm{E}_r^{p+r,q-r+1},$$

and that in particular on ${}^I\mathrm{E}_1^{\bullet,\bullet}$ the differential is induced by ${}^I\!d$ and on ${}^{II}\mathrm{E}_1^{\bullet,\bullet}$ by ${}^{II}\!d$.

After the manner of usual Hochschild-Serre spectral sequences, we shall see that the first spectral sequence converges rather trivially (Proposition 12.2.1), its only benefit being to indicate us what the cohomology of the total complex $(TL^\bullet, D)$ is. The second spectral sequence ${}^{II}\mathrm{E}_\bullet^{\bullet,\bullet}$ will really be of interest, and its interpretation takes all its sense since we know a priori that it must also abut to the cohomology of the total complex [**60**, III.7.9]. This second spectral sequence is properly called the Hochschild-Serre spectral sequence, whilst ${}^{I}\mathrm{E}_{\bullet}^{\bullet,\bullet}$ is rather an auxiliary spectral sequence.

The following well known "abstract nonsense" (see [**60**, III.7]) will be the pattern guiding us through the proof of Theorem 12.0.2.

LEMMA 12.1.1.— *Let $\mathrm{E}_{\bullet}^{\bullet,\bullet}$ be a first quadrant spectral sequence. For any $r \geq p+1, q+2$ one has $\mathrm{E}_{\infty}^{p,q} = \mathrm{E}_{r}^{p,q}$ and hence in particular for all $s \geq 1$ the differential $\mathrm{E}_{s}^{0,s-1} \to \mathrm{E}_{s}^{s,0}$ fits into the exact sequence*

$$0 \longrightarrow \mathrm{E}_{\infty}^{0,s-1} \longrightarrow \mathrm{E}_{s}^{0,s-1} \longrightarrow \mathrm{E}_{s}^{s,0} \longrightarrow \mathrm{E}_{\infty}^{s,0} \longrightarrow 0. \qquad \square$$

With the standard notation $\mathrm{E}_{\infty}^{n} = \bigoplus_{p+q=n} \mathrm{E}_{\infty}^{p,q}$, this implies immediately the

LEMMA 12.1.2.— *Let $\mathrm{E}_{\bullet}^{\bullet,\bullet}$ be a first quadrant spectral sequence.*

(i) *There is a canonical exact sequence*

$$0 \longrightarrow \mathrm{E}_{\infty}^{1,0} \longrightarrow \mathrm{E}_{\infty}^{1} \longrightarrow \mathrm{E}_{2}^{0,1} \longrightarrow \mathrm{E}_{2}^{2,0} \longrightarrow \mathrm{E}_{\infty}^{2}.$$

(ii) *If $\mathrm{E}_{1}^{\bullet,1} = 0$, then there is a canonical exact sequence*

$$0 \longrightarrow \mathrm{E}_{\infty}^{2,0} \longrightarrow \mathrm{E}_{\infty}^{2} \longrightarrow \mathrm{E}_{3}^{0,2} \longrightarrow \mathrm{E}_{3}^{3,0} \longrightarrow \mathrm{E}_{\infty}^{3}.$$

PROOF. For (i), we consider first the canonical inclusion $\mathrm{E}_{\infty}^{1,0} \to \mathrm{E}_{\infty}^{1}$, which yields the exact sequence

$$0 \longrightarrow \mathrm{E}_{\infty}^{1,0} \longrightarrow \mathrm{E}_{\infty}^{1} \longrightarrow \mathrm{E}_{\infty}^{0,1} \longrightarrow 0.$$

The case $s = 2$ of Lemma 12.1.1 yields the exact sequence

$$0 \longrightarrow \mathrm{E}_{\infty}^{0,1} \longrightarrow \mathrm{E}_{2}^{0,1} \longrightarrow \mathrm{E}_{2}^{2,0} \longrightarrow \mathrm{E}_{\infty}^{2,0} \longrightarrow 0.$$

Finally, we have the canonical inclusion $0 \to \mathrm{E}_{\infty}^{2,0} \to \mathrm{E}_{\infty}^{2}$. Connecting the three exact sequences yields the statement of (i).

As for (ii), the assumption implies $\mathrm{E}_{\infty}^{1,1} = 0$, so that the canonical injection $\mathrm{E}_{\infty}^{2,0} \to \mathrm{E}_{\infty}^{2}$ fits into the exact sequence

$$0 \longrightarrow \mathrm{E}_{\infty}^{2,0} \longrightarrow \mathrm{E}_{\infty}^{2} \longrightarrow \mathrm{E}_{\infty}^{0,2} \longrightarrow 0.$$

Setting now $s = 3$ in Lemma 12.1.1, we have

$$0 \longrightarrow \mathrm{E}_{\infty}^{0,2} \longrightarrow \mathrm{E}_{3}^{0,2} \longrightarrow \mathrm{E}_{3}^{3,0} \longrightarrow \mathrm{E}_{\infty}^{3,0} \longrightarrow 0.$$

As before, we connect these sequences with the canonical inclusion $0 \to \mathrm{E}_{\infty}^{3,0} \to \mathrm{E}_{\infty}^{3}$ and obtain the statement. $\square$

12.2. The first tableaux. With the tools of Section 7, the convergence of the first spectral sequence ${}^{I}\mathrm{E}_{\bullet}^{\bullet,\bullet}$ can easily be investigated.

PROPOSITION 12.2.1.— *The first spectral sequence ${}^{I}\mathrm{E}_{\bullet}^{\bullet,\bullet}$ collapses in the first tableau, converging in the second to the continuous bounded cohomology of G with coefficients in F :*

$$ {}^{I}\mathrm{E}_{\infty}^{n} = {}^{I}\mathrm{E}_{2}^{n,0} \cong \mathrm{H}_{\mathrm{cb}}^{\bullet}(G,F). $$

PROOF. Since N acts trivially on T, we have the identification

$$ L^{p,q} \cong L_{\mathrm{w}*}^{\infty}\Big(T^{q+1}, L_{\mathrm{w}*}^{\infty}(S^{p+1},F)\Big)^{G} \cong L_{\mathrm{w}*}^{\infty}\Big(T^{q+1}, L_{\mathrm{w}*}^{\infty}(S^{p+1},F)^{N}\Big)^{Q}. $$

Therefore, since T is an amenable regular Q-space, we apply may apply Theorem 7.5.3 and deduce that there is a canonical isomorphism

$$ {}^{I}\mathrm{E}_{1}^{p,q} \cong \mathrm{H}_{\mathrm{cb}}^{q}\Big(Q, L_{\mathrm{w}*}^{\infty}(S^{p+1},F)^{N}\Big). $$

Moreover, by Theorem 5.7.1, the coefficient module $L_{\mathrm{w}*}^{\infty}(S^{p+1},F)$ is G-relatively injective. Therefore, by Proposition 4.3.4, the submodule $L_{\mathrm{w}*}^{\infty}(S^{p+1},F)^{N}$ is Q-relatively injective. Now the Proposition 7.4.1 implies that ${}^{I}\mathrm{E}_{1}^{p,q} = 0$ unless $q=0$, proving that ${}^{I}\mathrm{E}_{1}^{\bullet,\bullet}$ collapses. Hence this spectral sequence is stationary from the second tableau on. Thus it remains to identify

$$ \bigoplus_{p+q=n} {}^{I}\mathrm{E}_{\infty}^{p,q} = {}^{I}\mathrm{E}_{\infty}^{n,0} = {}^{I}\mathrm{E}_{2}^{n,0}. $$

To this end, observe that

$$ {}^{I}\mathrm{E}_{1}^{n,0} \cong \mathrm{H}_{\mathrm{cb}}^{0}\Big(Q, L_{\mathrm{w}*}^{\infty}(S^{n+1},F)^{N}\Big) = \Big(L_{\mathrm{w}*}^{\infty}(S^{n+1},F)^{N}\Big)^{Q} = L_{\mathrm{w}*}^{\infty}(S^{n+1},F)^{G} $$

and that the differential ${}^{I}\mathrm{E}_{1}^{n,0} \to {}^{I}\mathrm{E}_{1}^{n+1,0}$ is induced by ${}^{I}d$, so that we may apply again Theorem 7.5.3 and conclude

$$ {}^{I}\mathrm{E}_{2}^{n,0} \cong \mathrm{H}_{\mathrm{cb}}^{\bullet}(G,F), $$

as claimed. □

Turning now our attention to the second spectral sequence, our next goal is to understand the second tableau ${}^{II}\mathrm{E}_{2}^{\bullet,\bullet}$. At first sight, since we have

$$ {}^{II}\mathrm{E}_{1}^{p,q} = \mathrm{H}^{q,p}(L^{\bullet,p}, {}^{I}d), $$

the identification

$$ L^{q,p} \cong L_{\mathrm{w}*}^{\infty}\Big(S^{q+1}, L_{\mathrm{w}*}^{\infty}(T^{p+1},F)\Big) $$

allows us to apply the Theorem 7.5.3 which yields a canonical isomorphism

$$ {}^{II}\mathrm{E}_{1}^{p,q} \cong \mathrm{H}_{\mathrm{cb}}^{q}\Big(G, L_{\mathrm{w}*}^{\infty}(T^{p+1},F)\Big) $$

under which the differential ${}^{II}\mathrm{E}_{1}^{p,q} \to {}^{II}\mathrm{E}_{1}^{p+1,q}$ is the map induced in cohomology by the usual homogeneous differential on $L_{\mathrm{w}*}^{\infty}(T^{p+1},F)$. However, we seek another realization of this space ${}^{II}\mathrm{E}_{1}^{p,q}$: namely, since we will have to iterate the process of taking cohomology, we want roughly speaking to switch $\mathrm{H}_{\mathrm{cb}}^{\bullet}$ with

L^∞_{w*}. This can be done thanks to the lifting statement of Corollary 8.2.11. Since there is no advantage here in working with a general amenable regular Q-space T, we shall continue now with the choice $T = Q$, so that actually we use just Lemma 8.2.5.

PROPOSITION 12.2.2.— *Let* $T = Q$.

(i) *There are canonical isomorphisms*

$$ {}^{II}\mathrm{E}_2^{p,0} \cong \mathrm{H}^p_{\mathrm{cb}}(Q, F^N) \quad \textit{and} \quad {}^{II}\mathrm{E}_2^{0,q} \cong \mathrm{H}^q_{\mathrm{cb}}(N,F)^Q. \qquad (p,q \geq 0)$$

(ii) *If for some* q *the space* $\mathrm{H}^q_{\mathrm{cb}}(N,F)$ *is Hausdorff, then there is a canonical isomorphism*

$$ {}^{II}\mathrm{E}_2^{p,q} \cong \mathrm{H}^p_{\mathrm{cb}}\Big(Q, \mathrm{H}^q_{\mathrm{cb}}(N,F)\Big). \qquad (p \geq 0)$$

(iii) *If* $\mathrm{H}^1_{\mathrm{cb}}(N,F) = 0$, *then* ${}^{II}\mathrm{E}_1^{\bullet,1} = 0$ *and there is a canonical isomorphism*

$$ {}^{II}\mathrm{E}_3^{p,0} \cong \mathrm{H}^p_{\mathrm{cb}}(Q, F^N). \qquad (p \geq 0)$$

As we have already mentioned, the Hausdorff restriction of point (ii) will actually not enter the statements of the exact sequence theorems. In the expression $\mathrm{H}^p_{\mathrm{cb}}(Q, \mathrm{H}^q_{\mathrm{cb}}(N,F))$, the space $\mathrm{H}^q_{\mathrm{cb}}(N,F)$ is endowed with its natural Banach Q-module structure, as explained in Section 8.8.

Here is a technical ingredient of Proposition 12.2.2.

LEMMA 12.2.3.— *Let* $A \xrightarrow{\alpha} B \xrightarrow{\beta} C$ *be an adjoint sequence of* Q*-morphisms of coefficient* Q*-modules with* $\beta\alpha = 0$. *If* $\alpha(A)$ *is closed, then the homology of*

$$L^\infty_{w*}(Q^{n+1}, A)^Q \xrightarrow{\alpha_*} L^\infty_{w*}(Q^{n+1}, B)^Q \xrightarrow{\beta_*} L^\infty_{w*}(Q^{n+1}, C)^Q \qquad (*)$$

is canonically isomorphic to $L^\infty_{w*}(Q^{n+1}, \mathrm{Ker}\beta/\alpha(A))^Q$ *for all* $n \geq 0$.

PROOF OF THE LEMMA. By the closed range theorem, the image of α is weak-* closed in B. Therefore, the exact sequence

$$0 \longrightarrow \alpha(A) \longrightarrow \mathrm{Ker}\beta \longrightarrow \mathrm{Ker}\beta/\alpha(A) \longrightarrow 0$$

is adjoint. Hence, by Corollary 8.2.11, it induces the exact sequence

$$0 \longrightarrow L^\infty_{w*}\big(Q^{n+1}, \alpha(A)\big)^Q \longrightarrow L^\infty_{w*}\big(Q^{n+1}, \mathrm{Ker}\beta\big)^Q \longrightarrow$$
$$\longrightarrow L^\infty_{w*}\big(Q^{n+1}, \mathrm{Ker}\beta/\alpha(A)\big)^Q \longrightarrow 0.$$

The first term here identifies with the image of α_* in the sequence $(*)$ and the second with the kernel of β_* in $(*)$. This proves the lemma. □

REMARK 12.2.4. More than just a technical point, the Lemma 12.2.3 is really on the cutting edge between the docile features that continuous bounded cohomology shares meekly with Eilenberg-MacLane cohomology, and its own pathologies.

Indeed, the assumption that $\alpha(A)$ be closed is not just intended to allow a well posed definition of $L^\infty_{w*}(Q^{n+1}, \mathrm{Ker}\beta/\alpha(A))$: let us see what happens if

$\alpha(A)$ is not closed. If for instance Q is discrete, then one can consistently define the ℓ^∞ space on Q^{n+1} of bounded maps in the semi-normed space $\mathrm{Ker}\beta/\alpha(A)$. Well, as soon as Q is infinite and $n \geq 1$, the natural inclusion

$$\alpha_*\Big(\ell^\infty(Q^{n+1}, A)^Q\Big) \hookrightarrow \ell^\infty\big(Q^{n+1}, \alpha(A)\big)^Q$$

is proper, so that the statement analogous to Lemma 12.2.3 fails in this situation.

PROOF OF PROPOSITION 12.2.2. Point (i) : the case of ${}^{II}\mathrm{E}_1^{p,0}$ is contained in (ii) because $\mathrm{H}^0_{\mathrm{cb}}(N,F) = F^N$ is Hausdorff. The term ${}^{II}\mathrm{E}_1^{0,q}$ is defined by

$$\cdots \xrightarrow{{}^I d_*} L^\infty_{\mathrm{w*}}\Big(Q, L^\infty_{\mathrm{w*}}(S^{q+1}, F)^N\Big)^Q \xrightarrow{{}^I d_*} \cdots$$

which is intertwined with

$$L^\infty_{\mathrm{w*}}(S^q, F)^N \xrightarrow{{}^I d} L^\infty_{\mathrm{w*}}(S^{q+1}, F)^N \xrightarrow{{}^I d} L^\infty_{\mathrm{w*}}(S^{q+2}, F)^N,$$

by the isomorphism

$$U^0 : L^\infty_{\mathrm{w*}}(S^\bullet, F)^N \longrightarrow L^\infty_{\mathrm{w*}}\Big(Q, L^\infty_{\mathrm{w*}}(S^\bullet, F)^N\Big)^Q$$

defined almost everywhere by $U^0 f(x) = \lambda_\pi(x) f$. This is indeed an isomorphism of Banach spaces since it is nothing else than the case $G = Q$ and $E = L^\infty_{\mathrm{w*}}(S^\bullet, F)^N$ of the isomorphism U^0 in the proof of Proposition 7.5.15. Hence, applying Theorem 7.5.3, we have the canonical isomorphism ${}^{II}\mathrm{E}_1^{0,q} \cong \mathrm{H}^q_{\mathrm{cb}}(N, F)$.

Now the isomorphism U^1 taken as above from the proof of Proposition 7.5.15 intertwines the sequence

$$\cdots \xrightarrow{{}^I d_*} L^\infty_{\mathrm{w*}}\Big(Q^2, L^\infty_{\mathrm{w*}}(S^{q+1}, F)^N\Big)^Q \xrightarrow{{}^I d_*} \cdots$$

defining ${}^{II}\mathrm{E}_1^{1,q}$ with

$$\cdots \xrightarrow{{}^I d_*} L^\infty_{\mathrm{w*}}\Big(Q, L^\infty_{\mathrm{w*}}(S^{q+1}, F)^N\Big) \xrightarrow{{}^I d_*} \cdots$$

in such a way that for a cochain $f \in L^\infty_{\mathrm{w*}}(S^{q+1}, F)^N$ the class of df in ${}^{II}\mathrm{E}_1^{1,q}$ is represented by the inhomogeneous coboundary $q \mapsto \lambda_\pi(q) f - f$. This implies, using the characterization of the action on the cohomology given in Corollary 8.8.3, that df vanishes in ${}^{II}\mathrm{E}_1^{1,q}$ if and only if the class of f in $\mathrm{H}^q_{\mathrm{cb}}(N, F)$ is Q-invariant. In conclusion, we have indeed ${}^{II}\mathrm{E}_2^{0,q} = \mathrm{H}^q_{\mathrm{cb}}(N, F)^Q$.

Point (ii) : the term ${}^{II}\mathrm{E}_1^{p,q}$ is defined by

$$\cdots \xrightarrow{{}^I d_*} L^\infty_{\mathrm{w*}}\Big(Q^{p+1}, L^\infty_{\mathrm{w*}}(S^{q+1}, F)^N\Big)^Q \xrightarrow{{}^I d_*} \cdots$$

so by Lemma 12.2.3 and Theorem 7.5.3 we have a canonical isomorphism ${}^{II}\mathrm{E}_1^{p,q} \cong L^\infty_{\mathrm{w*}}\big(Q^{p+1}, \mathrm{H}^q_{\mathrm{cb}}(N, F)\big)^Q$. This isomorphism intertwines

$$\cdots \xrightarrow{{}^{II} d} L^\infty_{\mathrm{w*}}\Big(Q^{p+1}, \mathrm{H}^q_{\mathrm{cb}}(N, F)\Big)^Q \xrightarrow{{}^{II} d} \cdots$$

with ${}^{II}\mathrm{E}_1^{p-1,q} \to {}^{II}\mathrm{E}_1^{p,q} \to {}^{II}\mathrm{E}_1^{p+1,q}$. Applying Theorem 7.5.3 once again we conclude ${}^{II}\mathrm{E}_2^{p,q} \cong \mathrm{H}^p_{\mathrm{cb}}(Q, \mathrm{H}^q_{\mathrm{cb}}(N,F))$.

Point (iii) : assume now $\mathrm{H}^1_{\mathrm{cb}}(N,F) = 0$. The above consideration gives ${}^{II}\mathrm{E}_1^{p,1} = 0$ whence ${}^{II}\mathrm{E}_\bullet^{p,1} = 0$. Since by definition ${}^{II}\mathrm{E}_3^{p,0}$ is the (algebraic) cokernel of ${}^{II}\mathrm{E}_2^{p-2,1} \to {}^{II}\mathrm{E}_2^{p,0}$ we have ${}^{II}\mathrm{E}_3^{p,0} = {}^{II}\mathrm{E}_2^{p,0}$. □

12.3. Exact sequences. In this section, we derive from the above some exact sequences modelled on, and generalizing, the Lyndon-Hochschild-Serre exact sequence. The first is really an analogue of the usual Lyndon-Hochschild-Serre sequence :

PROPOSITION 12.3.1.— *Let G be a locally compact second countable group, $N \lhd G$ a closed subgroup and $Q = G/N$ the quotient. Let (π, F) be a coefficient G-module. Then inflation and restriction fit into a natural sequence*

$$0 \longrightarrow \mathrm{H}^1_{\mathrm{cb}}(Q, F^N) \xrightarrow{\inf} \mathrm{H}^1_{\mathrm{cb}}(G,F) \xrightarrow{\mathrm{res}} \mathrm{H}^1_{\mathrm{cb}}(N,F)^Q \longrightarrow$$
$$\longrightarrow \mathrm{H}^2_{\mathrm{cb}}(Q, F^N) \xrightarrow{\inf} \mathrm{H}^2_{\mathrm{cb}}(G,F).$$

PROOF. Since (by Lemma 12.1.1) ${}^{II}\mathrm{E}_\infty^{1,0} = {}^{II}\mathrm{E}_2^{1,0}$, we have only to interpret the exact sequence

$$0 \longrightarrow {}^{II}\mathrm{E}_2^{1,0} \longrightarrow {}^{II}\mathrm{E}_\infty^{1} \longrightarrow {}^{II}\mathrm{E}_2^{0,1} \longrightarrow {}^{II}\mathrm{E}_2^{2,0} \longrightarrow {}^{II}\mathrm{E}_\infty^{2}$$

arising from Lemma 12.1.2. The second and last terms are limit terms giving the cohomology of the total complex $TL^\bullet$, hence they are given also by the first spectral sequence ${}^{I}\mathrm{E}_\bullet^{\bullet,\bullet}$. Thus the Proposition 12.2.1 identifies them with $\mathrm{H}^1_{\mathrm{cb}}(G,F)$ and $\mathrm{H}^2_{\mathrm{cb}}(G,F)$, respectively. All remaining terms are identified in the point (i) of Proposition 12.2.2. Thus we have an exact sequence of the required type ; unravelling the identifications, we check that except for ${}^{II}\mathrm{E}_2^{0,1} \to {}^{II}\mathrm{E}_2^{2,0}$, the maps come from inflation and restriction. □

There is the following particular case of the above proposition :

COROLLARY 12.3.2.— *Suppose $G = N \rtimes Q$ is a topological semi-direct product of the locally compact second countable groups N, Q. Let (π, F) be a coefficient G-module. Then we have the exact sequence*

$$0 \longrightarrow \mathrm{H}^1_{\mathrm{cb}}(Q, F^N) \xrightarrow{\inf} \mathrm{H}^1_{\mathrm{cb}}(G,F) \xrightarrow{\mathrm{res}} \mathrm{H}^1_{\mathrm{cb}}(N,F)^Q \longrightarrow 0.$$

PROOF. There is a topological group homomorphism $\sigma : Q \to G$ with $p\sigma = Id$, where p is the canonical map $G \to Q$. Since the inflation is nothing but $\mathrm{H}^\bullet_{\mathrm{cb}}(p)$, we deduce from $\mathrm{H}^\bullet_{\mathrm{cb}}(\sigma)\mathrm{H}^\bullet_{\mathrm{cb}}(p) = Id$ that the inflation is injective (in any degree). By exactness at $\mathrm{H}^2_{\mathrm{cb}}(Q, F^N)$ in Proposition 12.3.1, we deduce that the map

$$\mathrm{H}^1_{\mathrm{cb}}(N,F)^Q \longrightarrow \mathrm{H}^2_{\mathrm{cb}}(Q, F^N)$$

vanishes, whence the statement. □

PROOF OF THEOREM 12.0.2. By the last point of Proposition 12.2.2, we have ${}^{II}\mathrm{E}_1^{\bullet,1} = 0$, so that we can use the second exact sequence of Lemma 12.1.2 in a way similar to the proof of Proposition 12.3.1 above : since ${}^{II}\mathrm{E}_\infty^{2,0} = {}^{II}\mathrm{E}_3^{2,0}$ (Lemma 12.1.1), we have to interpret the sequence

$$0 \longrightarrow {}^{II}\mathrm{E}_3^{2,0} \longrightarrow {}^{II}\mathrm{E}_\infty^{2} \longrightarrow {}^{II}\mathrm{E}_3^{0,2} \longrightarrow {}^{II}\mathrm{E}_3^{3,0} \longrightarrow {}^{II}\mathrm{E}_\infty^{3}.$$

The second and last terms coincide with the corresponding terms in ${}^{I}\mathrm{E}_\infty^{\bullet}$, hence Proposition 12.2.1 identifies them with $\mathrm{H}^2_{\mathrm{cb}}(G, F)$ and $\mathrm{H}^3_{\mathrm{cb}}(G, F)$, respectively. The first and fourth terms are identified in the point (iii) of Proposition 12.2.2, so that only the term ${}^{II}\mathrm{E}_3^{0,2}$ remains. The latter is given as the cohomology of

$$ {}^{II}\mathrm{E}_2^{-2,3} \longrightarrow {}^{II}\mathrm{E}_2^{0,2} \longrightarrow {}^{II}\mathrm{E}_2^{2,1}. \qquad (*)$$

The first term here vanishes. On the other hand, $\mathrm{H}^1_{\mathrm{cb}}(N, F) = 0$ is indeed Hausdorff, so Proposition 12.2.2 point (ii) identifies ${}^{II}\mathrm{E}_2^{2,1}$ as $\mathrm{H}^2_{\mathrm{cb}}(Q, 0) = 0$. Thus the sequence $(*)$ degenerates and we have ${}^{II}\mathrm{E}_3^{0,2} = {}^{II}\mathrm{E}_2^{0,2}$, which is now identified by the first point of Proposition 12.2.2. Thus we have an exact sequence of the required type ; unravelling the identifications, it is a routine verification to check that except for ${}^{II}\mathrm{E}_3^{0,2} \to {}^{II}\mathrm{E}_3^{3,0}$ the maps come from inflation and restriction. □

Again, we may consider the particular case of semi-direct products :

COROLLARY 12.3.3.— *Suppose $G = N \rtimes Q$ is a topological semi-direct product of the locally compact second countable groups N, Q. Let (π, F) be a coefficient G-module.*

If $\mathrm{H}^1_{\mathrm{cb}}(N, F) = 0$, then we have the exact sequence

$$0 \longrightarrow \mathrm{H}^2_{\mathrm{cb}}(Q, F^N) \xrightarrow{\ \mathrm{inf}\ } \mathrm{H}^2_{\mathrm{cb}}(G, F) \xrightarrow{\ \mathrm{res}\ } \mathrm{H}^2_{\mathrm{cb}}(N, F)^Q \longrightarrow 0.$$

PROOF. As explained in the proof of Corollary 12.3.2, the inflation

$$\mathrm{H}^\bullet_{\mathrm{cb}}(Q, F^N) \longrightarrow \mathrm{H}^\bullet_{\mathrm{cb}}(G, F)$$

is injective in any degree. By exactness at $\mathrm{H}^3_{\mathrm{cb}}(Q, F^N)$ in Theorem 12.0.2, we deduce that the map

$$\mathrm{H}^2_{\mathrm{cb}}(N, F)^Q \longrightarrow \mathrm{H}^3_{\mathrm{cb}}(Q, F^N)$$

vanishes, whence the statement. □

PROOF OF THEOREM 12.0.3. Applying Theorem 11.4.4 to $H = N$, we obtain the canonical identification $\mathrm{H}^2_{\mathrm{cb}}(N, F^{Z_G(N)})^G \cong \mathrm{H}^2_{\mathrm{cb}}(N, E)^G$. Recalling that the G-action on cohomology factors through Q (Corollary 8.7.3), we plug this identification in the exact sequence of Theorem 12.0.2 and thus obtain the statement of Theorem 12.0.3. □

The conjunction of Theorem 12.0.3 and of Corollary 12.3.3 gives :

COROLLARY 12.3.4.— *Suppose $G = N \rtimes Q$ is a topological semi-direct product of the locally compact second countable groups N, Q with N compactly generated. Let (π, F) be a separable coefficient G-module. Then we have the exact* *sequence*

$$0 \longrightarrow \mathrm{H}^2_{\mathrm{cb}}(Q, F^N) \xrightarrow{\ \inf\ } \mathrm{H}^2_{\mathrm{cb}}(G, F) \xrightarrow{\ \mathrm{res}\ } \mathrm{H}^2_{\mathrm{cb}}(N, F^{Z_G(N)})^Q \longrightarrow 0. \qquad \square$$

PROOF OF COROLLARY 12.0.4. Due to the symmetry of the statement, we may assume that the only possibly non compactly generated factor is G_n. Arguing by induction, it is enough to consider the case $n = 2$, where G_1 is compactly generated. Since G_1 is contained in $Z_G(G_2)$, the Corollary 11.4.6 shows that the sequence of inclusions

$$F^{Z_G(G_2)} \subset F^{G_1} \subset F$$

forces the natural map $\mathrm{H}^2_{\mathrm{cb}}(N, F^{G_1}) \to \mathrm{H}^2_{\mathrm{cb}}(N, F)^{G_1}$ to be an isomorphism. Thus the exact sequence of Corollary 12.3.4 becomes :

$$0 \longrightarrow \mathrm{H}^2_{\mathrm{cb}}(G_1, F^{G_2}) \xrightarrow{\ \inf_1\ } \mathrm{H}^2_{\mathrm{cb}}(G, F) \xrightarrow{\ \mathrm{res}_2\ } \mathrm{H}^2_{\mathrm{cb}}(G_2, F^{G_1}) \longrightarrow 0,$$

where the indices remind which inflation and restriction are considered. We cannot use the analogous sequence with switched indices because G_2 needs not be compactly generated. However, the functoriality of contravariance implies the four relations

$$\mathrm{res}_1 \circ \inf_1 = Id, \qquad \mathrm{res}_2 \circ \inf_2 = Id,$$
$$\mathrm{res}_1 \circ \inf_2 = 0, \qquad \mathrm{res}_1 \circ \inf_2 = 0.$$

Hence we have already

$$\big(\mathrm{res}_1 \oplus \mathrm{res}_2\big)\big(\inf_1 + \inf_2\big) = Id.$$

Now, in order to conclude the proof of the claimed isomorphism, it remains to check that the above short exact sequence together with the four relations implies

$$\big(\inf_1 + \inf_2\big)\big(\mathrm{res}_1 \oplus \mathrm{res}_2\big) = Id.$$

Thus, for ω in $\mathrm{H}^2_{\mathrm{cb}}(G, F)$, write $\overline{\omega} = \inf_1 \mathrm{res}_1 \omega + \inf_2 \mathrm{res}_2 \omega$. Now we have

$$\mathrm{res}_2(\omega - \overline{\omega}) = \mathrm{res}_2(\omega) - \mathrm{res}_2(\omega) = 0$$

and thus by the exact sequence there is α in $\mathrm{H}^2_{\mathrm{cb}}(G_1, F^{G_2})$ with $\omega - \overline{\omega} = \inf_1 \alpha$. Applying res_1 we find $\mathrm{res}_1 \omega - \mathrm{res}_1 \omega = \alpha$, hence α and also ω vanish. $\square$

We still have to justify the Example 12.0.5, that is, we have to prove the

LEMMA 12.3.5.— *Keep the notation of Example 12.0.5 (one of the G_j might not be compactly generated). Then $F_0 = \sum_{j=0}^{n} F^{G'_j}$ is weak-* closed in F, so that it is again a coefficient G-module.*

PROOF. Again, we may suppose that G_j is compactly generated for $j \leq n-1$; moreover the case $n = 1$ is void. Pick v in the weak-* closure of F_0 and take sequences $(v_j^k)_{k \in \mathbf{N}}$ of $F^{G'_j}$ such that $v^k = \sum_{j=0}^{n} v_j^k$ converges weak-* to v. For any $g \in G_1$, we have $\pi(g)v^k - v^k = \pi(g)v_1^k - v_1^k$, which is in $F^{G'_1}$ and yet converges to $\pi(g)v - v$. Since $F^{G'_1}$ is weak-* closed, we conclude that for every $g \in G_1$ the difference $\pi(g)v - v$ is in $F^{G'_1}$. This yields a 1-cocycle for $\mathrm{H}^1_{\mathrm{cb}}(G_1, F^{G'_1})$. This cohomology group vanishes by Corollary 11.4.2, so that there is $u_1 \in F^{G'_1}$ with $\pi(g)v - v = \pi(g)u_1 - u_1$ for all $g \in G_1$, and therefore $v - u_1 \in F^{G_1}$.

We may now repeat the argument with $G_2 \times \cdots \times G_n$ instead of G, F^{G_1} instead of F and $v - u_1$ replacing v. This involves the vanishing of $\mathrm{H}^1_{\mathrm{cb}}(G_2, -)$ with coefficients in $F^{G_1} \cap F^{G_3 \times \cdots \times G_n} = F^{G'_2}$, granted by Corollary 11.4.2. This way, using the vanishing of $\mathrm{H}^1_{\mathrm{cb}}(G_j, F^{G'_j})$, we obtain by induction that there are $u_j \in F^{G'_j}$ for $j = 1, \ldots, n-1$ such that

$$v - u_1 - u_2 \ldots - u_{n-1} \in F^{G_1} \cap F^{G_2} \cap \ldots \cap F^{G_{n-1}} = F^{G'_n},$$

and hence v is in F_0, finishing the proof. Observe that we have not needed to know whether any $\mathrm{H}^1_{\mathrm{cb}}(G_n, -)$ vanishes for the possibly not compactly generated group G_n. □

12.4. Exact sequences for discrete groups. The results of the preceding section apply of course also to the special case of finitely generated "abstract" groups :

COROLLARY 12.4.1.— *Let $1 \to \Delta \to \Gamma \to \Lambda \to 1$ be an exact sequence of countable groups with Δ finitely generated, and let F be a separable coefficient Γ-module.*

Then we have the exact sequence

$$0 \longrightarrow \mathrm{H}^2_{\mathrm{b}}(\Lambda, F^{\Delta}) \xrightarrow{\text{inf}} \mathrm{H}^2_{\mathrm{b}}(\Gamma, F) \xrightarrow{\text{res}} \mathrm{H}^2_{\mathrm{b}}(\Delta, F^{Z_\Gamma(\Delta)})^{\Lambda} \longrightarrow \\ \longrightarrow \mathrm{H}^3_{\mathrm{b}}(\Lambda, F^{\Delta}) \xrightarrow{\text{inf}} \mathrm{H}^3_{\mathrm{b}}(\Gamma, F).$$

□

However, since the theory of bounded cohomology for discrete groups is much simpler than for general topological groups, the arguments can be simplified in order to give a statement for general Banach modules ; the countability condition can also be removed. The price to pay to the absence of double ergodicity is that one does not have the extra precision on the $\mathrm{H}^2_{\mathrm{b}}(\Delta, -)$ term :

THEOREM 12.4.2.— *Let $1 \to \Delta \to \Gamma \to \Lambda \to 1$ be an exact sequence of groups and let F be a Banach Γ-module. Then we have the exact sequence*

$$0 \longrightarrow \mathrm{H}^1_{\mathrm{b}}(\Lambda, F^{\Delta}) \xrightarrow{\text{inf}} \mathrm{H}^1_{\mathrm{b}}(\Gamma, F) \xrightarrow{\text{res}} \mathrm{H}^1_{\mathrm{b}}(\Delta, F)^{\Lambda} \longrightarrow \\ \longrightarrow \mathrm{H}^2_{\mathrm{b}}(\Lambda, F^{\Delta}) \xrightarrow{\text{inf}} \mathrm{H}^2_{\mathrm{b}}(\Gamma, F).$$

If moreover $\mathrm{H}^1_{\mathrm{b}}(N, F) = 0$, e.g. if F is reflexive, then we have the exact sequence

$$0 \longrightarrow \mathrm{H}^2_{\mathrm{b}}(\Lambda, F^{\Delta}) \xrightarrow{\text{inf}} \mathrm{H}^2_{\mathrm{b}}(\Gamma, F) \xrightarrow{\text{res}} \mathrm{H}^2_{\mathrm{b}}(\Delta, F)^{\Lambda} \longrightarrow \\ \longrightarrow \mathrm{H}^3_{\mathrm{b}}(\Lambda, F^{\Delta}) \xrightarrow{\text{inf}} \mathrm{H}^3_{\mathrm{b}}(\Gamma, F).$$

EXAMPLE 12.4.3. Let $1 \to \Delta \to \Gamma \to \Lambda \to 1$ be an exact sequence of groups. Then we have the exact sequence

$$0 \longrightarrow \mathrm{H}^2_{\mathrm{b}}(\Lambda) \longrightarrow \mathrm{H}^2_{\mathrm{b}}(\Gamma) \longrightarrow \mathrm{H}^2_{\mathrm{b}}(\Delta)^{\Lambda} \longrightarrow \mathrm{H}^3_{\mathrm{b}}(\Lambda) \longrightarrow \mathrm{H}^3_{\mathrm{b}}(\Gamma).$$

Indeed, we have $\mathrm{H}^1_\mathrm{b}(\Delta) = 0$ (Section 6.2) and hence we may apply Theorem 12.4.2.

The fact that a quotient map $\Gamma \to \Lambda$ induces an injection $\mathrm{H}^2_\mathrm{b}(\Lambda) \longrightarrow \mathrm{H}^2_\mathrm{b}(\Gamma)$ has been observed already at the very beginnings of bounded cohomology.

The fact that the second statement of the above theorem applies to the reflexive case follows from the Ryll-Nardzewski fixed point theorem, see Proposition 6.2.1.

PROOF OF THEOREM 12.4.2. We define a first quadrant spectral sequence in total analogy with Section 12.1, except that the double complex is given by the general term

$$L^{p,q} = \ell^\infty(\Gamma^{p+1} \times \Lambda^{q+1}, F)^\Gamma \cong \ell^\infty\Big(\Lambda^{q+1}, \ell^\infty(\Gamma^{p+1}, F)^\Delta\Big)^\Lambda.$$

All remaining arguments go through without change, except that whenever amenable actions were used to establish relative injectivity, one has to use the fact that for every Banach Γ-module A, be it a coefficient module or not, the Banach module $\ell^\infty(\Gamma^{p+1}, A)$ is relatively injective ; in particular, all

$$\ell^\infty\Big(\Lambda^{q+1}, \ell^\infty(\Gamma^{p+1}, B)^\Delta\Big)$$

are relatively Λ-injective, for all Banach Λ-modules B. Likewise, $\ell^\infty(\Gamma^{p+1}, F)$ is Γ-relatively injective so that $\ell^\infty(\Gamma^{p+1}, F)^\Delta$ is Λ-relatively injective. Both statements are special cases of Proposition 4.4.1, wherein one has to take for G the relevant discrete group. □

We conclude our exploration of the Lyndon-Hochschild-Serre exact sequences by a brief glance at the topological setting. As this is meant as an illustration, we do not consider local coefficient systems.

Let $\mathcal{F} = \big(E \xrightarrow{p} B\big)$ be a fibration of topological spaces and denote by F the homotopy type of the fibres. There is classically a Hochschild-Serre spectral sequence converging to the usual (singular) cohomology of the total space E, leading to an exact sequence in low degree involving the cohomology of the base B and of F. In this classical setting, the whole matter is more involved than for group extensions, since the spaces in presence have a priori higher homotopy groups, all of which are linked by the homotopy long exact sequence corresponding to the fibration $\mathcal{F}$. Therefore the Lyndon-Hochschild-Serre exact sequence for groups can be considered as a special case of its topological counterpart, namely the case where all spaces in presence are Eilenberg-MacLane spaces.

In singular bounded cohomology, however, a fundamental theorem of Gromov's (page 40 in [**69**], stated in the proof below) establishes a canonical isometric isomorphism between the bounded cohomology of a space and the bounded cohomology of its fundamental group. Thus, taking into account the amenability of the higher homotopy groups appearing in the homotopy exact sequence, we may take the opposite viewpoint and deduce the topological statement from the group-theoretic result that we established in Theorem 12.0.2.

This yields the following statement, in which we suppose the base connected for simplicity :

COROLLARY 12.4.4.— *Let $\mathcal{F} = (E \xrightarrow{p} B)$ be a fibration with arc-wise connected base B and fibre homotopy type F. Then we have the following exact sequence in singular bounded cohomology :*

$$0 \longrightarrow \mathrm{H}^2_\mathrm{b}(B) \xrightarrow{p^*} \mathrm{H}^2_\mathrm{b}(E) \xrightarrow{\iota^*} \mathrm{H}^2_\mathrm{b}(F)^{\pi_1(B)} \longrightarrow \mathrm{H}^3_\mathrm{b}(B) \xrightarrow{p^*} \mathrm{H}^3_\mathrm{b}(E),$$

where ι^* is induced by the fibre inclusions.

PROOF. Since B is arc-wise connected, the fibration $\mathcal{F}$ induces the following homotopy exact sequence :

$$\cdots \longrightarrow \pi_2(E) \xrightarrow{p_*} \pi_2(B) \longrightarrow \pi_1(F) \xrightarrow{\iota_*} \pi_1(E) \xrightarrow{p_*} \pi_1(B) \longrightarrow 1$$

(see e.g. Section 4.7 in [**135**]). We write A for the image of $\pi_2(B)$ in $\pi_1(F)$ and set moreover $\Gamma = \pi_1(E)$, $\Lambda = \pi_1(B)$, $\Delta = \pi_1(F)/\mathrm{A}$. Thus we have an exact sequence

$$1 \longrightarrow \Delta \xrightarrow{\iota_*} \Gamma \xrightarrow{p_*} \Lambda \longrightarrow 1.$$

As mentionned in Example 12.4.3 above, Theorem 12.4.2 yields the exact sequence

$$0 \longrightarrow \mathrm{H}^2_\mathrm{b}(\Lambda) \xrightarrow{\mathrm{H}^2_\mathrm{b}(p_*)} \mathrm{H}^2_\mathrm{b}(\Gamma) \xrightarrow{\mathrm{H}^2_\mathrm{b}(\iota_*)} \mathrm{H}^2_\mathrm{b}(\Delta)^\Lambda \longrightarrow \mathrm{H}^3_\mathrm{b}(\Lambda) \xrightarrow{\mathrm{H}^3_\mathrm{b}(p_*)} \mathrm{H}^3_\mathrm{b}(\Gamma).$$

Since $\pi_2(B)$ is Abelian, A is amenable and hence by Remark 8.5.4 the inflation

$$\inf : \mathrm{H}^\bullet_\mathrm{b}(\Delta) \longrightarrow \mathrm{H}^\bullet_\mathrm{b}(\pi_1(F)) \qquad (*)$$

is an isometric isomorphism. We recall now that Gromov's theorem (3.1, Corollary (A) page 40 in [**69**]) states that for a topological space X the classifying map $\chi_X : X \to K(\pi_1(X), 1)$ induces an isometric isomorphism χ_X^* in bounded cohomology. If we compose this isomorphism with the inflation $(*)$, the exact sequence attached above to Γ, Δ, Λ is transformed isometrically into the sequence of the statement. □

CHAPTER V

Towards applications

The aim of this chapter is, firstly, to show what kind of applications pertinent to various fields can be derived from the theory of continuous bounded cohomology, and secondly to establish such results for concrete groups.

All this is concerned merely with bounded cohomology in degree two, for various coefficients. An example of very concrete computations in degree three based heavily on the cohomological techniques is given for $\mathrm{SL}_3(\mathbf{R})$ and more general SL_n by M. Burger and the author in [**38, 113**].

Moreover, an example of the benefit that one can derive from the metric information encapsulated in the whole theory, as emphasized in the head of Section 4, is the following result that we quote from [**38**] :

THEOREM.— *Let $M = X_n/\Gamma$ be a quotient of the Siegel upper half space X_n endowed with the Hermitian metric with holomorphic sectional curvature between -1 and $-1/n$. Suppose M has sub-exponential volume growth (e.g. M has finite volume).*

Then the Gromov norm of the Kähler class $\Omega \in \mathrm{H}^2_{\mathrm{b}}(M)$ is

$$\|\Omega\|_\infty = \pi n.$$

ON THE PROOF. The idea of the proof that we give in [**38**] is as follows. The connected component of the group of isometries of X_n is $G = \mathrm{Sp}(2n, \mathbf{R})$, hence Γ is a discrete co-Følner subgroup of G. Let $\omega \in \Omega^2(X_n)^G$ be the corresponding Kähler form and

$$c_M : \mathcal{L}_n \times \mathcal{L}_n \times \mathcal{L}_n \longrightarrow \mathbf{Z}$$

the Maslov cocycle on the Grassmannian space $\mathcal{L}_n$ of Lagrangians (for background see [**7, 51, 96**]). Given a minimal parabolic P of G, there is an G-equivariant quotient map $G/P \to \mathcal{L}_n$. Thus, pulling back c_M to G/P, we get a bounded measurable G-invariant alternating 2-cocycle. Since P is amenable, we deduce from Corollary 7.5.9 that the supremum norm $\|c_M\|_\infty$ is already the norm of the corresponding class $[c_M]$ in $\mathrm{H}^2_{\mathrm{cb}}(G)$. Then we show that under the Dupont isomorphism one has $\mathcal{D}\omega = \pi[c_M]$ (the Dupont isomorphism $\mathcal{D}$ was defined in Section 9.3 above). Finally, the set of values of c_M consists by definition of all integers between $-n$ and n. □

This generalizes the main result of the paper [**45**] of A. Domic and D. Toledo, which are bound to the compact case because of their use of Thurston's proportionality principle for the simplicial norm of the fundamental class of M.

13. Interpretations of $\mathrm{EH}^2_{\mathrm{cb}}$

13.1. Rough actions. In this section, all Banach spaces are over the field of reals $\mathbf{R}$.

Let E be a Banach space and G a group. We denote by $\mathrm{Isom}(E)$ the group of all isometries of E.

DEFINITION 13.1.1. A *rough action* of G on E is a map $\varrho : G \to \mathrm{Isom}(E)$ such that the expression

$$D(\varrho) = \sup_{x,y \in G} \sup_{v \in E} \big\| \varrho(x)\varrho(y)v - \varrho(xy)v \big\|_E$$

is finite.

By the Mazur-Ulam theorem [**106**], one has the semi-direct product decomposition

$$\mathrm{Isom}(E) = \mathcal{O}(E) \ltimes E,$$

where $\mathcal{O}(E)$ is the group of linear (over $\mathbf{R}$) isometries and E acts by translations. Therefore, one has for each $x \in G$ a unique decomposition

$$\varrho(x) = \pi(x) + h(x)$$

with $\pi(x) \in \mathcal{O}(E)$ and $h(x) \in E$.

LEMMA 13.1.2.—

(i) *The map $\pi : G \to \mathcal{O}(E)$ is a group homomorphism.*

(ii) *The inhomogeneous coboundary $\delta_\pi h : G^2 \to E$ is bounded and satisfies $\|\delta_\pi h\|_\infty = D(\varrho)$.*

PROOF. Let $x, y \in G$ and $v \in E$. For every $r \in \mathbf{R}$ we compute

$$\begin{aligned} \varrho(x)\varrho(y)rv - \varrho(xy)rv &= \\ &= h(x) + \pi(x)\big(\pi(y)rv + h(y)\big) - \pi(xy)rv - h(xy) = \\ &= r\Big(\pi(x)\pi(y)v - \pi(xy)v\Big) + \delta_\pi h(x,y). \end{aligned}$$

Since this expression is bounded in norm by $D(\varrho)$ independently of r, we conclude

$$\pi(x)\pi(y)v - \pi(xy)v = 0,$$

whence (i). But now (ii) follows too. □

REMARK 13.1.3. Of course, a rough action ϱ with $D(\varrho) = 0$ is but a usual action by isometries ; moreover, the relation $\delta_\pi h = 0$ following in this case from point (ii) above is the well known cocycle relation for such actions.

The first point of Lemma 13.1.2 justifies the following definition.

DEFINITION 13.1.4. We call the homomorphism π the *linear part* of ϱ and endow E with the Banach G-module structure (π, E). We call h the *translation part* of ϱ.

We suppose now in addition that G has a topology turning it into a topological group and investigate continuity requirements.

LEMMA 13.1.5.— *The following assertions are equivalent :*

(i) *The orbit of some $v \in CE$ under ϱ varies continuously over G.*

(ii) *The orbit of every $v \in CE$ under ϱ varies continuously over G.*

(iii) *The map $h : G \to E$ is continuous.*

PROOF. Observe that (iii) amounts to say that the orbit of zero under ϱ varies continuously over G ; thus we have already (ii)$\Leftrightarrow$(iii) in view of $\varrho = \pi + h$. As (ii)$\Rightarrow$(i) is obvious, we are left with (i)$\Rightarrow$(iii). Since under the assumption (i) both maps $x \mapsto \varrho(x)v$ and $x \mapsto \pi(x)v$ are continuous, their difference h is also continuous. □

DEFINITION 13.1.6. The rough action ϱ is *semi-continuous* if the translation part h ranges in CE and moreover satisfies the equivalent conditions of Lemma 13.1.5.

The rough action is *continuous* if in addition $CE = E$, that is, if in addition (π, E) is a continuous Banach G-module.

We use the same terminology for the special case $D(\varrho) = 0$ of actions.

If we want to exhibit rough actions, an obvious thing to do is to perturb usual actions :

LEMMA 13.1.7.— *Let ϱ_0 be an action of G by isometries of E and let $b : G \to E$ be a bounded map. Then $\varrho = \varrho_0 + b$ is a rough action.*

If moreover ϱ_0 is semi-continuous and b is a continuous bounded map $G \to CE$, then ϱ is semi-continuous.

PROOF. Let $\varrho_0 = \pi + h$ be the decomposition in linear and translation parts, so that $\varrho = \pi + (h + b)$. A computation already performed above shows that

$$\varrho(x)\varrho(y)v - \varrho(xy)v = \delta_\pi(h + b)(x, y). \qquad (x, y \in G, v \in E)$$

In the present case, this is $\delta_\pi b(x, y)$ because $\delta_\pi h = 0$. Therefore, $D(\varrho) = \|\delta_\pi b\|_\infty$ which is bounded by $3\|b\|_\infty$. The continuity statement follows also from the decomposition $\varrho = \varrho_0 + (h + b)$. □

DEFINITION 13.1.8. A rough action ϱ as in Lemma 13.1.7 is called a *perturbed action*. If the continuity assumptions are fulfilled, ϱ is a perturbed semi-continuous, or continuous if $CE = E$, action.

Now, at last, we are faced with the natural rigidity question : are all rough actions perturbed actions ?

The answer is described in the following proposition. In order to formulate it, we observe that the set of all rough actions with a given linear part π has a natural structure of affine space ; if we fix the linear action $\varrho - \pi$ as origin, we obtain a vector space in which the perturbed actions form a subspace. The same holds for semi-continuous, respectively continuous, rough actions.

PROPOSITION 13.1.9.— *Let G be a locally compact group and (π, E) a Banach G-module. There is a canonical isomorphism of vector spaces*

$$\mathrm{EH}^2_{\mathrm{cb}}(G,E) \cong \frac{\Big\{\text{semi-continuous rough actions with linear part } \pi\Big\}}{\Big\{\text{perturbed semi-continuous actions with linear part } \pi\Big\}}.$$

Two special cases are of course :

COROLLARY 13.1.10.— *(i) Let G be a group and (π, E) a Banach G-module. There is a canonical isomorphism of vector spaces*

$$\mathrm{EH}^2_{\mathrm{b}}(G,E) \cong \frac{\Big\{\text{rough actions with linear part } \pi\Big\}}{\Big\{\text{perturbed actions with linear part } \pi\Big\}}.$$

(ii) Let G be a locally compact group and (π, E) a continuous Banach G-module. There is a canonical isomorphism of vector spaces

$$\mathrm{EH}^2_{\mathrm{cb}}(G,E) \cong \frac{\Big\{\text{continuous rough actions with linear part } \pi\Big\}}{\Big\{\text{perturbed continuous actions with linear part } \pi\Big\}}.$$

□

PROOF OF THE PROPOSITION. According to Remark 9.2.6, the kernel

$$\mathrm{EH}^2_{\mathrm{cb}}(G,E)$$

can be read off in the diagram

$$\begin{array}{ccccccccc}
0 \longrightarrow & \mathcal{C}E & \xrightarrow{\delta_\pi} & \mathrm{C_b}(G,\mathcal{C}E) & \xrightarrow{\delta_\pi} & \mathrm{C_b}(G^2,\mathcal{C}E) & \xrightarrow{\delta_\pi} & \mathrm{C_b}(G^3,\mathcal{C}E) & \xrightarrow{\delta_\pi} \cdots \\
 & \| & & \downarrow & & \downarrow & & \downarrow & \\
0 \longrightarrow & \mathcal{C}E & \xrightarrow{\delta_\pi} & \mathrm{C}(G,\mathcal{C}E) & \xrightarrow{\delta_\pi} & \mathrm{C}(G^2,\mathcal{C}E) & \xrightarrow{\delta_\pi} & \mathrm{C}(G^3,\mathcal{C}E) & \xrightarrow{\delta_\pi} \cdots
\end{array}$$

If h is the translation part of a semi-continuous rough action $\varrho = \pi + h$, then by Lemmata 13.1.2 and 13.1.5 the map $\delta_\pi h$ is in $\mathrm{C_b}(G^2, \mathcal{C}E)$. Now $\delta_\pi h$ defines an inhomogeneous cocycle in $\mathrm{EH}^2_{\mathrm{cb}}(G,E)$ since $h \in \mathrm{C}(G,\mathcal{C}E)$ trivializes its image in $\mathrm{H}^2_{\mathrm{c}}(G,E)$. This cocycle represents a trivial class in $\mathrm{EH}^2_{\mathrm{cb}}(G,E)$ if and only if $\delta_\pi h = \delta_\pi h'$ for some $h' \in \mathrm{C_b}(G, \mathcal{C}E)$. In this case, $\varrho_0 = \pi + (h - h')$ is a semi-continuous action and ϱ is a perturbation of ϱ_0. Conversely, if $f \in \mathrm{C_b}(G^2, \mathcal{C}E)$ is a cocycle such that $f = \delta_\pi h$ for some $h \in \mathrm{C}(G, \mathcal{C}E)$, then $\varrho = \pi + h$ is a semi-continuous rough action ; these two correspondences are inverse to each other. □

13.2. Property (TT). The interpretation of $\mathrm{EH}^2_{\mathrm{cb}}$ in terms of rough actions allows us to define, study and in certain cases establish a strengthening of Kazhdan's property (T) : the property (TT).

Amongst the numerous characterizations of Kazhdan's property we recall the formulation which is best suited to our analogy :

A locally compact group has property (T) if and only if any continuous action by isometries on a Hilbert space has bounded orbits.

For many more characterizations, see the book [**78**] by P. de la Harpe and A. Valette or Lubotzky's book [**97**]. Usually, one requires in the above characterization the group to have a fixed point, but this is equivalent to having bounded orbits by a well known lemma of F. Bruhat and J. Tits : the circumcentre of a bounded orbit is a fixed point.

DEFINITION 13.2.1. A locally compact group has *property (TT)* if any continuous rough action on a Hilbert space has bounded orbits.

REMARK 13.2.2. We ask loosely for "bounded orbits" since all orbits are bounded as soon as any one of them is bounded.

The very formulation of Definition 13.2.1 yields

LEMMA 13.2.3.— *Property (TT) implies property (T).* □

We shall see below that the converse is by no means true, so that the question of the rigidity of isometric actions on Hilbert spaces is really of interest. But first we take a look at the cohomological side ; a well known characterization of Kazhdan's property is

PROPOSITION 13.2.4.— *Let G be a locally compact group. The following assertions are equivalent :*

(i) *G has property (T).*

(ii) $\mathrm{H}^1_{\mathrm{c}}(G,\mathfrak{H}) = 0$ *for every continuous orthogonal representation* $(\pi,\mathfrak{H})$ *of* G.

ON THE PROOF. This follows e.g. from Théorème 7 in [**78**, chap. 4]. □

In parallel to this, we have :

PROPOSITION 13.2.5.— *Let G be a locally compact group. The following assertions are equivalent :*

(i) *G has property (TT).*

(ii) $\mathrm{H}^1_{\mathrm{c}}(G,\mathfrak{H}) = 0$ *and* $\mathrm{EH}^2_{\mathrm{cb}}(G,\mathfrak{H}) = 0$ *for every continuous orthogonal representation* $(\pi,\mathfrak{H})$ *of* G.

PROOF. "(i)⇒(ii)" : By Lemma 13.2.3 and Proposition 13.2.4, we have only to show that $\mathrm{EH}^2_{\mathrm{cb}}(G,\mathfrak{H})$ vanishes. In view of Corollary 13.1.10, it is enough to show that any continuous rough action $\varrho = \pi + h$ on $\mathfrak{H}$ is a perturbed continuous action. But since the orbit of $0 \in \mathfrak{H}$ is bounded, h is bounded and hence ϱ is a perturbation of π.

"(ii)⇒(i)" : Let $\varrho = \pi + h$ be a continuous rough action on $\mathfrak{H}$. Since $\mathrm{EH}^2_{\mathrm{cb}}(G,\mathfrak{H}) = 0$, the Corollary 13.1.10 implies that $\varrho = \varrho_0 + h'$, where h' is bounded and ϱ_0 is a continuous action by isometries. Since $\mathrm{H}^1_{\mathrm{c}}(G,\mathfrak{H}) = 0$, the Proposition 13.2.4 shows that ϱ_0 has bounded orbits (and even a fixed point). Therefore ϱ has also bounded orbits since h' is bounded. □

The behaviour of property (TT) with respect to uniform lattices and more generally co-compact subgroups with finite invariant measure on the quotient is simple :

PROPOSITION 13.2.6.— *Let G be a locally compact second countable group and $H < G$ a co-compact subgroup such that $H\backslash G$ has finite invariant measure. Then the following assertions are equivalent :*

(i) *G has property (TT).*

(ii) *H has property (TT).*

PROOF. We use the cohomological interpretation : since property (T) for G and H are equivalent (chapitres 3.a and 3.c in [**78**]), we are reduced to study $\mathrm{EH}^2_{\mathrm{cb}}$. Suppose G has property (TT) and let $\mathfrak{H}$ be a continuous orthogonal representation of H ; there is no loss of generality in assuming $\mathfrak{H}$ separable because H is second countable. We have the following commutative diagram

$$\begin{array}{ccc} \mathrm{H}^2_{\mathrm{cb}}(H,\mathfrak{H}) & \xrightarrow{L^2\mathbf{i}} & \mathrm{H}^2_{\mathrm{cb}}(G, L^2\mathrm{I}_H^G\mathfrak{H}) \\ {\scriptstyle \Psi^2_H}\downarrow & & \downarrow{\scriptstyle \Psi^2_G} \\ \mathrm{H}^2_{\mathrm{c}}(H,\mathfrak{H}) & \xrightarrow{\mathbf{i}} & \mathrm{H}^2_{\mathrm{c}}(G, L^2\mathrm{I}_H^G\mathfrak{H}) \end{array}$$

The L^2 induction $L^2\mathbf{i}$ is injective by Corollary 11.1.5, and Ψ^2_G is injective because G has property (TT) ; thus Ψ^2_H is injective.

Conversely, suppose H has property (TT) and let $\mathfrak{H}$ be a continuous orthogonal representation of G, again assumed separable. This time we consider the commutative diagram

$$\begin{array}{ccc} \mathrm{H}^2_{\mathrm{cb}}(G,\mathfrak{H}) & \xrightarrow{\mathrm{res}} & \mathrm{H}^2_{\mathrm{cb}}(H,\mathfrak{H}) \\ {\scriptstyle \Psi^2_G}\downarrow & & \downarrow{\scriptstyle \Psi^2_H} \\ \mathrm{H}^2_{\mathrm{c}}(G,\mathfrak{H}) & \xrightarrow{\mathrm{res}} & \mathrm{H}^2_{\mathrm{c}}(H,\mathfrak{H}) \end{array}$$

The upper restriction is injective by Proposition 8.6.2 and Ψ^2_H is injective by assumption. Thus Ψ^2_G is injective. □

In the next section, we shall see that almost simple algebraic groups of higher rank have property (TT). As a way of emphasizing this result, we shall now show that this property is much scarcely fulfilled than Kazhdan's property (T). We begin with a geometric construction :

PROPOSITION 13.2.7.— *Let X be a complete simply connected Riemannian manifold of pinched negative curvature. Then the group of isometries $G = \mathrm{Isom}(X)$ does not have property (TT) unless it is compact.*

The proposition applies notably to quaternionic hyperbolic spaces, and in this case G has property (T). The proof consists basically in a construction due to Z. Sela [**128**] and exposed by M. Gromov in Section 7.E_1 of [**70**].

PROOF OF THE PROPOSITION. For $\varepsilon > 0$ small enough, there is for all $x, y \in X$ a normal equivariant projection from the ε-neighbourhood $N_\varepsilon([x,y])$ of the geodesic segment $[x,y]$ onto $[x,y]$. Pulling back the oriented length one-form from $[x,y]$ to $N_\varepsilon([x,y])$, one obtains a continuous G-equivariant map

$$\alpha : X \times X \longrightarrow L^2\Omega^1(X)$$

ranging in the space $L^2\Omega^1(X)$ of square integrable one-forms on X, endowed with the natural continuous unitary representation π of G. We fix now $x_0 \in X$ and define a continuous map $h : G \to L^2\Omega^1(X)$ by $h(g) = \alpha(x_0, gx_0)$ for $g \in G$. The homogeneous coboundary $d\alpha$ is bounded because geodesic triangles are uniformly thin in X. Therefore the inhomogeneous coboundary $\delta_\pi h$ is bounded and thus $\varrho = \pi + h$ gives a continuous rough action of G on the Hilbert space $L^2\Omega^1(X)$. This rough action does not have bounded orbits unless the orbits of G in X are bounded ; in this case, however, the group G would be compact. □

On the more combinatorial side, we observe the following

PROPOSITION 13.2.8.— *There is no infinite Gromov-hyperbolic group with property (TT).*

PROOF. The case of an *elementary* hyperbolic group is immediate, since such a group is virtually cyclic, hence amenable : infinite amenable groups are not Kazhdan.

Suppose now that Γ is a non-elementary Gromov-hyperbolic group. Then, by a result of D. Epstein and K. Fujiwara [**53**], the space $\mathrm{EH}^2_{\mathrm{b}}(\Gamma)$ is infinite dimensional. Hence, by Proposition 13.2.5, the group Γ cannot have property (TT). □

Using M. Gromov's notion of *generic* finitely presented groups (see [**70**] and [**143**]), we can draw the consequence that "most" groups are counter-examples to the converse of Lemma 13.2.3 :

COROLLARY 13.2.9.— *A generic group presentation with relator density* $1/3 < d < 1/2$ *defines a group with property (T) but without property (TT).*

PROOF. Gromov shows [**70**] that a generic presentation with density $d < 1/2$ defines an infinite hyperbolic group (whilst the generic group for $d > 1/2$ is finite) ; a proof of this fact can also be found in [**119**]. Therefore, by the above Proposition 13.2.8, such groups fail to satisfy property (TT). On the other hand, A. Żuk shows in [**143**, Theorem 9] that a generic group presentation with density $1/3 < d < 1/2$ determines a Kazhdan group. □

13.3. Quasi-morphisms. We turn now our attention to a particular case of rough actions of a group G, namely rough actions on the Banach space $\mathbf{R}$ with trivial linear part. Such a rough action ϱ is reduced to its translation part h which is characterized by

$$D(\varrho) = \|\delta h\|_\infty = \sup_{x,y \in G} \big|h(x) + h(y) - h(xy)\big| < \infty.$$

DEFINITION 13.3.1. We call such a function $h : G \to \mathbf{R}$ a *quasi-morphism*. The vector space of quasi-morphisms is denoted by $\mathrm{QM}(G)$.

Suppose now G is moreover a topological group. Since the linear part is trivial, continuity and semi-continuity of ϱ are one and the same thing and amount to the continuity of h. We denote by $\mathrm{QM}_{\mathrm{c}}(G)$ the vector space of continuous quasi-morphisms.

Now the Proposition 13.1.9 gives the

COROLLARY 13.3.2.— *Let G be a locally compact group. There is a canonical isomorphism of vector spaces*

$$\mathrm{EH}^2_{\mathrm{cb}}(G, \mathbf{R}) \cong \frac{\mathrm{QM}_{\mathrm{c}}(G)}{\mathrm{C_b}(G, \mathbf{R}) \oplus \mathrm{Hom}_{\mathrm{c}}(G)},$$

where $\mathrm{Hom}_{\mathrm{c}}(G)$ denotes the space of continuous characters $G \to \mathbf{R}$. □

Thus we see that the kernel $\mathrm{EH}^2_{\mathrm{cb}}(G, \mathbf{R})$ of the comparison map classifies the obstruction to the rigidity of continuous characters under bounded perturbations. An example of this kind of rigidity will be given below : Corollary 14.2.9.

Ch. Bavard has given in [**9**] another interpretation of the vanishing of $\mathrm{EH}^2_{\mathrm{b}}(\Gamma, \mathbf{R})$ for a finitely generated group Γ. We recall that the ***stable length*** of an element γ of the commutator subgroup $[\Gamma, \Gamma]$ is the limit

$$\ell(\gamma) = \lim_{n \to \infty} \frac{\|\gamma^n\|}{n}$$

of the non-increasing sequence $\|\gamma^n\|/n$, where $\|\cdot\|$ denotes the word metric associated to the generally infinite set

$$\{[\gamma, \eta] : \gamma, \eta \in \Gamma\}$$

generating $[\Gamma, \Gamma]$.

THEOREM 13.3.3 (Ch. Bavard).— *For a finitely generated group Γ, the following assertions are equivalent :*

(i) Every quasi-morphism is at bounded distance of a homomorphism, that is, $\mathrm{EH}^2_{\mathrm{b}}(\Gamma, \mathbf{R}) = 0$.

(ii) The stable length ℓ vanishes on the commutator subgroup $[\Gamma, \Gamma]$.

ON THE PROOF. This is the Théorème on page 110 in [**9**]. □

This result shows in particular that the stable length vanishes for any group with property (TT).

Since we have now introduced quasimorphisms, we can bring as an example the following lemma, which was Lemma 6.1 in [**35**] ; it justifies the Examples 9.3.3 and 9.3.5 given above.

LEMMA 13.3.4.— *Let $G = \mathbf{G}(k)$, where $\mathbf{G}$ is a connected, simply connected, almost simple k-isotropic group and k a local field.*

Then every continuous quasi-morphism $h : G \to \mathbf{R}$ is bounded.

PROOF. One can write G as a product $G = N_1 \cdots N_\ell$ of unipotent subgroups such that for each N_j there is a semi-simple element $s_j \in G$ whose action as inner automorphism contracts N_j to the neutral element $e \in G$:

$$\lim_{k\to\infty} s_j^{-k} u s_j^k = e. \qquad (\forall\, u \in N_j)$$

It is enough to show that $h|_{N_j}$ is bounded for every $1 \leq j \leq \ell$ because of the estimate

$$\|h\|_\infty \;\leq\; \sum_{j=1}^{\ell} \|h|_{N_j}\|_\infty \;+ (\ell-1)\|\delta h\|_\infty.$$

For s_j as above we have for every $u \in N_j$ and every $k \in \mathbf{N}$

$$\begin{aligned}|h(u)| &\leq \big|h(s_j^k) + h(s_j^{-k} u s_j^k) + h(s_j^{-k})\big| + 2\|\delta h\|_\infty \\ &\leq |h(s_j^{-k} u s_j^k)| + |h(e)| + 3\|\delta h\|_\infty.\end{aligned}$$

Therefore, h being continuous, the limit statement above shows that

$$\|h|_{N_j}\|_\infty \;\leq\; 2|h(e)| + 3\|\delta h\|_\infty.$$

□

Another easy lemma of the sort will be useful later for certain automorphism groups of trees ; the assumption below is an analogue of a rank one Cartan decomposition.

LEMMA 13.3.5.— *Let G be a topological group. Suppose that there is $a \in G$ and a compact subset $K \subset G$ such that*

$$G = K \cdot \{a^n | n \geq 0\} \cdot K.$$

Then every continuous quasi-morphism of G is bounded.

PROOF. Let f be a continuous quasi-morphism of G. Since K is compact, there is a bound L for f on K. The assumption implies that for every $r \in \mathbf{Z}$ there is $n_r \geq 0$ and $k, k' \in K$ such that $a^{-r} = k a^{n_r} k'$. Now we have

$$\big|f(a^{-r}) - f(a^{n_r})\big| \leq |f(k)| + |f(k')| + 2\|\delta f\|_\infty \leq 2L + 2\|\delta f\|_\infty.$$

On the other hand,

$$\big|f(a^{r}) - f(a^{-r})\big| \leq |f(e)| + \|\delta f\|_\infty.$$

Thus for all r there is $n_r \geq 0$ such that

$$|f(a^{r+n_r})| \leq |f(e)| + 2L + 3\|\delta f\|_\infty$$

and in particular f is bounded on arbitrarily high positive powers of a. However, the group A generated by a is Abelian, so $\mathrm{H}^2_{\mathrm{b}}(A) = 0$ and therefore $f|_A$ is at bounded distance of a character of A. Since a character which is bounded on arbitrarily high positive powers vanishes, $f|_A$ is bounded and therefore f is bounded too. □

EXAMPLE 13.3.6. The above argument appeared in Lemma 7.1 of our joint work [**35**] with M. Burger in the following context. Let $\mathcal{T}$ be a locally finite regular (or bi-regular) tree and let $G < \mathrm{Aut}(\mathcal{T})$ be a closed subgroup acting transitively on the boundary $\partial_\infty\mathcal{T}$. Then any continuous quasi-morphism of G is bounded.

Indeed, let a be any hyperbolic element of G and $x, y \in \mathcal{T}$ be adjacent vertices on the axis of a. Denote by K_1 and K_2 the stabilizers of x_1 and x_2 in G. Then M. Burger and Sh. Mozes show in Chapter 4 of [**42**] that

$$G = K_1 \cdot A^+ \cdot K_1 \qquad \text{if } G \text{ is vertex transitive,}$$
$$G = K_1 \cdot A^+ \cdot (K_1 \cup K_2) \quad \text{otherwise.}$$

Therefore one may apply Lemma 13.3.5 and conclude.

13.4. Algebraic groups and their lattices. Higher rank algebraic groups have property (TT), as we have established with M. Burger in [**35**].

THEOREM 13.4.1.— *Let* $\mathbf{G}$ *be a connected, simply connected, almost simple* k*-isotropic group over a local field* k *(of any characteristic).*

Then $G = \mathbf{G}(k)$ *has property (TT) if and only if* $\mathrm{rank}_k(\mathbf{G}) \geq 2$.

This is essentially contained in [**35**], but for the reader's convenience we give the proof. We isolate the following elementary lemma of independent interest, which is Lemma 5.1 in [**35**].

LEMMA 13.4.2.— *Let* $(\pi, \mathfrak{H})$ *be a continuous unitary representation of a locally compact group* G *and let* $H < G$ *be a second countable closed subgroup. Let* $\beta : G \to \mathfrak{H}$ *be a locally bounded map with bounded inhomogeneous coboundary* $\eta = \delta_\pi\beta$. *If* $\pi|_H$ *has no almost invariant vectors, then* $\beta|_{Z_G(H)}$ *is bounded.*

PROOF OF THE LEMMA. It is well known that the assumption on $\pi|_H$ is equivalent to the existence of a non negative compactly supported continuous function $f : H \to \mathbf{R}$ of integral one with $\|\pi(f)\| < 1$, see e.g. Proposition III.1.3 in [**103**] (the reference given in [**35**] was mistaken). Recall that by definition of δ_π we have

$$\eta(x, y) = \pi(x)\beta(y) - \beta(xy) + \beta(x). \qquad x, y \in G$$

In particular, for all $z \in Z_G(H)$ and $h \in H$, both

$$\pi(h)\beta(zh^{-1}) - \beta(hzh^{-1}) + \beta(h)$$
$$\text{and} \quad \pi(z)\beta(h^{-1}) - \beta(zh^{-1}) + \beta(z)$$

are bounded in norm by $\|\eta\|_\infty$ independently of z and h ; note $hzh^{-1} = z$. Applying $\pi(h)$ to the second expression before adding it to the first yields that

$$\pi(h)\beta(z) - \beta(z) + \pi(h)\pi(z)\beta(h^{-1}) + \beta(h)$$

is bounded independently of z and h, hence

$$\Big\|\Big(\mathrm{Id} - \pi(h)\Big)\beta(z)\Big\| \leq \|\beta(h^{-1})\| + \|\beta(h)\| + C$$

for some $C < \infty$. Now we have

$$\left\| \Big(\mathrm{Id} - \pi(f)\Big)\beta(z) \right\| \leq \sup_{h \in \mathrm{Supp}(f)} \Big(\|\beta(h^{-1})\| + \|\beta(h)\| + C\Big)$$

for all z in $Z_G(H)$. Since $\mathrm{Id} - \pi(f)$ is invertible, we conclude that $\beta|_{Z_G(H)}$ is bounded. □

PROOF OF THEOREM 13.4.1. We begin with the easier direction : assume that $\mathbf{G}$ has k-rank one. If k is non-Archimedean or if G is locally $\mathrm{SO}(n,1)$ or $\mathrm{SU}(n,1)$, then G does not have property (T), thus a minori fails to satisfy property (TT). In the remaining cases, the associated symmetric space is either a quaternionic hyperbolic space or the hyperbolic plane over Cayley's octonions. We are then in a very special case of the above Proposition 13.2.7, which shows that G does not have property (TT).

For the converse implication, suppose the rank is at least two. Then G has property (T) (see e.g. Chapitre 2 in [**78**]). Thus one has to show that $\mathrm{EH}^2_{\mathrm{cb}}(G, \mathfrak{H}) = 0$ for every continuous orthogonal representation in a separable Hilbert space $\mathfrak{H}$. Upon complexification, we may assume that we have a unitary representation. We decompose $\mathfrak{H} = \mathfrak{H}^G \oplus \mathfrak{H}_1$, where $\mathfrak{H}_1$ is the orthogonal complement of the invariants. By the Corollary 8.2.10 on direct sums, one can handle these components separately, and the vanishing of $\mathrm{EH}^2_{\mathrm{cb}}(G, \mathfrak{H}^G)$ follows immediately from Lemma 13.3.4. We are therefore left with the following problem : let $\omega : G^3 \to \mathfrak{H}_1$ be a continuous bounded G-invariant cocycle and $\alpha : G^2 \to \mathfrak{H}_1$ be a continuous G-equivariant[1] map with $d\alpha = \omega$. We have to show that ω vanishes in $\mathrm{H}^2_{\mathrm{cb}}(G, \mathfrak{H}_1)$. Now we make the stronger claim that actually α is bounded :

Let $G = KSK$ be a Cartan decomposition, where K is a maximal compact subgroup and $S = \mathbf{S}(k)$ is a maximal k-split torus. By the k-rank assumption, $\mathbf{S}$ can be written as a product of singular tori $\mathbf{S}_1, \ldots, \mathbf{S}_r$ such that the centralizer $H_j = Z_G\big(\mathbf{S}_j(k)\big)$ is non amenable for all $1 \leq j \leq r$. Since $\mathfrak{H}_1^G = 0$, the classical arguments for Kazhdan's property (given in Sections III.4 and III.5 of [**103**]) show that $\pi|_{H_j}$ has no almost invariant vectors in $\mathfrak{H}_1$. Since $\mathbf{S}_j(k) < Z_G(H_j)$, Lemma 13.4.2 applied to the inhomogeneous representative β of α implies that $\alpha|_{\mathbf{S}_j(k)}$ is bounded. Thus α is bounded on S and hence on G, as claimed. □

In particular, applying Proposition 13.2.6, we deduce that uniform lattices in these algebraic groups have property (TT). In this particular setting, however, the powerful theories available allow a more general statement : one can get rid of the co-compactness assumption (see also [**37**]).

THEOREM 13.4.3.— *Let Γ be a lattice in $G = \mathbf{G}(k)$, where $\mathbf{G}$ is a connected, simply connected, almost simple k-isotropic group and k a local field.*

Then Γ has property (TT) if and only if $\mathrm{rank}_k(\mathbf{G}) \geq 2$.

PROOF. Suppose first that the rank is one. Then Theorem 13.4.1 tells us that G does not have property (TT). Thus either G is not Kazhdan, in which

[1] we recall from Remark 1.2.8 that we allows ourselves this liberty in the terminology.

case Γ is not Kazhdan (see chapitre 3.c in [**78**]), or $\mathrm{EH}^2_{\mathrm{cb}}(G,\mathfrak{H}) \neq 0$ for some continuous orthogonal G-representation in a separable Hilbert space $\mathfrak{H}$. In the latter case, Proposition 8.6.2 ensures that by restriction we have also $\mathrm{EH}^2_{\mathrm{b}}(\Gamma,\mathfrak{H}) \neq 0$.

The difficulty lies in the converse. Suppose $\mathrm{rank}_k(\mathbf{G}) \geq 2$. Then Γ is Kazhdan by Corollaire 5 in [**78**, chap 3], but we have still to prove that $\mathrm{EH}^2_{\mathrm{b}}(\Gamma,\mathfrak{H})$ vanishes for every orthogonal Γ-representation in a separable Hilbert space $\mathfrak{H}$; we may take $\mathfrak{H}$ unitary. We consider the commutative diagram

$$\begin{array}{ccc} \mathrm{H}^2_{\mathrm{b}}(\Gamma,\mathfrak{H}) & \xrightarrow{L^2\mathbf{i}} & \mathrm{H}^2_{\mathrm{cb}}(G, L^2\mathrm{I}_\Gamma^G\mathfrak{H}) \\ & & \searrow{\scriptstyle \Psi^2_G} \\ \Big\downarrow {\scriptstyle \Psi^2_\Gamma} & & \qquad \mathrm{H}^2_{\mathrm{c}}(G, L^2\mathrm{I}_\Gamma^G\mathfrak{H}) \\ & & \swarrow \\ \mathrm{H}^2(\Gamma,\mathfrak{H}) & \xrightarrow{\mathbf{i}} & \mathrm{H}^2_{\mathrm{c}}(G, L^2_{\mathrm{loc}}(G,\mathfrak{H})^\Gamma) \end{array}$$

Here the lower map $\mathbf{i}$ is the induction towards the Fréchet module $L^2_{\mathrm{loc}}(G,\mathfrak{H})^\Gamma$ considered e.g. by Ph. Blanc [**13**] and alluded to in Section 10.2.

The L^2 induction $L^2\mathbf{i}$ is injective by Corollary 11.1.5 and Ψ^2_G is injective because G has property (TT) by Theorem 13.4.1. Therefore, it is enough to show that $L^2\mathbf{i}$ maps $\mathrm{EH}^2_{\mathrm{b}}(\Gamma,\mathfrak{H})$ into $\mathrm{EH}^2_{\mathrm{cb}}(G, L^2\mathrm{I}_\Gamma^G\mathfrak{H})$. This is taken care of in Proposition 13.4.4 below. □

The remaining step, very similar to an argument of Y. Shalom [**130**, Sec. 2], relies notably on the results of Lubotzky, Mozes and Raghunathan [**99**] about the word metrics of the lattices under consideration here. The following statement, in the unitary setting, is presented by M. Burger and the author in [**37**]. Since it will be of later use, we consider the non almost simple case as well.

PROPOSITION 13.4.4.— *Let $\Gamma < G = \prod_{\alpha\in\mathrm{A}} \mathbf{G}_\alpha(k_\alpha)$ be an irreducible lattice, where $(k_\alpha)_{\alpha\in\mathrm{A}}$ is a finite family of local fields and the $\mathbf{G}_\alpha$ are connected simply connected k_α-almost simple groups. Let (π, E) be a separable coefficient Γ-module. If*

$$\sum_{\alpha\in\mathrm{A}} \mathrm{rank}_{k_\alpha} \mathbf{G}_\alpha(k_\alpha) \geq 2,$$

then the L^2 induction

$$\mathrm{H}^2_{\mathrm{b}}(\Gamma, E) \to \mathrm{H}^2_{\mathrm{cb}}(G, L^2\mathrm{I}_\Gamma^G E)$$

maps $\mathrm{EH}^2_{\mathrm{b}}(\Gamma, E)$ to $\mathrm{EH}^2_{\mathrm{cb}}(G, L^2\mathrm{I}_\Gamma^G E)$.

Note that the proof of Theorem 13.4.3 is only concerned with the almost simple case $|\mathrm{A}| = 1$; there, irreducibility of Γ is a void assumption.

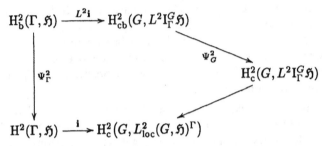

PROOF OF THE PROPOSITION. A class in the kernel $\mathrm{EH}^2_{\mathrm{b}}(\Gamma, E)$ is given by a Γ-equivariant bounded cocycle $\omega : \Gamma^3 \to E$ such that there is a Γ-equivariant map $\alpha : \Gamma^2 \to E$ with $\omega = d\alpha$. Now we want to look at induction as in the second point of Remarks 10.2.5. That is, we fix a left Γ-equivariant Borelian map $\sigma : G \to \Gamma$; in particular $\mathcal{F} = \sigma^{-1}(e)$ is a Borelian fundamental domain. We have observed in Remarks 10.2.5 that the induced cocycle $\mathbf{i}\sigma^*\omega$ is defined by

$$\mathbf{i}\sigma^*\omega(g_0, g_1, g_2)(g) \;=\; \omega\big(\sigma(gg_0), \sigma(gg_1), \sigma(gg_2)\big).$$

Define now

$$\mathbf{i}\sigma^*\alpha : G^2 \longrightarrow L^2_{\mathrm{loc}}(G, E)^\Gamma, \qquad \mathbf{i}\sigma^*\alpha(g_0, g_1)(g) = \alpha\big(\sigma(gg_0), \sigma(gg_1)\big),$$

where $L^2_{\mathrm{loc}}(G, E)^\Gamma$ is the Fréchet space of locally square-summable function classes considered at the very end of Section 10.2. We know that $d\mathbf{i}\sigma^*\alpha$ ranges in $L^2\mathrm{I}_\Gamma^G E$ since it coincides with $\mathbf{i}\sigma^*\omega$. Therefore, what we have to show is that $\mathbf{i}\sigma^*\alpha$ actually ranges also in $L^2\mathrm{I}_\Gamma^G E$, that is :

$$\int_{\mathcal{F}} \Big\|\alpha\big(\sigma(gg_0), \sigma(gg_1)\big)\Big\|^2 dm(g) < \infty$$

for all g_0, g_1, wherein m is a left Haar measure on G. Equivalently, setting $\psi(\gamma) = \alpha(\gamma, e)$ and $\kappa(g, g') = \sigma(g)^{-1}\sigma(gg')$, we must show

$$\int_{\mathcal{F}} \big\|\psi\big(\kappa(g, g')\big)\big\|^2 dm(g) < \infty \qquad\qquad (\forall g' \in G)$$

(note that κ is Borelian). By the conclusion of Section IX.3 in Margulis' book [**103**], Γ is finitely generated ; we fix a finite generating set S and denote by ℓ the corresponding word length on Γ. Now we set

$$C = \max_{s \in S} \|\psi(s)\| + \|\omega\|_\infty$$

and claim that $\|\psi(\gamma)\| \le C\ell(\gamma)$ holds for all $\gamma \in \Gamma$. Indeed, we have for $\gamma_1, \gamma_2 \in \Gamma$ the estimate

$$\begin{aligned}\|\psi(\gamma_1\gamma_2)\| &\le \|\psi(\gamma_1) + \pi(\gamma_1)\psi(\gamma_2)\| + \|\omega\|_\infty \\ &\le \|\psi(\gamma_1))\| + \|\psi(\gamma_2)\| + \|\omega\|_\infty.\end{aligned}$$

Since γ can be written

$$\gamma = s_1 s_2 \cdots s_{\ell(\gamma)}, \qquad\qquad (s_j \in S)$$

the claim follows by induction on $\ell(\gamma)$.

Therefore the above integral is dominated by

$$C \int_{\mathcal{F}} \ell\big(\kappa(g, g')\big)^2 dm(g)$$

(observe that this expression does not depend on the coefficient module E anymore). Such integrals are known to be finite for lattices as in the statement of the proposition : this is shown explicitely in Section 2 of [**130**], pages 16-18.

(The reason is that the integral over $\mathcal{F}$ can be replaced by an integral over a generalised Dirichlet fundamental domain ; but then, one can invoque the result [**99**] of A. Lubotzky, S. Mozes and M.S. Raghunathan, where ℓ is dominated in terms of a metric on the ambient group.) □

We observe that in the almost simple case needed for Theorem 13.4.3, the fact that Γ is finitely generated is a bit less hard to obtain since it follows from property (T).

13.5. Actions on the circle. É. Ghys' result [**63**] about actions on the circle has opened a domain of application with, in our opinion, striking consequences. Let us recall the necessary terminology following [**63**].

We consider an action of a group Γ on the circle $\mathbf{S}^1$ by orientation preserving homeomorphism as a homomorphism $\phi : \Gamma \to \mathrm{Homeo}^0(\mathbf{S}^1)$. Two such actions ϕ_1, ϕ_2 are *semi-conjugated* if there is a non-decreasing map $\bar{h} : \mathbf{R} \to \mathbf{R}$ with $\bar{h}(x+1) = \bar{h}(x) + 1$ for all $x \in \mathbf{R}$ such that the induced map $h : \mathbf{S}^1 \to \mathbf{S}^1$ satisfies

$$\phi_1(\gamma) \circ h = h \circ \phi_2(\gamma). \qquad (\forall \gamma \in \Gamma)$$

The fact that h is allowed to be non continuous has the consequence that semi-conjugacy is indeed an equivalence relation (Proposition 2-1 in [**63**]).

An orientation preserving action ϕ gives rise to a *bounded Euler class*

$$\mathcal{E}_{\phi,\mathbf{Z}} \in \mathrm{H_b}(\Gamma, \mathbf{Z}).$$

É. Ghys shows that this class is a complete invariant of semi-conjugacy :

THEOREM 13.5.1 (É. Ghys).— *Two Γ-actions ϕ_1, ϕ_2 on the circle by orientation preserving homeomorphism are semi-conjugated if and only if $\mathcal{E}_{\phi_1,\mathbf{Z}} = \mathcal{E}_{\phi_2,\mathbf{Z}}$.*

ON THE PROOF. This is contained in the Téorème A in [**63**]. □

Now we denote by $\mathcal{E}_{\phi,\mathbf{R}}$ the image of $\mathcal{E}_{\phi,\mathbf{Z}}$ in $\mathrm{H}^2_\mathrm{b}(\Gamma, \mathbf{R})$ under the natural map $\mathrm{H}^2_\mathrm{b}(\Gamma, \mathbf{Z}) \to \mathrm{H}^2_\mathrm{b}(\Gamma, \mathbf{R})$. The above theorem has an immediate consequence :

COROLLARY 13.5.2.— *Let ϕ be an action of a group Γ on the circle by orientation preserving homeomorphisms. Then ϕ is semi-conjugated to an action by rotations if and only if $\mathcal{E}_{\phi,\mathbf{R}} = 0$.*

PROOF. Consider the standard isomorphism $\mathbf{S}^1 \cong \mathbf{R}/\mathbf{Z}$; the statement follows from Theorem 13.5.1 in view of the first terms

$$0 \longrightarrow \mathrm{H}^1(\Gamma, \mathbf{R}/\mathbf{Z}) \longrightarrow \mathrm{H}^2_\mathrm{cb}(\Gamma, \mathbf{Z}) \longrightarrow \mathrm{H}^2_\mathrm{cb}(\Gamma, \mathbf{R}) \longrightarrow \cdots$$

of the long exact sequence of Proposition 8.2.12. □

We can further deduce the

COROLLARY 13.5.3.— *Let Γ be a group with $\mathrm{H}^2_\mathrm{b}(\Gamma, \mathbf{R}) = 0$ and with finite Abelianization $\Gamma_\mathrm{Ab} = \Gamma/[\Gamma, \Gamma]$.*

Then any Γ-action on the circle by orientation preserving homeomorphisms is semi-conjugated to an action by a finite group of rotations. In particular, Γ has a finite orbit in $\mathbf{S}^1$.

This applies clearly to any group with property (TT) and trivial H^2. Indeed, the cohomological characterization of property (TT) given in Proposition 13.2.5 implies that $\mathrm{EH}^2_\mathrm{b}(\Gamma, \mathbf{R})$ vanishes, thus $\mathrm{H}^2_\mathrm{b}(\Gamma, \mathbf{R}) = 0$. By Lemma 13.2.3, the group Γ is Kazhdan and therefore Γ_Ab is finite (Proposition 7 in [**78**, chap. 1]).

PROOF OF THE COROLLARY. By Corollary 13.5.2, our action ϕ_1 is semi-conjugated to an action by rotations ϕ_2. As an action by rotations, ϕ_2 must factor

$$\Gamma \longrightarrow \Gamma_{\mathrm{Ab}} \longrightarrow \mathbf{S}^1 \longrightarrow \mathrm{Homeo}^0(\mathbf{S}^1).$$

It remains to show that Γ has a finite orbit under ϕ_1. Let h be as in the definition of semi-conjugacy with $\phi_1(\cdot)h = h\phi_2(\cdot)$; then any point in the image of h has finite orbit under ϕ_1. □

Even though the action of the above statement is semi-conjugated to an action by a finite group, we can not deduce that the original action itself factors through a finite group, because the map h used in the proof needs not be surjective.

Nevertheless, it has been independently observed by several authors, e.g. by D. Witte in [**138**], that for C^1 actions one can use W.P. Thurston's stability theorem (Theorem 3 in [**136**]) and deduce that the action is finite, provided Γ'_{Ab} is finite for any finite index subgroup $\Gamma' < \Gamma$.

Indeed, since we have a finite orbit, there is a finite index subgroup $\Gamma^{(1)} < \Gamma$ with a fixed point. The C^1 assumption allows us to differentiate the $\Gamma^{(1)}$-action at this point. This yields a one dimensional linear representation, which must be finite by the assumption on $\Gamma^{(1)}_{\mathrm{Ab}}$. Upon passing again to a finite index subgroup $\Gamma^{(2)}$, we can assume the representation trivial. Now Thurston's stability theorem asserts that either the $\Gamma^{(2)}$-action is trivial or there is a non trivial character of $\Gamma^{(2)}$ associated to it. The later possibility is excluded since $\Gamma^{(2)}_{\mathrm{Ab}}$ is finite, whence ce claim.

The proviso about Γ'_{Ab} is of course satisfied for Kazhdan groups, since Kazhdan's property is stable under passing to finite index subgroups. Thus we deduce for instance :

COROLLARY 13.5.4.— *Let Γ be a group with property (TT) such that* $\mathrm{H}^2(\Gamma, \mathbf{R}) = 0$.

Then any Γ-action on the circle by orientation preserving C^1 diffeomorphisms factors through a finite group. □

Since we have established in 13.4.3 the property (TT) for higher rank lattices, we deduce

COROLLARY 13.5.5.— *Let Γ be a lattice in $G = \mathbf{G}(k)$, where $\mathbf{G}$ is a connected, simply connected, almost simple k-isotropic group and k a local field.*

If $\mathrm{H}^2(\Gamma, \mathbf{R}) = 0$ *and* $\mathrm{rank}_k(\mathbf{G}) \geq 2$, *then any Γ-action on the circle by orientation preserving C^1 diffeomorphisms factors through a finite group.* □

The essential consequences of property (TT) used here are : the vanishing of $\mathrm{EH}^2_{\mathrm{b}}(\Gamma, \mathbf{R})$ and the finiteness of Γ'_{Ab} for any finite index subgroup $\Gamma' < \Gamma$. We will see that these properties can also be established for a number of interesting groups that fail to have property (TT), namely certain irreducible lattices (Corollary 14.2.4 below).

13.6. Mapping class group and $\mathrm{PSU}(n,1)$**.** In short, the previous section was dealing with É. Ghys' idea to use $\mathrm{H}^2_{\mathrm{b}}(\Gamma)$ in order to deduce rigidity statements for representations

$$\phi: \Gamma \longrightarrow \mathrm{Homeo}^0(\mathbf{S}^1).$$

We shall now indicate two further situations in which very recent research has connected rigidity issues to $\mathrm{H}^2_{\mathrm{b}}(\Gamma)$.

Let Σ be a compact orientable surface, possibly with boundary. Let $\mathrm{Mod}(\Sigma)$ denote the mapping class group of Σ. Recall that $\mathrm{Mod}(\Sigma)$ has a natural action on the curve complex of Σ (see [**76, 77, 79**]). By a theorem of H. Masur and Y. Minsky [**104**], the curve complex is Gromov-hyperbolic. Mind however that $\mathrm{Mod}(\Sigma)$ is not a Gromov-hyperbolic group.

Now this is the place to recall that Gromov-hyperbolicity has a tendency to create a lot of bounded cohomology in degree two. We shall not go into the early history of the related results, but only mention that K. Fujiwara [**56, 57**] has improved on his original joint work with D. Epstein [**53**] on Gromov-hyperbolic groups in such a way that he could finally show the following in collaboration with M. Bestvina [**12**] :

THEOREM 13.6.1 (M. Bestvina, K. Fujiwara).— *Let Λ be a subgroup of* $\mathrm{Mod}(\Sigma)$. *If Λ is not virtually Abelian, then* $\mathrm{EH}^2_{\mathrm{b}}(\Lambda)$ *is infinite dimensional.*

ON THE PROOF. This is Theorem 12 in [**12**]. The authors use the classification [**107**] of such subgroups to show that their action on the curve complex satisfies a fairly weak properness condition, which is still enough to deduce the result. □

The above theorem shows that vanishing results for $\mathrm{EH}^2_{\mathrm{cb}}$ of a group Γ, such as the results that will be given in Section 14, will force any representation

$$\phi: \Gamma \longrightarrow \mathrm{Mod}(\Sigma)$$

to have virtually Abelian image. This is indeed a strong statement on rigidity, considering that for instance higher rank arithmetic lattices Γ have finite Abelianization.

One can use vanishing results on $\mathrm{EH}^2_{\mathrm{b}}$ in order to obtain fixed point theorems for Teichmüller spaces. Indeed, $\mathrm{Mod}(\Sigma)$ identifies with the group of isometries of the Teichmüller space $\mathrm{Teich}(\Sigma)$, whether this space is endowed with the Weil-Petersson or Teichmüller metric (see [**124**]). On the other hand, S.P. Kerckhoff has shown (Theorem 4 in [**93**]) that every *finite* subgroup of $\mathrm{Mod}(\Sigma)$ fixes a point in $\mathrm{Teich}(\Sigma)$. Thus, using the injectivity of inflation at the $\mathrm{H}^2_{\mathrm{b}}$ level (see Example 12.4.3), one has the following statement, which we record for later reference :

COROLLARY 13.6.2.— *Let Γ be a group with finite Abelianization* $\Gamma_{\mathrm{Ab}} = \Gamma/[\Gamma,\Gamma]$ *and finite dimensional* $\mathrm{H}^2_{\mathrm{b}}(\Gamma)$. *Then any representation*

$$\phi: \Gamma \longrightarrow \mathrm{Mod}(\Sigma)$$

has finite image and any Γ-action by isometries of $\mathrm{Teich}(\Sigma)$ *has a fixed point.* □

We come now to the second example of this section. Let

$$\phi : \Gamma \longrightarrow \mathrm{PSU}(n,1)$$

be a representation of a group Γ in the connected component $\mathrm{PSU}(n,1)$ of the group of isometries of the complex hyperbolic space $\mathbf{H}^n_{\mathbf{C}}$. Denote by $\partial_\infty \mathbf{H}^n_{\mathbf{C}}$ the boundary at infinity and recall that a *totally real* subspace is an isometrically embedded copy of a real hyperbolic space $\mathbf{H}^k_{\mathbf{R}}$ for some k.

Recall from Example 9.3.3 that the comparison map

$$\Psi^2 : \mathrm{H}^2_{\mathrm{cb}}(\mathrm{PSU}(n,1)) \longrightarrow \mathrm{H}^2_{\mathrm{c}}(\mathrm{PSU}(n,1))$$

is an isomorphism and that both spaces are one dimensional. Let $\omega \in \mathrm{H}^2_{\mathrm{cb}}(\mathrm{PSU}(n,1))$ be the class corresponding to the Kähler form. By contravariance, we obtain $\pi^*\omega \in \mathrm{H}^2_{\mathrm{b}}(\Gamma)$.

THEOREM 13.6.3 (M. Burger, A. Iozzi).— *For Γ finitely generated, the following assertions are equivalent :*

(i) $\pi^*\omega = 0$.

(ii) $\pi(\Gamma)$ *either fixes a point in $\partial_\infty \mathbf{H}^n_{\mathbf{C}}$ or leaves a totally real subspace invariant.*

ON THE PROOF. This statement is proved in [**34**]. The authors realize ω by the Cartan angle. Since Γ is finitely generated, they may take advantage of the amenable doubly ergodic Γ-space constructed in Section 11 above or in [**37**] and pull the Cartan angle back to this space as cocycle for $\pi^*\omega$. Then they show that the vanishing of the latter forces the former to be constant on the limit set of $\pi(\Gamma)$. □

14. GENERAL IRREDUCIBLE LATTICES

In this section we shall study the second bounded cohomology of certain lattices in products of arbitrary (compactly generated) locally compact groups ; the main result is Theorem 14.2.2, which gives a quite general super-rigidity statement. The key notion here is *irreducibility* ; for the questions relevant here, we can prove our results in the more general setting of *non elementary* lattices, see Definition 14.1.4 below.

14.1. Irreducibility. Suppose that Γ is a lattice in a direct product of locally compact groups :

$$\Gamma < G = G_1 \times \cdots \times G_n.$$

A degenerate example of this situation would be that Γ itself essentially splits as a product $\Gamma = \Gamma' \times \Gamma''$ of lattices in subproducts of G, say for instance $\Gamma' < G_1$ and $\Gamma'' < G_2 \times \cdots \times G_n$. One does *not* want to consider here that kind of examples which can be reduced readily to the study of Γ' and Γ''.

Within the theory of lattices in algebraic groups, this observation led to consider the following notion.

DEFINITION 14.1.1 (Algebraic case). Let $(\mathbf{G}_\alpha)_{\alpha\in\mathrm{A}}$ be a finite family of connected simply connected k_α-almost simple groups, where $(k_\alpha)_{\alpha\in\mathrm{A}}$ is a finite family of local fields. Set $G = \prod_{\alpha\in\mathrm{A}} \mathbf{G}_\alpha(k_\alpha)$ and for $\mathrm{B}\subset\mathrm{A}$ denote by

$$\mathrm{pr}_{\mathrm{B}} : G \longrightarrow G_{\mathrm{B}} = \prod_{\beta\in\mathrm{B}} \mathbf{G}_\beta(k_\beta)$$

the canonical projection.

A lattice $\Gamma < G$ is called *reducible* if there is a non-empty proper subset $\mathrm{B}\subset\mathrm{A}$ such that $\mathrm{pr}_{\mathrm{B}}(\Gamma)\times\mathrm{pr}_{\mathrm{A}\setminus\mathrm{B}}(\Gamma)$ is of finite index in Γ.

Otherwise, we say that Γ is *irreducible*.

There is a characterization of irreducible lattices which is of basic importance for the related super-rigidity issues ; in this generality, it is due to G.A. Margulis :

THEOREM 14.1.2 (G.A. Margulis).— *Let G be as in Definition 14.1.1 and suppose that all $\mathbf{G}_\alpha$ are k_α-isotropic.*

A lattice Γ is irreducible if and only if $\mathrm{pr}_{\mathrm{B}}(\Gamma)$ is dense in G_B for every non-empty proper subset $\mathrm{B}\subset\mathrm{A}$.

Observe that the assumption that the groups $\mathbf{G}_\alpha$ are isotropic is obviously necessary since otherwise $\mathbf{G}_\alpha(k_\alpha)$ is compact.

ON THE PROOF. The density is proved by G.A. Margulis in [**103**], Theorem II.6.7 point (b) ; for the definition of irreducibility, see also his Section III.5.9. The converse is obvious since in the reducible case there is a projection with discrete image. □

Let us now come back to the general setting of a lattice in a product of locally compact groups. We shall here take the characterization of the above theorem as *definiens* :

DEFINITION 14.1.3 (Topological case). Let $(G_\alpha)_{\alpha\in\mathrm{A}}$ be a finite family of locally compact groups, set $G = \prod_{\alpha\in\mathrm{A}} G_\alpha$ and for $\mathrm{B}\subset\mathrm{A}$ denote by

$$\mathrm{pr}_{\mathrm{B}} : G \longrightarrow G_{\mathrm{B}} = \prod_{\beta\in\mathrm{B}} G_\beta$$

the canonical projection.

A lattice $\Gamma < G$ is *irreducible* if $\mathrm{pr}_{\mathrm{B}}(\Gamma)$ is dense in G_B for every non-empty proper subset $\mathrm{B}\subset\mathrm{A}$.

A first observation is that this definition is indeed compatible with the notion of irreducibility of Definition 14.1.1 because of Theorem 14.1.2. As second observation, we point out that we do not bother to speak of reducible lattices in Definition 14.1.3 because in general there is not such a striking dichotomy as the one that Theorem 14.1.2 states about the lattices of Definition 14.1.1.

On the other hand, the more general Definition 14.1.3 captures the appropriate information for further super-rigidity statements : one cannot expect to relate certain subtle properties of Γ to a larger group than the "hull" obtained by taking the closure of all projections to the factors.

A relevant family of examples here is given by the work [**40, 41, 42**] of M. Burger and Sh. Mozes, where very interesting lattices

$$\Gamma < \mathrm{Aut}(\mathcal{T}_1) \times \mathrm{Aut}(\mathcal{T}_2)$$

in the product of automorphism groups of two simplicial trees $\mathcal{T}_1, \mathcal{T}_2$ are constructed. However, in these examples, the projection of Γ is never dense in $\mathrm{Aut}(\mathcal{T}_j)$; the fact that Γ does not split as direct product is merely reflected by the non-discreteness of its projections on the factors. Accordingly, the authors study rather Γ as a lattice in its natural "hull"

$$\Gamma < \overline{\mathrm{pr}_1(\Gamma)} \times \overline{\mathrm{pr}_2(\Gamma)}.$$

When considering lattices in the product of strictly more than two factors, one can set a weaker requirement then irreducibility :

DEFINITION 14.1.4. Let $G_1, \ldots, G_n$ be a family of locally compact groups, set $G = \prod_{j=1}^n G_j$ and for $1 \leq j \leq n$ denote by pr_j the canonical projection $G \to G_j$.

A lattice $\Gamma < G$ is *non elementary* if $\mathrm{pr}_j(\Gamma)$ is dense in G_j for every $1 \leq j \leq n$.

REMARKS 14.1.5. The relation between being non elementary and irreducibility in the generality of Definition 14.1.3 is the following.

(i) If $n = 2$, the two notions coincide.

(ii) If $n \geq 2$, then clearly irreducible implies non elementary. Examples of non elementary lattices that are not irreducible are given by the direct product of lattices which are themselves irreducible in a product. For instance, the embedding of

$$\Gamma = \mathrm{SL}_2(\mathbf{Z}[1/p]) \times \mathrm{SL}_2(\mathbf{Z}[\sqrt{d}])$$

into

$$G = \mathrm{SL}_2(\mathbf{R}) \times \mathrm{SL}_2(\mathbf{Q}_p) \times \mathrm{SL}_2(\mathbf{R}) \times \mathrm{SL}_2(\mathbf{R})$$

for p prime and $d \neq 0, 1$ square-free. In the algebraic group setting a non dense projection of a lattice on a subproduct is again a lattice (compare Sections III.5.9 and II.6 in [**103**]). We deduce that a non elementary lattice has always a finite index subgroup which is the direct product of irreducible lattices in subproducts ; each subproduct has at least two factors since the lattice is non elementary. In particular :

(iii) If $n = 3$, the two notions coincide.

(iv) If $n = 1$ there is a slight formal difference : whilst the irreducibility condition is void, making thus any lattice irreducible, we see that the only way for a lattice $\Gamma < G = G_1$ to be non elementary is to have $G = \Gamma$ discrete. The latter case will however be completely trivial for the statements that we are to establish below. As for the former, the only situation in which we make a statement regards almost simple algebraic group of rank at least two.

This more general notion of non elementary lattices suits the statements that we want to prove. This setting is also the framework of the super-rigidity for characters proved by Y. Shalom in [**130**], wherein he generalizes results known for irreducible lattices in semi-simple groups. Indeed the setup (0.1) in [**130**] is consistent with our Definition 14.1.4. (In Shalom's paper, there are also rigidity results for which the lattices must be irreducible.)

14.2. The general results and applications. Using induction, our results on double ergodicity and the Hochschild-Serre sequences allow us to relate the bounded cohomology in degree two of a non elementary lattice to the continuous bounded cohomology of the factors in the product. As outcome, we get super-rigidity statements for bounded 2-cocycles. The applications of Sections 13.5 and 13.6 yield then interesting rigidity statements.

Actually, as far as cohomological statements are concerned, we can also work with more general subgroups that are not necessarily lattices, therefore we agree that

DEFINITION 14.2.1. A subgroup $H < G_1 \times \cdots \times G_n$ is said *non elementary* if for all $1 \leq j \leq n$ the projection $\mathrm{pr}_j(H)$ is dense in G_j.

Before stating the result, we recall from Section 12 the notation $G'_j = \prod_{i \neq j} G_i$. Moreover, for any coefficient H-module (π, E) we denote by E_j the maximal coefficient H-submodule of E such that the restriction $\pi|_{E_j}$ extends continuously to G, factoring through $G \twoheadrightarrow G_j$; this is well-defined because $\overline{\mathrm{pr}_j(H)} = G_j$. Thus we have a G-action on the sum $\sum_{j=1}^n E_j$; we shall see (Lemma 14.2.10) that the latter space is again a coefficient G-module. In this setting we have

THEOREM 14.2.2.— *Let $G_1, \ldots, G_n$ be compactly generated locally compact second countable groups and let $H < G = G_1 \times \cdots \times G_n$ be a closed non elementary subgroup such that $H \backslash G$ has finite invariant measure (e.g. an irreducible lattice with $n \geq 2$). Let (π, E) be a separable coefficient H-module. Then there are topological isomorphisms*

$$\mathrm{H}^2_{\mathrm{cb}}(H, E) \cong \bigoplus_{j=1}^{n} \mathrm{H}^2_{\mathrm{cb}}(G_j, E_j) \cong \mathrm{H}^2_{\mathrm{cb}}\big(G, \textstyle\sum_{j=1}^n E_j\big).$$

The result still holds true if one of the groups G_j is not compactly generated.

(Proof below.)

If we apply this theorem to the case of lattices and combine it with Corollary 13.5.2, we have

COROLLARY 14.2.3.— *Let $G_1, \ldots, G_n$ be compactly generated locally compact second countable groups and let $\Gamma < G_1 \times \cdots \times G_n$ be a non elementary (e.g. irreducible) lattice. Assume $\mathrm{H}^2_{\mathrm{cb}}(G_j) = 0$ for $1 \leq j \leq n$.*

Then any Γ-action by orientation-preserving homeomorphisms of the circle is semi-conjugated to a Γ-action by rotations.

The result still holds true if one of the groups G_j is not compactly generated.

□

Using a result of Y. Shalom, we can conclude :

COROLLARY 14.2.4.— *Let $G_1, \ldots, G_n$ be compactly generated locally compact second countable groups and let $\Gamma < G_1 \times \cdots \times G_n$ be a non elementary (e.g. irreducible) co-compact lattice. Assume $\mathrm{H}^2_{\mathrm{cb}}(G_j) = 0$ and $\mathrm{Hom_c}(G_j) = 0$ for $1 \leq j \leq n$.*

Then any Γ-action by orientation-preserving homeomorphisms of the circle has a finite orbit. Moreover, if the action is by C^1 diffeomorphisms, then it factors through a finite quotient of Γ.

This time, we have to assume all factors compactly generated because this is an assumption of Shalom's result.

PROOF OF COROLLARY 14.2.4. The discussion of Section 13.5 results in the following conclusion : in order to deduce Corollary 14.2.4 from Corollary 14.2.3, it is enough to show that for any finite index subgroup $\Gamma' < \Gamma$, the Abelianization Γ'_{Ab} is finite. Shalom's super-rigidity for characters (Theorem 0.8 in [**130**]) allows us to deduce it from the condition $\mathrm{Hom_c}(G_j) = 0$ for $1 \leq j \leq n$. □

The next consequence of Theorem 14.2.2 is immediate given the result of M. Bestvina and K. Fujiwara presented above as Theorem 13.6.1.

COROLLARY 14.2.5.— *Let $G_1, \ldots, G_n$ be compactly generated locally compact second countable groups and let $\Gamma < G_1 \times \cdots \times G_n$ be a non elementary (e.g. irreducible) lattice. Assume $\dim \mathrm{H}^2_{\mathrm{cb}}(G_j) < \infty$ for $1 \leq j \leq n$.*

For every compact orientable surface Σ, the image of any representation $\Gamma \to \mathrm{Mod}(\Sigma)$ has virtually Abelian image.

The result still holds true if one of the groups G_j is not compactly generated.

□

Replacing Theorem 13.6.1 with Corollary 13.6.2, one deduces from Theorem 14.2.2 the following fixed point statement :

COROLLARY 14.2.6.— *Let $G_1, \ldots, G_n$ be compactly generated locally compact second countable groups and let $\Gamma < G_1 \times \cdots \times G_n$ be a non elementary (e.g. irreducible) lattice. Assume $\dim \mathrm{H}^2_{\mathrm{cb}}(G_j) < \infty$ for $1 \leq j \leq n$. Suppose Γ_{Ab} finite, which happens[2] e.g. if Γ is co-compact and $\mathrm{Hom_c}(G_j) = 0$ for $1 \leq j \leq n$.*

Then any Γ-action by isometries of a Teichmüller space Teich(Σ) *has a fixed point.* □

[2] by the already mentioned character super-rigidity of Y. Shalom, Theorem 0.8 in [**130**].

We emphasize the following particular case of Corollary 14.2.5 in which the assumptions are strikingly general :

COROLLARY 14.2.7.— *Let $G_1, \ldots, G_n$ be connected locally compact groups and let $\Gamma < G_1 \times \cdots \times G_n$ be a non elementary (e.g. irreducible) lattice.*

For every compact orientable surface Σ, the image of any representation $\Gamma \to \mathrm{Mod}(\Sigma)$ has virtually Abelian image.

We point out that the proof given below also applies if each G_j has finitely many connected components. These Corollaries generalize the main result of the paper [**55**] by B. Farb and H. Masur.

PROOF OF COROLLARY 14.2.7. We recall first that connected locally compact groups are compactly generated (by any compact neighbourhood of e). Moreover, we may assume each G_j to be second countable :

Indeed, G_j is σ-compact and thus, as already mentioned a couple of times, there is a compact normal subgroup $K_j \lhd G_j$ such that $H_j = G_j/K_j$ is second countable (Satz 6 in [**90**]). As Γ is canonically a non-elementary lattice in the connected group $H_1 \times \cdots \times H_n$, we see that there is no loss of generality in assuming G_j second countable.

Now, in view of Corollary 14.2.5, we have only to show that for all $1 \leq j \leq n$ the dimension of $\mathrm{H}^2_{\mathrm{cb}}(G_j)$ is finite. We define $A(G_j) \lhd G_j$ to be the unique maximal amenable closed normal subgroup, as we did in Section 11.3. Let $L_j = G_j/A(G_j)$. By Corollary 8.5.2, the inflation

$$\inf: \ \mathrm{H}^2_{\mathrm{cb}}(L_j) \longrightarrow \mathrm{H}^2_{\mathrm{cb}}(G_j)$$

is an isomorphism. On the other hand, we have seen in the proof of Theorem 11.3.4 that the solution of Hilbert's fifth problem (Theorem 4.6 in [**114**]) forces L_j^0, which here is L_j, to be a connected semi-simple adjoint Lie group without compact factors. Such a group has finite dimensional second continuous bounded cohomology $\mathrm{H}^2_{\mathrm{cb}}(L_j)$, as explained in Remark 9.3.4 above. □

Given the result of M. Burger and A. Iozzi presented above as Theorem 13.6.3, Theorem 14.2.2 implies :

COROLLARY 14.2.8.— *Let $G_1, \ldots, G_n$ be compactly generated locally compact second countable groups and let $\Gamma < G_1 \times \cdots \times G_n$ be a non elementary (e.g. irreducible) finitely generated lattice. Assume $\mathrm{H}^2_{\mathrm{cb}}(G_j) = 0$ for $1 \leq j \leq n$.*

Then the image of any representation $\Gamma \to \mathrm{PSU}(n,1)$ either fixes a point in $\partial_\infty \mathbf{H}^n_{\mathbf{C}}$ or leaves a totally real subspace invariant.

The result still holds true if one of the groups G_j is not compactly generated.
□

Another consequence, bringing in Shalom's aforementioned result, regards super-rigidity of quasi-morphisms ; we shall prove below :

COROLLARY 14.2.9.— *Let $G_1, \ldots, G_n$ be compactly generated locally compact second countable groups and let $\Gamma < G = G_1 \times \cdots \times G_n$ be a non elementary (e.g. irreducible) co-compact lattice. Then any quasi-morphism $f : \Gamma \to \mathbf{R}$ extends to a continuous quasi-morphism $f_{\mathrm{ext}} : G \to \mathbf{R}$.*

Now, in view of the proof of Theorem 14.2.2, we regularize the case of the space $\sum_{j=1}^{n} E_j$ with the help of Lemma 12.3.5 :

LEMMA 14.2.10.— *The sum $\sum_{j=0}^{n} E_j$ is weak-* closed in E, so that it is a coefficient G-module extending the H-action.*

PROOF. The weak-* continuous G-action on the E_j extends to $\sum_{j=0}^{n} E_j$ and hence to its weak-* closure that we shall denote by E_∞. Applying Lemma 12.3.5 to E_∞ yields the statement since we have precisely $(E_\infty)^{G'_j} = E_j$. □

Now we are going to use L^p induction in order to reduce the Theorem 14.2.2 to our Lyndon-Hochschild-Serre exact sequence, more precisely to Corollary 12.0.4 and Example 12.0.5. To this end we need the

LEMMA 14.2.11.— *For every $1 < p < \infty$ there is a natural isometric isomorphism of G-modules*

$$\alpha_j : \; E_j \cong \big(L^p\mathrm{I}_H^G E\big)^{G'_j}.$$

PROOF. Define the map $\alpha_j : E_j \to L^p\mathrm{I}_H^G E$ by $\alpha_j(v)(g) = \pi_j(g)v$, so that $\alpha_j(v)$ is indeed H-invariant. Since π_j factors through G_j, the map $\alpha(v)$ is moreover G'_j-invariant under right translation. Moreover, the map α_j is G-equivariant and it preserves the norm since π_j is isometric and the invariant measure on $H\backslash G$ is normalized. It remains thus to show surjectivity onto the G'_j-invariants. If $f : G \to F$ is in $\big(L^p\mathrm{I}_H^G E\big)^{G'_j}$, then by Fubini-Lebesgue it is represented by a $\mathrm{pr}_j(H)$-equivariant map $G_j \to E$, which has to be of the form $\alpha_j(v)$ for some $v \in E_j$ by the density of $\mathrm{pr}_j(H)$, since E is continuous by Proposition 3.3.2. □

We can now complete the

PROOF OF THEOREM 14.2.2. Let $1 < p < \infty$ and take α_j as in (the proof of) Lemma 14.2.11. We consider the following diagram :

$$\begin{array}{ccccc}
\mathrm{H}^2_{\mathrm{cb}}(H,E) & \xrightarrow{L^p\mathrm{i}} & \mathrm{H}^2_{\mathrm{cb}}(G, L^p\mathrm{I}_H^G E) & & \\
\uparrow & & \updownarrow & \searrow\!\!\!\nwarrow & \\
 & & & & \bigoplus_{j=1}^{n} \mathrm{H}^2_{\mathrm{cb}}\big(G, (L^p\mathrm{I}_H^G E)^{G'_j}\big) \\
 & & & \swarrow\!\!\!\nearrow & \updownarrow \\
 & & \mathrm{H}^2_{\mathrm{cb}}\big(G, \sum_{j=1}^{n} (L^p\mathrm{I}_H^G E)^{G'_j}\big) & & \oplus_{j=1}^{n}(\alpha_j)_* \\
 & & \updownarrow (\sum_{j=1}^{n}\alpha_j)_* & & \\
\mathrm{H}^2_{\mathrm{cb}}\big(H, \sum_{j=1}^{n} E_j\big) & \xleftarrow{\mathrm{res}} & \mathrm{H}^2_{\mathrm{cb}}\big(G, \sum_{j=1}^{n} E_j\big) & & \bigoplus_{j=1}^{n} \mathrm{H}^2_{\mathrm{cb}}(G, E_j)
\end{array}$$

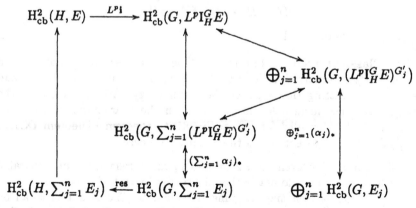

On the top right we have a commutative triangle of topological isomorphisms because this is an instance of the Lyndon-Hochschild-Serre sequence exactly as considered in Corollary 12.0.4 and Example 12.0.5, but applied to the induced coefficient G-module $L^p\mathrm{I}_H^G E$. The maps induced by the α_j are topological isomorphisms because of Lemma 14.2.11. The left square is commutative because the formula for α_j coincides with the composition of restriction and induction. Thus the L^p induction $L^p\mathbf{i}$ on the top is surjective ; on the other hand it is injective by Corollary 11.1.5. It remains only to show that this isomorphism $L^p\mathbf{i}$ is a topological isomorphism. It is continuous by the first point in Remarks 10.2.5, and since its range is a Banach space because of Corollary 11.4.2, we can apply the open mapping theorem and conclude. □

We have still to give the

PROOF OF COROLLARY 14.2.9. Consider the commutative diagram

$$\begin{array}{ccc}
\mathrm{H}^2_{\mathrm{b}}(\Gamma,\mathbf{R}) & \xleftarrow[\cong]{\text{res}} & \mathrm{H}^2_{\mathrm{cb}}(G,\mathbf{R}) \\
\big\downarrow & & \big\downarrow \\
\mathrm{H}^2(\Gamma,\mathbf{R}) & \xleftarrow{\text{res}} & \mathrm{H}^2_{\mathrm{c}}(G,\mathbf{R})
\end{array}$$

The upper restriction map is an isomorphism by Theorem 14.2.2 with $H = \Gamma$ and $E = \mathbf{R}$ trivial. It is well known that the lower restriction map is injective because Γ is co-compact. If now $f : \Gamma \to \mathbf{R}$ is a quasimorphism, it follows from this diagram that there is a continuous quasimorphism $F : G \to \mathbf{R}$ and $h \in \ell^\infty(\Gamma, \mathbf{R})$ such that $\delta(f + h) = \delta F|_{\Gamma\times\Gamma}$. In particular, $\chi = f + h - F|_\Gamma$ is in $\mathrm{Hom}(\Gamma)$. By Y. Shalom's result (Theorem 0.8 in [**130**]), χ extends to a continuous homomorphism $\mathrm{X} : G \to \mathbf{R}$. Pick now any $H \in \mathrm{C}_{\mathrm{b}}(G, \mathbf{R})$ with $H|_\Gamma = h$. We claim then $f_{\mathrm{ext}} = F - H + \mathrm{X}$ is the desired extension. Indeed, we have

$$f_{\mathrm{ext}}|_\Gamma = (F - H + \mathrm{X})|_\Gamma = F|_\Gamma - h + \chi = f,$$

so that f_{ext} extends f. Finally f_{ext} is continuous by construction and it is a quasi-morphism because

$$\delta(F - H + \mathrm{X}) = \delta F - \delta H$$

and both terms are bounded. □

14.3. Classical irreducible lattices. The most classical examples of non elementary lattices are irreducible lattices in semi-simple algebraic groups, and indeed it is this setting that inspires the terminology. We consider also the situations where the local field varies, and even the characteristic is allowed to vary. By G.A. Margulis' famous arithmeticity theorem (Theorem IX.1.11 in [**103**]), we are thus essentially in the S-arithmetic case.

In this context, Theorem 14.2.2 applies in full strength for all separable coefficient modules, so we do not bother the reader by restating it.

As for the comparison map, arguments similar to Theorem 14.2.2 can be used to derive a substitute for property (TT) of Theorem 13.4.3 for irreducible lattices. The statement below generalizes the main result of [**35**], which was to be amended by [**36**], see also [**37**]. We need the

DEFINITION 14.3.1. A unitary representation $(\pi, \mathfrak{H})$ of a lattice Γ in a product G of algebraic groups is *degenerate* if the unitarily induced representation of G contains a subrepresentation which factors non trivially through a rank one factor.

THEOREM 14.3.2.— *Let $\Gamma < G = \prod_{\alpha\in A} \mathbf{G}_\alpha(k_\alpha)$ be a non elementary (e.g. irreducible with $|A| \geq 2$) lattice, where $(k_\alpha)_{\alpha\in A}$ is a finite family of local fields and the $\mathbf{G}_\alpha$ are connected simply connected k_α-almost simple groups of positive k_α-rank. Then the comparison map*

$$\mathrm{H}^2_{\mathrm{b}}(\Gamma, \mathfrak{H}) \longrightarrow \mathrm{H}^2(\Gamma, \mathfrak{H})$$

is injective for every non degenerate unitary representation of Γ.

In Definition 14.3.1, "non trivially" just means that the subrepresentation should not consist of G-invariant vectors. The non degeneracy assumption of the theorem is clearly necessary in view of the fact that any rank one simple Lie group G admits a unitary representation with non vanishing $\mathrm{EH}^2_{\mathrm{cb}}$.

Before we present the proof of Theorem 14.3.2, we consider in more detail the important particular case of trivial coefficients :

COROLLARY 14.3.3.— *Let $\Gamma < G = \prod_{\alpha\in A} \mathbf{G}_\alpha(k_\alpha)$ be a non elementary (e.g. irreducible with $|A| \geq 2$) lattice, where $(k_\alpha)_{\alpha\in A}$ is a finite family of local fields and the $\mathbf{G}_\alpha$ are connected simply connected k_α-almost simple groups of positive k_α-rank.*

Then the comparison map from bounded to ordinary cohomology induces an isomorphism

$$\mathrm{H}^2_{\mathrm{b}}(\Gamma) \longrightarrow \mathrm{H}^2(\Gamma)^{\mathrm{inv}},$$

where the latter is the image in $\mathrm{H}^2(\Gamma)$ under restriction of the continuous cohomology $\mathrm{H}^2_{\mathrm{c}}(G)$. Both spaces have the dimension of the number of Hermitian factors of G.

One may of course not expect Γ to have property (TT) here, for it may happen that some of the factors $\mathbf{G}_\alpha(k_\alpha)$ fail to be Kazhdan groups.

Due to the numerous applications connected with $\mathrm{H}^2_{\mathrm{b}}(\Gamma)$, we can deduce :

COROLLARY 14.3.4.— *Let $\Gamma < G = \prod_{\alpha\in A} \mathbf{G}_\alpha(k_\alpha)$ be a non elementary (e.g. irreducible with $|A| \geq 2$) lattice, where $(k_\alpha)_{\alpha\in A}$ is a finite family of local fields and the $\mathbf{G}_\alpha$ are connected simply connected k_α-almost simple groups of positive k_α-rank. Then*

(i) *The commutator length vanishes on the commutator subgroup $[\Gamma, \Gamma]$.*

(ii) *For every compact orientable surface Σ, the image of any representation $\Gamma \to \mathrm{Mod}(\Sigma)$ has finite image ; thus any Γ-action by isometries of the Teichmüller space $\mathrm{Teich}(\Sigma)$ has a fixed point.*

Suppose moreover $\mathrm{H}^2(\Gamma) = 0$. Then

(iii) *Any Γ-action on the circle by orientation preserving homeomorphisms is semi-conjugated to an action by a finite group of rotations, and hence has a finite orbit.*

(iv) *Any Γ-action on the circle by orientation preserving C^1 diffeomorphisms factors through a finite group.*

(v) *The image of any representation $\Gamma \to \mathrm{PSU}(n,1)$ either fixes a point in $\partial_\infty \mathbf{H}^n_{\mathbf{C}}$ or leaves a totally real subspace invariant.*

PROOF OF COROLLARY 14.3.4. The first point is but the juxtaposition of Corollary 14.3.3 and Ch. Bavard's criterion Theorem 13.3.3. For point (ii), we observe that Corollary 14.3.3 identifies $\mathrm{H}^2_{\mathrm{b}}(\Gamma)$ with a finite dimensional space ; therefore, one may apply Theorem 13.6.1 and Corollary 14.2.5.

In the remaining cases, we have $\mathrm{H}^2_{\mathrm{b}}(\Gamma) = 0$. In point (iii), Corollary 13.5.2 shows therefore that the action is semi-conjugated to an action by rotations. It remains to show that the corresponding Abelian quotient is finite. For Γ irreducible, this follows from Margulis' normal subgroup theorem (IV.4.10 in [**103**]). If Γ is not irreducible, it contains a finite index subgroup that splits into irreducible lattices in subproduct ; it is enough to show that none of these lattices admit infinite Abelian quotients. But since Γ is non elementary, each subproduct has at least two factors, so that one can again apply Margulis' theorem. The same argument yields (iv) using W.P. Thurston's stability theorem [**136**], as explained in the discussion preceding Corollary 13.5.4.

The last point is the bare combination of Corollary 14.3.3 and Theorem 13.6.3. □

We give now an example wich shows that the second option in point (v) can still give quite complicated representations :

EXAMPLE 14.3.5. Fix a totally real subspace $\mathbf{H}^k_{\mathbf{R}} \subset \mathbf{H}^n_{\mathbf{C}}$ and denote by $G_1 = G_2 = \mathrm{SO}(k,1)$ the corresponding group of isometries. Choose an irreducible lattice

$$\Gamma < G = G_1 \times G_2.$$

Since $\mathbf{H}^k_{\mathbf{R}}$ is not of Hermitian type, we know from Example 9.3.3 that $\mathrm{H}^2_{\mathrm{cb}}(G_j) = 0$. Therefore, the lattice Γ satisfies all assumptions of point (v) in Corollary 14.3.4. Now here is a representation of Γ in $\mathrm{PSU}(n,1)$:

$$\Gamma \xrightarrow{\mathrm{pr}_1} G_1 = \mathrm{SO}(k,1) \subset \mathrm{PSU}(n,1).$$

PROOF OF THEOREM 14.3.2. The proof is very much in the spirit of the proof of Theorem 13.4.3, except that now the ambient group does not have property (TT) in general ; however the Künneth type result of Corollary 12.0.4 will make up for that.

Recall that by Corollary 11.1.5 the L^2 induction

$$\mathrm{H}^2_{\mathrm{b}}(\Gamma, \mathfrak{H}) \longrightarrow \mathrm{H}^2_{\mathrm{cb}}(G, L^2\mathrm{I}_\Gamma^G \mathfrak{H})$$

is injective. We have observed in the second point of Remarks 14.1.5 that Γ is, up to finite index, a direct product of irreducible lattices in subproducts of at least two factors. Therefore we may apply Proposition 13.4.4, and it remains to show that the comparison map

$$\mathrm{H}^2_{\mathrm{cb}}(G, L^2\mathrm{I}_\Gamma^G\mathfrak{H}) \longrightarrow \mathrm{H}^2_{\mathrm{c}}(G, L^2\mathrm{I}_\Gamma^G\mathfrak{H}) \qquad (*)$$

is injective. We split $L^2\mathrm{I}_\Gamma^G\mathfrak{H}$ as $\big(L^2\mathrm{I}_\Gamma^G\mathfrak{H}\big)^G \oplus \mathfrak{L}$, where by definition $\mathfrak{L}$ is the orthogonal complement to the G-invariants. Appealing to Corollary 8.2.10, me may handle the summands of this decomposition separately. The injectivity with the G-invariant part follows at once from the fact that the factors of G have no continuous quasi-morphisms, proved in Lemma 13.3.4.

As for $\mathfrak{L}$, we write for all $\alpha \in \mathrm{A}$

$$\mathfrak{L}_\alpha = \mathfrak{L}^{\prod_{\beta\neq\alpha}\mathbf{G}_\beta(k_\beta)}.$$

According to Corollary 12.0.4, we have

$$\mathrm{H}^2_{\mathrm{cb}}(G, \mathfrak{L}) \cong \bigoplus_{\alpha\in\mathrm{A}} \mathrm{H}^2_{\mathrm{cb}}(G, \mathfrak{L}_\alpha).$$

By assumption, we have $\mathfrak{L}_\alpha = 0$ whenever the k_α-rank of $\mathbf{G}_\alpha$ is one. Therefore, since all other factors have property (TT) by Theorem 13.4.1, we conclude that the map

$$\oplus_\alpha \Psi^2 : \bigoplus_{\alpha\in\mathrm{A}} \mathrm{H}^2_{\mathrm{cb}}(G, \mathfrak{L}_\alpha) \longrightarrow \bigoplus_{\alpha\in\mathrm{A}} \mathrm{H}^2_{\mathrm{c}}(G, \mathfrak{L}_\alpha)$$

is injective. Therefore the injectivity of $(*)$ follows from naturality and the injectivity of

$$\bigoplus_{\alpha\in\mathrm{A}} \mathrm{H}^2_{\mathrm{c}}(G, \mathfrak{L}_\alpha) \longrightarrow \mathrm{H}^2_{\mathrm{c}}(G, \mathfrak{L}).$$

The later can be verified e.g. by remarking that the the spaces $\mathfrak{L}_\alpha$ intersect trivially and combine as a topologically direct sum by Lemma 12.3.5. □

Proof of Corollary 14.3.3. The comparison map

$$\mathrm{H}^2_{\mathrm{b}}(\Gamma) \longrightarrow \mathrm{H}^2(\Gamma)$$

is injective by Theorem 14.3.2, so that it remains only to show that its image is indeed the image $\mathrm{H}^2(\Gamma)^{\mathrm{inv}}$ of the restriction map

$$\mathrm{res}: \ \mathrm{H}^2_{\mathrm{c}}(G) \longrightarrow \mathrm{H}^2(\Gamma).$$

This follows from the commutativity of the diagram

$$\begin{array}{ccc}
\mathrm{H}^2_{\mathrm{c}}(G) & \xrightarrow{\ \cong\ } & \mathrm{H}^2_{\mathrm{cb}}(G) \\
\Big\downarrow{\scriptstyle \Psi} & & \Big\downarrow{\scriptstyle \Psi} \\
\mathrm{H}^2(\Gamma) & \xleftarrow{\ \mathrm{res}\ } & \mathrm{H}^2_{\mathrm{c}}(G)
\end{array}$$

because the upper arrow is an isomorphism by Theorem 14.2.2 and the right comparison map is an isomorphism by Example 9.3.3, Remark 9.3.4 and Example 9.3.5. □

REMARK 14.3.6. If we consider the diagram of the above proof, we see that we have moreover shown that in ordinary [continuous] cohomology the restriction map

$$\mathrm{res}: \ \mathrm{H}^2_{\mathrm{c}}(G) \longrightarrow \mathrm{H}^2(\Gamma).$$

is injective.

14.4. Lattices in products of trees. We shall now cast a glance at lattices in products of automorphism groups of trees. Upon passing to closed subgroups of the automorphism groups, such lattices give rise to highly interesting examples of non elementary lattices in products. Recall for instance that M. Burger and Sh. Mozes [**40, 42**] have constructed in this way the first known examples of torsion free finitely presented simple groups.

REMARK 14.4.1. These groups being in particular both finitely generated and non residually finite, they cannot be linear. On the other hand, the setting of lattices in products of trees encompasses also many arithmetic examples. For instance, one may work with the Bruhat-Tits trees associated to rank one p-adic groups.

Theorem 14.2.2 has the following consequence (see also [**37**]).

COROLLARY 14.4.2.— *Let $\mathcal{T}_1, \ldots, \mathcal{T}_n$ be locally finite regular (or bi-regular) trees and let*

$$\Gamma < \mathrm{Aut}(\mathcal{T}_1) \times \cdots \times \mathrm{Aut}(\mathcal{T}_n)$$

be a lattice such that the closure $\overline{\mathrm{pr}_j(\Gamma)}$ acts transitively on $\partial_\infty \mathcal{T}_j$ for all $1 \leq j \leq n$. Then we have

$$\mathrm{H}^2_{\mathrm{b}}(\Gamma) = 0.$$

PROOF. Define $G_j = \overline{\mathrm{pr}_j(\Gamma)}$ and view Γ as a non elementary lattice

$$\Gamma < G = G_1 \times \cdots \times G_n.$$

We have seen in Example 13.3.6 that the assumption on G_j allows us to apply Lemma 13.3.5 and deduce that every continuous quasi-morphism of G_j is bounded. Therefore the comparison map

$$\mathrm{H}^2_{\mathrm{cb}}(G_j) \longrightarrow \mathrm{H}^2_{\mathrm{c}}(G_j)$$

is injective for all $1 \leq j \leq n$. On the other hand, G_j acts properly simplicially on $\mathcal{T}_j$, which is a one dimensional acyclic simplicial complex ; this implies that $\mathrm{H}^n_{\mathrm{c}}(G_j)$ vanishes for all $n > 1$ (see for instance Lemma X.1.12 in [**19**]).

Therefore $\mathrm{H}^2_{\mathrm{cb}}(G_j) = 0$ for all $1 \leq j \leq n$ and thus the corollary is indeed a particular case of Theorem 14.2.2. □

Again, we can take advantage of some applications of the vanishing of $\mathrm{H}^2_{\mathrm{b}}(\Gamma)$ and deduce :

COROLLARY 14.4.3.— *Let $\mathcal{T}_1, \ldots, \mathcal{T}_n$ be locally finite regular (or bi-regular) trees and let*

$$\Gamma < \mathrm{Aut}(\mathcal{T}_1) \times \cdots \times \mathrm{Aut}(\mathcal{T}_n)$$

be a lattice such that the closure $\overline{\mathrm{pr}_j(\Gamma)}$ acts transitively on $\partial_\infty \mathcal{T}_j$ for all $1 \leq j \leq n$. Then

(i) *The commutator length vanishes on the commutator subgroup* $[\Gamma,\Gamma]$.

(ii) *For every compact orientable surface* Σ*, the image of any representation* $\Gamma \to \mathrm{Mod}(\Sigma)$ *has virtually Abelian image.*

(iii) *Any* Γ*-action on the circle by orientation preserving homeomorphisms is semi-conjugated to an action by rotations.*

(iv) *The image of any representation* $\Gamma \to \mathrm{PSU}(n,1)$ *either fixes a point in* $\partial_\infty \mathbf{H}^n_{\mathbb{C}}$ *or leaves a totally real subspace invariant.*

PROOF. Given the above Corollary 14.4.2, point (i) follows from Theorem 13.3.3, point (ii) from Theorem 13.6.1, point (iii) from Corollary 13.5.2 and point (iv) from Theorem 13.6.3. □

14.5. The adélic case. The embedding of a global field into the corresponding ring of adèles gives rise, for algebraic groups, to interesting examples of irreducible lattices in the sense of Definition 14.1.3. Intuitively speaking, this situation can be thought of as the limiting case of S-arithmetic lattices when the number of places goes to infinity. Accordingly, it is not surprising that we are are led away from the compactly generated case. Quite conveniently, the relevant results of Sections 12 and 14.2 allow exactly for that amount of generality.

Additional ingredients will be the strong approximation theorem and Kolmogorov's zero-one law.

THEOREM 14.5.1.— *Let* K *be a global field and* $\mathbf{G}$ *a simply connected semi-simple linear algebraic group over* K*. Denote by* $\mathcal{V}_\infty$ *the collection of Archimedean places of* K*.*

There are canonical topological isomorphisms

$$\mathrm{H}^2_{\mathrm{b}}(\mathbf{G}(K)) \cong \bigoplus_{v\in\mathcal{V}_\infty} \mathrm{H}^2_{\mathrm{cb}}(\mathbf{G}(K_v)) \cong \bigoplus_{v\in\mathcal{V}_\infty} \mathrm{H}^2_{\mathrm{c}}(\mathbf{G}(K_v)),$$

realized notably by the restriction morphisms.

(Proof below ; see also [**37**].)

Notice that since $\mathcal{V}_\infty = \varnothing$ when K has positive characteristic, the Theorem 14.5.1 implies immediately

COROLLARY 14.5.2.— *Let* K *be a global field of positive characteristic and* $\mathbf{G}$ *a simply connected semi-simple linear algebraic group over* K*. Then*

$$\mathrm{H}^2_{\mathrm{b}}(\mathbf{G}(K)) = 0.$$ □

REMARK 14.5.3. Suppose that K is a number field, i.e. $\mathrm{char}(K) = 0$. In ordinary cohomology, A. Borel and J. Yang prove the statement corresponding to Theorem 14.5.1 for any positive degree (Theorem 2.1 in [**20**]). Therefore, the rightmost term in the statement of Theorem 14.5.1 is in return isomorphic to $\mathrm{H}^2(\mathbf{G}(K))$ under the restriction. Since by functoriality the corresponding diagram

$$\begin{array}{ccc} \mathrm{H}^2_{\mathrm{b}}(\mathbf{G}(K)) & \xleftarrow{\text{res}} & \bigoplus_{v\in\mathcal{V}_\infty} \mathrm{H}^2_{\mathrm{cb}}(\mathbf{G}(K_v)) \\ \downarrow & & \downarrow \\ \mathrm{H}^2(\mathbf{G}(K)) & \xleftarrow{\text{res}} & \bigoplus_{v\in\mathcal{V}_\infty} \mathrm{H}^2_{\mathrm{c}}(\mathbf{G}(K_v)) \end{array}$$

commutes, we conclude that the comparison map

$$\mathrm{H}^2_{\mathrm{b}}(\mathbf{G}(K)) \longrightarrow \mathrm{H}^2(\mathbf{G}(K))$$

is an isomorphism.

This remark was for $\mathrm{char}(K) = 0$; together with the statement of Corollary 14.5.2 for $\mathrm{char}(K) \neq 0$, it implies :

COROLLARY 14.5.4.— *Let K be a global field and $\mathbf{G}$ a simply connected semi-simple linear algebraic group over K. The comparison map*

$$\mathrm{H}^2_{\mathrm{b}}(\mathbf{G}(K)) \longrightarrow \mathrm{H}^2(\mathbf{G}(K))$$

is injective. □

By Bavard's criterion (Theorem 13.3.3), the above corollary is equivalent to the vanishing of the stable length ℓ on the commutator subgroup of $\mathbf{G}(K)$. While this is trivially verified for $\mathbf{G} = \mathrm{SL}_n$ $(n \geq 2)$, we do not know the status of the question for a general group $\mathbf{G}$.

Let us consider the following particular case of Theorem 14.5.1.

COROLLARY 14.5.5.— *Let $\mathbf{G}$ be a simply connected semi-simple linear group defined over $\mathbf{Q}$. Then the restriction map*

$$\mathrm{H}^2_{\mathrm{cb}}(\mathbf{G}(\mathbf{R})) \longrightarrow \mathrm{H}^2_{\mathrm{b}}(\mathbf{G}(\mathbf{Q}))$$

is an isomorphism. In particular (Corollary 14.3.3) the dimension of $\mathrm{H}^2_{\mathrm{b}}(\mathbf{G}(\mathbf{Q}))$ is exactly the number of factors of Hermitian type in $\mathbf{G}(\mathbf{R})$. □

SCHOLIUM 14.5.6. In the setting of this corollary, we recall from Corollary 7.5.9 that if P is a minimal parabolic subgroup of $G = \mathbf{G}(\mathbf{R})$, then $\mathrm{H}^\bullet_{\mathrm{cb}}(G)$ is realized by the G-invariants of

$$0 \longrightarrow L^\infty_{\mathrm{w*}}(G/P) \longrightarrow L^\infty_{\mathrm{w*}}((G/P)^2) \longrightarrow L^\infty_{\mathrm{w*}}((G/P)^3) \longrightarrow \cdots \qquad (*)$$

If now the $\mathbf{G}(\mathbf{Q})$-action on G/P were amenable, then one could deduce immediately the statement of Corollary 14.5.5 without appealing to Theorem 14.5.1 and in any degree. Indeed, one could by Theorem 7.5.3 realize $\mathrm{H}^\bullet_{\mathrm{b}}(\mathbf{G}(\mathbf{Q}))$ by the $\mathbf{G}(\mathbf{Q})$-invariants of the complex $(*)$; however, since the G-action is weak-* continuous and $\mathbf{G}(\mathbf{Q})$ is dense in G, the subcomplex of $\mathbf{G}(\mathbf{Q})$-invariants coincides with the G-invariants.

This argument is however not applicable, because the $\mathbf{G}(\mathbf{Q})$-action on the Furstenberg boundary G/P of G is non amenable according to a result of R. Zimmer [**142**].

Changing the number field, we can give an example in which the effects of Zimmer's result become obvious in the light of Theorem 14.5.1 : if $d \in \mathbf{N}$ is not a square, then the latter theorem yields an isomorphism

$$\mathrm{H}^2_{\mathrm{b}}(\mathrm{SL}_2(\mathbf{Q}[\sqrt{d}])) \cong \mathrm{H}^2_{\mathrm{c}}(\mathrm{SL}_2(\mathbf{R})) \oplus \mathrm{H}^2_{\mathrm{c}}(\mathrm{SL}_2(\mathbf{R})).$$

This is two dimensional according to Example 9.3.3. Now if the $\mathrm{SL}_2(\mathbf{Q}[\sqrt{d}])$-action on the Furstenberg boundary of $\mathrm{SL}_2(\mathbf{R})$ (i.e. the circle) were amenable, the same argument as above would yields an isomorphism between $\mathrm{H}^2_{\mathrm{b}}(\mathrm{SL}_2(\mathbf{Q}[\sqrt{d}]))$ and the one dimensional space $\mathrm{H}^2_{\mathrm{b}}(\mathrm{SL}_2(\mathbf{R}))$, a contradiction.

We prepare now the proof of Theorem 14.5.1, recalling the notation — which we base on Margulis' book [**103**], to which we refer for further information.

To begin with, the product formula of Corollary 12.0.4 shows that we may assume that $\mathbf{G}$ is K-almost simple ; thus we make this assumption.

We denote by $\mathcal{V}$ the set of places of K, by K_v the completion of K at $v \in \mathcal{V}$ and by $\mathbf{A}_K$ the ring of adèles ; we write $\mathbf{G}(\mathbf{A}_K)$ for the corresponding adélic group ; recall that $\mathbf{G}(\mathbf{A}_K)$ is a locally compact group, but in general not compactly generated. For any set of places $\mathcal{U} \subset \mathcal{V}$ let $G_{\mathcal{U}}$ be the group of all elements of $\mathbf{G}(\mathbf{A}_K)$ which are trivial outside $\mathcal{U}$. In particular, given any subsets $\mathcal{U}' \subset \mathcal{U} \subset \mathcal{V}$, the product map yields an isomorphism

$$\prod_{v\in\mathcal{U}'} \mathbf{G}(K_v) \times G_{\mathcal{U}\setminus\mathcal{U}'} \cong G_{\mathcal{U}}$$

if and only if $\mathcal{U}'$ is *finite*.

Recall that $\mathcal{V}_\infty$ is the finite set of Archimedean places of K. We write $\mathcal{A}$ for the set of places at which $\mathbf{G}$ is anisotropic ; let $\mathcal{I} = \mathcal{V} \setminus \mathcal{A}$ be the set of places where $\mathbf{G}$ is isotropic. Recall that $\mathcal{A}$ is finite and that $\mathbf{G}(K_v)$ is compact if and only if $v \in \mathcal{A}$. In particular, the decomposition

$$\mathbf{G}(\mathbf{A}_K) \cong G_{\mathcal{A}} \times G_{\mathcal{I}}$$

has compact first factor.

Next we recall that the embedding $K \to \mathbf{A}_K$ that takes $x \in K$ to the *principal* adèle $(x)_{v\in\mathcal{V}}$ realizes $\mathbf{G}(K)$ as a lattice in $\mathbf{G}(\mathbf{A}_K)$ (see Theorem 3.2.2 in [**103**, chap. I]). Thus we can also view $\mathbf{G}(K)$ as a lattice in $G_{\mathcal{I}}$. One has the celebrated

THEOREM 14.5.7 (Strong Approximation).— *The image of $\mathbf{G}(K)$ in $G_{\mathcal{U}}$ is dense for every proper subset $\mathcal{U} \subset \mathcal{I}$.*

ON THE PROOF. This is the result of Section II.6.8 in [**103**]. In the latter reference, $\mathbf{G}$ is any simply connected K-almost simple linear group. □

In view of the irreducibility concept introduced in Definition 14.1.3, the above theorem implies immediately the

COROLLARY 14.5.8.— *Let $\mathcal{U} \subset \mathcal{I}$ be a finite non empty set of isotropic places. Then the lattice*

$$\mathbf{G}(K) < \prod_{v\in\mathcal{U}} \mathbf{G}(K_v) \times G_{\mathcal{I}\setminus\mathcal{U}} \cong G_{\mathcal{I}}$$

is irreducible. □

At this point, we bring the additional ingredient needed to deduce Theorem 14.5.1 from Corollary 12.0.4 : it is the following exhaustion principle based on Kolmogorov's zero-one law. The point is, of course, that the set $\mathcal{B}$ below will be taken infinite.

PROPOSITION 14.5.9.— *Let $\mathcal{B} \subset \mathcal{I}$ be a set of places with $\mathcal{B} \cap \mathcal{V}_\infty = \varnothing$. Then $\mathrm{H}^2_{\mathrm{cb}}(G_\mathcal{B}) = 0$.*

PROOF. Let $\mathbf{P}$ be a minimal K-parabolic subgroup of $\mathbf{G}$ and define for $\mathcal{U} \subset \mathcal{V}$ the product

$$S_\mathcal{U} = \prod_{v \in \mathcal{U}} \mathbf{G}(K_v)/\mathbf{P}(K_v).$$

We insist that contrary to the definition of $\mathbf{G}(\mathbf{A}_K)$ and thus of $G_\mathcal{U}$, the product space $S_\mathcal{U}$ is an unrestricted product. The $G_\mathcal{B}$-action on $S_\mathcal{B}$ is transitive because $G_\mathcal{B}$ contains the unrestricted product of a choice of maximal compact subgroups in each of the $\mathbf{G}(K_v)$ as v ranges over $\mathcal{B}$. Thus $S_\mathcal{B}$ is a homogeneous $G_\mathcal{B}$-space with amenable isotropy groups, and hence the action is amenable by Theorem 5.3.9. Therefore, Theorem 7.5.3 tells us that any class in $\mathrm{H}^2_{\mathrm{cb}}(G_\mathcal{B})$ is given by an alternating measurable bounded $G_\mathcal{B}$-invariant cocycle

$$\omega : S_\mathcal{B} \times S_\mathcal{B} \times S_\mathcal{B} \longrightarrow \mathbf{R}.$$

For every *finite* subset $\mathcal{F} \subset \mathcal{B}$, we have $\mathrm{H}^2_{\mathrm{cb}}(G_\mathcal{F}) = 0$. Indeed, the product formula of Corollary 12.0.4 decomposes this space as the direct sum of the local terms $\mathrm{H}^2_{\mathrm{cb}}(\mathbf{G}(K_v))$ over $v \in \mathcal{F}$, and we have seen in Example 9.3.5 that the latter terms vanish since $\mathcal{B} \cap \mathcal{V}_\infty = \varnothing$.

By the functoriality statement of Proposition 8.4.2, we may realize the restriction map

$$\mathrm{H}^2_{\mathrm{cb}}(G_\mathcal{B}) \longrightarrow \mathrm{H}^2_{\mathrm{cb}}(G_\mathcal{F}) = 0$$

associated to $G_\mathcal{F} \to G_\mathcal{B}$ by the inclusion

$$L^\infty_{\mathrm{alt}}(S^3_\mathcal{B})^{G_\mathcal{B}} \longrightarrow L^\infty_{\mathrm{alt}}(S^3_\mathcal{B})^{G_\mathcal{F}},$$

so that there is a bounded $G_\mathcal{F}$-invariant measurable function

$$\alpha_\mathcal{F} : S_\mathcal{B} \times S_\mathcal{B} \longrightarrow \mathbf{R}$$

with $d\alpha_\mathcal{F} = \omega$. We claim that $\alpha_\mathcal{F}$ does not depend on the first factor in the decomposition

$$S^2_\mathcal{B} \cong S^2_\mathcal{F} \times S^2_{\mathcal{B}\setminus F}.$$

Indeed, the diagonal $G_\mathcal{F}$-action on $S^2_\mathcal{F}$ is ergodic because, due to the Bruhat decomposition (see Section 11.2), each $\mathbf{G}(K_v)$ has an orbit of full measure in

$$\big(\mathbf{G}(K_v)/\mathbf{P}(K_v)\big) \times \big(\mathbf{G}(K_v)/\mathbf{P}(K_v)\big).$$

We conclude that whenever $\mathcal{F} \subset \mathcal{B}$ is finite, ω is independent of the factor $S^3_\mathcal{F}$ of $S^3_\mathcal{B}$. In other words, ω is invariant under the cofinality equivalence relation. The Kolmogorov zero-one law states that this equivalence relation is ergodic ; therefore, the cocycle ω must be constant and hence $\omega = 0$ by alternation. □

We can now present the

END OF PROOF OF THEOREM 14.5.1. The Corollary 14.5.8 of Strong Approximation precisely states that we are in situation to apply the general Corollary 12.0.4 on irreducible lattices to the product decomposition

$$G_{\mathcal{I}} \cong \prod_{v \in \mathcal{I} \cap \mathcal{V}_\infty} \mathbf{G}(K_v) \times G_{\mathcal{I} \setminus \mathcal{V}_\infty}$$

and deduce

$$\mathrm{H}^2_{\mathrm{b}}(\mathbf{G}(K)) \cong \bigoplus_{v \in \mathcal{I} \cap \mathcal{V}_\infty} \mathrm{H}^2_{\mathrm{cb}}(\mathbf{G}(K_v)) \oplus \mathrm{H}^2_{\mathrm{cb}}(G_{\mathcal{I} \setminus \mathcal{V}_\infty}). \qquad (*)$$

We recall here that Corollary 12.0.4 was allowing for the one non compactly generated factor $G_{\mathcal{I} \setminus \mathcal{V}_\infty}$.

Now Proposition 14.5.9 shows that the last term in (*) vanishes. Moreover, as we have seen in Examples 9.3.3 and 9.3.5, we have

$$\mathrm{H}^2_{\mathrm{cb}}(\mathbf{G}(K_v)) \cong \mathrm{H}^2_{\mathrm{c}}(\mathbf{G}(K_v))$$

and the latter vanishes if $v \notin \mathcal{V}_\infty$. This proves Theorem 14.5.1 up to terms $\mathrm{H}^2_{\mathrm{cb}}(\mathbf{G}(K_v))$ associated to places $v \in \mathcal{V}_\infty \cap \mathcal{A}$. However, for such places, $\mathbf{G}(K_v)$ is compact and hence both $\mathrm{H}^2_{\mathrm{cb}}$ and $\mathrm{H}^2_{\mathrm{c}}$ vanish. □

Bibliography

– A –

[1] S. Adams *Boundary amenability for word hyperbolic groups and an application to smooth dynamics of simple groups* Topology **33** n° 4 (1994) 765–783.

[2] S. Adams *Generalities on amenable actions* unpublished notes.

[3] S. Adams, G. A. Elliott, Th. Giordano *Amenable actions of groups* Trans. AMS **344** No. 2 (1994) 803–822.

[4] V. I. Arnold *Chapitres supplémentaires de la théorie des équations différentielles ordinaires* (translated from the Russian by D. Embarek) Éd. Mir, Moscou (1980).

[5] V. I. Arnold *Geometrical methods in the theory of ordinary differential equations* (translated from the Russian by J. Szücs) Grundlehren **250** Springer-Verlag (1988).

[6] R. Azencott *Espaces de Poisson des groupes localement compacts* Lecture Notes in Mathematics **148** Springer-Verlag (1970).

– B –

[7] J. Barge, É. Ghys *Cocycles d'Euler et de Maslov* Math. Ann. **294** No. 2 (1992) 235–265.

[8] R. G. Bartle, L. M. Graves *Mappings between function spaces* Trans. AMS **72** (1952) 400–413.

[9] Ch. Bavard *Longueur stable des commutateurs* L'Enseignement Math. **37** (1991) 109–150.

[10] M. E. B. Bekka *Complemented subspaces of $L^\infty(G)$, ideals of $L^1(G)$ and amenability* Monatsh. Math. **109** No. 3 (1990) 195–203.

[11] M. E. B. Bekka *On uniqueness of invariant means* Proc. AMS **126** N° 2 (1998) 507–514.

[12] M. Bestvina, K. Fujiwara *Bounded cohomology of subgroups of mapping class groups* preprint (2000).

[13] Ph. Blanc *Sur la cohomologie continue des groupes localement compacts* Ann. Sci. École Norm. Sup. (4) **12** No. 2 (1979) 137–168.

[14] S. Bloch *Applications of the dilogarithm function in algebraic K-theory and algebraic geometry* Proc. Int. Symp. on Algebraic Geometry (Kyoto 1977) 103–114 Kinokuniya Book Store, Tokyo (1978).

[15] A. Borel *Sur la cohomologie des espaces fibrés principaux et des espaces homogènes de groupes de Lie compacts* Ann. of Math. (2) **57** (1953) 115–207.

[16] A. Borel *Œuvres, volume III (1969-1982)* Springer-Verlag (1983).

[17] A. Borel *Stable real cohomology of arithmetic groups* Ann. Sci. École Norm. Sup. (4) **7** (1974) 235–272.

[18] A. Borel, J.-P. Serre *Corners and arithmetic groups* Comment. Math. Helv. **48** (1973, 436–491.

[19] A. Borel, N. R. Wallach *Continuous cohomology, discrete subgroups, and representations of reductive groups* Annals of Mathematics Studies **94** (1980).

[20] A. Borel, J. Yang *The rank conjecture for number fields* Math. Res. Lett. **1** n° 6 (1994) 689–699.

[21] R. Bott, L. W. Tu *Differential forms in algebraic topology* Graduate Texts in Mathematics **82** Springer-Verlag (1982).

[22] A. Bouarich *Suites exactes en cohomologie bornée réelle des groupes discrets* C. R. Acad. Sci. Paris Série I **320** No. 11 (1995) 1355–1359.

[23] A. Bouarich *Suite spectrale de Hochschild-Serre en cohomologie bornée* Thèse de l'Université Paul Sabatier, Toulouse.

[24] N. Bourbaki *Intégration I à IV* Éléments de mathématique Actualits Sci. Ind. **1175** Hermann, Paris (1952).

[25] N. Bourbaki *Intégration VI* Éléments de mathématique Actualits Sci. Ind. **1281** Hermann, Paris (1959).

[26] N. Bourbaki *Intégration VII et VIII* Éléments de mathématique Actualits Sci. Ind. **1306** Hermann, Paris (1963).

[27] N. Bourbaki *Topologie générale 1 à 4* Éléments de mathématique Hermann, Paris (1971).

[28] N. Bourbaki *Topologie générale 5 à 10 (nouvelle édition)* Éléments de mathématique Hermann, Paris (1974).

[29] N. Bourbaki *Algèbre 10 (algèbre homologique)* Éléments de mathématique Masson, Paris (1980).

[30] N. Bourbaki *Éspaces vectoriels topologiques (nouvelle édition)* Éléments de mathématique Masson, Paris (1981).

[31] R. Brooks *Some remarks on bounded cohomology* Riemann surfaces and related topics Proc. 1978 Stony Brook Conf., Ann. Math. Stud. **97** (1981) 53–63.

[32] K. S. Brown *Cohomology of groups* (second print) Graduate Texts in Mathematics **87** Springer-Verlag (1994).

[33] F. Bruhat *Sur les représentations induites des groupes de Lie* Bull. Soc. Math. de France **84** (1956) 97–205.

[34] M. Burger, A. Iozzi to appear (Springer-Verlag) in the proceedings of the EuroWorkshop on *Rigidity in Dynamics and Geometry* held at the Isaac Newton Institute for Mathematical Sciences, Cambridge (UK) in March 2000.

[35] M. Burger, N. Monod *Bounded cohomology of lattices in higher rank Lie groups* J. Eur. Math. Soc. **1** N° 2 (1999) 199–235.

[36] M. Burger, N. Monod Erratum to the above J. Eur. Math. Soc. **1** N° 3 (1999) 338.

[37] M. Burger, N. Monod *Continuous bounded cohomology and applications to rigidity theory* preprint (2000) to appear.

[38] M. Burger, N. Monod to appear (Springer-Verlag) in the proceedings of the EuroWorkshop on *Rigidity in Dynamics and Geometry* held at the Isaac Newton Institute for Mathematical Sciences, Cambridge (UK) in March 2000.

[39] M. Burger, Sh. Mozes *CAT(-1)-spaces, divergence groups and their commensurators* J. AMS **9** n° 1 (1996) 57–93.

[40] M. Burger, Sh. Mozes *Finitely presented simple groups and products of trees*. C. R. Acad. Sci., Paris, Ser. I **324** No. 7 (1997) 747–752.

[41] M. Burger, Sh. Mozes *Groups acting on trees : from local to global structure* preprint IHÉS M/99/15 (1999).

[42] M. Burger, Sh. Mozes *Lattices in products of trees* preprint IHÉS M/99/37 (1999).

– C –

[43] A. Colojoară *The exterior algebra of certain Banach spaces* Bull. Math. Soc. Sci. Math. Roumanie **36(84)** Nr. 1 (1992) 31–38.

– D –

[44] J. Diestel, J. Uhl *Vector measures* Mathematical Surveys of the AMS No. **15** (1977).

[45] A. Domic, D. Toledo *Gromov norm of the Kähler class of symmetric domains* Math. Ann. **276** (1987) 425–432.

[46] R. S. Doran, J. M. G. Fell *Representations of *-algebras, locally compact groups, and Banach *-algebraic bundles* Vol. 1 : Basic representation theory of groups and algebras Pure and Applied Mathematics **125** Academic Press, Boston (MA) (1988).

[47] R. S. Doran, J. Wichmann *Approximate identities and factorization in Banach modules* Lecture Notes in Mathematics **768** Springer-Verlag (1979).

[48] N. Dunford, J. T. Schwartz *Linear operators, I : General theory* Pure and Appl. Math. Vol. VII Interscience Publishers, New York, London (1958).

[49] J. L. Dupont *Bounds for characteristic numbers of flat bundles* In : Algebraic topology, Aarhus 1978 Lecture Notes in Mathematics **763** Springer-Verlag (1979).

[50] J. L. Dupont *The dilogarithm as a characteristic class for flat bundles* Proceedings of the Northwestern conference on cohomology of groups J. Pure Appl. Algebra **44** n° 1–3 (1987) 137–164.

[51] J. L. Dupont, A. Guichardet *À propos de l'article "sur la cohomologie réelle des groupes de Lie simples réels"* Ann. Sci. École Norm. Sup. (4) **11** (1978) 293–295.

– E –

[52] B. Eckmann *Cohomology of groups and transfer* Ann. of Math. **58** (1953) 481–493.

[53] D. B. A. Epstein, K. Fujiwara *The second bounded cohomology of word-hyperbolic groups* Topology **36** No. 6 (1997) 1275–1289.

[54] P. Eymard *Moyennes invariantes et représentations unitaires* Lecture Notes in Mathematics **300** Springer-Verlag (1972)

– F –

[55] B. Farb, H. Masur *Superrigidity and mapping class groups* Topology **37** n° 6 (1998) 1169–1176.

[56] K. Fujiwara *The second bounded cohomology of a group acting on a Gromov-hyperbolic space* Proc. Lond. Math. Soc., III **76** n° 1 (1998) 70–94.

[57] K. Fujiwara *The second bounded cohomology of an amalgamated free product of groups* Trans. AMS **352** n° 3 (2000) 1113–1129.

[58] H. Furstenberg *Random walks and discrete subgroups of Lie groups* In : Advances in Probability and Related Topics, Vol. 1 1–63 Dekker, New York (1971).

[59] H. Furstenberg *Boundary theory and stochastic processes on homogeneous spaces* In : Harmonic Analysis on homogeneous Spaces, Proc. Sympos. pure Math. 26, Williamstown (1972) 193–229.

– G –

[60] I. M. Gelfand, Yu. Manin *Methods of homological algebra* Springer-Verlag (1996).

[61] I. M. Gelfand, R. D. MacPherson *Geometry in Grassmannians and a generalization of the dilogarithm* Adv. Math. **44** (1982) 279–312.

[62] S. M. Gersten *Bounded cocycles and combings of groups* Internat. J. Algebra Comput. **2** n° 3 (1992) 307–326.

[63] É. Ghys *Groupes d'homéomorphismes du cercle et cohomologie bornée* In : The Lefschetz centennial conference, Contemporary math. **58** part III (1987) 81–106.

[64] É. Ghys *Actions de réseaux sur le cercle* Invent. Math. **137** No. 1 (1999) 199–231.

[65] I. Glicksberg, K. de Leeuw *The decomposition of certain group representations* J. d'analyse math. **15** (1965) 135–192.

[66] A. B. Goncharov *Chow polylogarithms and regulators* Math. Res. Lett. **2** No. 1 (1995) 95–112.

[67] A. B. Goncharov *Geometry of configurations, polylogarithms, and motivic cohomology* Adv. Math. **114** No. 2 (1995) 197–318.

[68] F. P. Greenleaf *Invariant means on topological groups.* Van Nostrand (1969).

[69] M. Gromov *Volume and bounded cohomology* Publ. math. de l'IHÉS **56** (1982) 5–100.

[70] M. Gromov *Asymptotic invariants of infinite groups* In : Geometric group theory (Vol. 2) London Math. Soc. Lecture Notes Series **182** Cambridge University Press, Cambridge (1993).

[71] M. Gromov *Spaces and questions* preprint (1999).

[72] A. Grothendieck *Produits tensoriels topologiques et espaces nucléaires* Mem. AMS **16** (1955).

[73] A. Grothendieck *Une caractérisation vectorielle-métrique des espaces L^1* Canad. J. Math. **7** (1955) 552–561.

[74] A. Guichardet *Cohomologie des groupes topologiques et des algèbres de Lie* Textes Mathématiques **2** Fernand Nathan, Paris (1980).

[75] A. Guichardet, D. Wigner *Sur la cohomologie réelle des groupes de Lie simples réels* Ann. Sci. École Norm. Sup. (4) **11** (1978) 277–292.

– H –

[76] J. L. Harer *Stability of the homology of the mapping class groups of orientable surfaces* Ann. of Math. (2) **121** n° 2 (1985) 215–249.

[77] J. L. Harer *The virtual cohomological dimension of the mapping class group of an orientable surface* Invent. Math. **84** n° 1 (1986) 157–176.

[78] P. de la Harpe, A. Valette *La propriété (T) de Kazhdan pour les groupes localement compacts* (Avec un appendice de M. Burger) Astérisque N° **175** (1989).

[79] W. J. Harvey *Boundary structure of the modular group* in : Riemann surfaces and related topics, Proceedings of the 1978 Stony Brook Conference 245–251 Ann. of Math. Stud.**97** 245–251 Princeton Univ. Press (1981).

[80] A. Ya. Helemskiĭ *The homology of Banach and topological algebras* (translated from the Russian by A. West) Mathematics and its Applications (Soviet Series) **41** Kluwer, Dordrecht (1989).

[81] A. Ya. Helemskiĭ *Banach and locally convex algebras* (translated from the Russian by A. West) Clarendon Press, Oxford (1993).

[82] G. Hochschild *Relative homological algebra* Trans. AMS **82** (1956) 246–269.

[83] G. Hochschild, G. D. Mostow *Cohomology of Lie groups* Illinois J. Math. **6** (1962) 367–401.

[84] E. Hewitt, K. A. Ross *Abstract harmonic analysis, Vol. I (second edition)* Grundlehren **115** Springer-Verlag (1979).

– I –

[85] A. Iozzi, A. Nevo *Algebraic hulls and the Foelner property* GAFA **6** No. 4 (1996) 666–688.

[86] N. V. Ivanov *Foundations of the theory of bounded cohomology* J. of Soviet Mathematics **37** No. 1 (1987) 1090–1115.

[87] N. V. Ivanov, V. G. Turaev *A canonical cocycle for the Euler class of a flat vector bundle* Soviet Math. Dokl. **26** No. 1 (1982) 78–81.

– J –

[88] H. Jarchow *Locally convex spaces* Math. Leitfäden, B. G. Teubner, Stuttgart (1981).

[89] B. E. Johnson *Cohomology in Banach algebras* Mem. AMS **127** (1972).

– K –

[90] Sh. Kakutani, K. Kodaira *Über das Haarsche Mass in der lokal bikompakten Gruppe* Proc. Imp. Acad. of Japan **20** N° 7 (1944) 444–450.

[91] J. L. Kelley *General topology* Van Nostrand, Princeton (1955).

[92] J. L. Kelley, T. P. Srinivasan *Measure and integral* Graduate Texts in Math. **116** Springer-Verlag (1988).

[93] S. P. Kerckhoff *The Nielsen realization problem* Ann. of Math. **117** n° 2 (1983) 235–265.

[94] J.-L. Koszul *Lectures on groups of transformations* Notes by R.R. Simha and R. Sridharan Tata Institute of Fundamental Research Lectures on Mathematics, No. **32** Tata Institute of Fundamental Research, Bombay (1965).

– L –

[95] L. Lewin *The evolution of the ladder concept* In : Structural properties of polylogarithms Mathematical Surveys and Monographs **37** (1991) 1–10.

[96] G. Lion, M. Vergne *The Weil representation, Maslov index and Theta series* Progress in Math. **6** Birkhäuser, Basel (1980).

[97] A. Lubotzky *Discrete groups, expanding graphs and invariant measures* (With an appendix by J.D. Rogawski) Progress in Math. **125** Birkhäuser, Basel (1994).

[98] A. Lubotzky, Sh. Mozes *Asymptotic properties of unitary representations of tree automorphisms* In : Harmonic analysis and discrete potential theory Plenum, New York (1992).

[99] A. Lubotzky, Sh. Mozes, M. S. Raghunathan *Cyclic subgroups of exponential growth and metrics on discrete groups* C. R. Acad. Sci., Paris, Ser. I **317** No. 8 (1993) 735–740.

[100] Yu. I. Lyubich *Introduction to the theory of Banach representations of groups* (translated from the Russian by A. Jacob) Operator Theory : Advances and Applications **30** Birkhäuser, Basel (1988).

– M –

[101] G. W. Mackey *Point realizations of transformation groups* Ill. J. Math. **6** (1962) 327–335.

[102] G. A. Margulis *Some remarks on invariant means* Monatsh. Math. **90** N° 3 (1980) 233–235.

[103] G. A. Margulis *Discrete subgroups of semisimple Lie groups* Ergebnisse der Mathematik und ihrer Grenzgebiete (3) **17** Springer-Verlag (1991).

[104] H. A. Masur, Y. N. Minsky *Geometry of the complex of curves. I. Hyperbolicity* Invent. Math. **138** n° 1 (1999) 103–149.

[105] S. Matsumoto, S. Morita *Bounded cohomology of certain groups of homeomorphisms* Proc. AMS **94** No. 3 (1985) 539–544.

[106] S. Mazur, S. Ulam *Sur les transformations isométriques d'espaces vectoriels normés* C. R. Acad. Sci. Paris **194** (1932) 946–948.

[107] J. McCarthy, A. Papadopoulos *Dynamics on Thurston's sphere of projective measured foliations* Comment. Math. Helv. **64** n° 1 (1989) 133–166.

[108] E. Michael *A note on paracompact spaces* Proc. AMS **4** (1953) 831–838.

[109] E. Michael *Continuous selections. I* Ann. of Math. **63** (1956) 361–382.

[110] E. Michael *Another note on paracompact spaces* Proc. AMS **8** (1957) 822–828.

[111] I. Mineyev *Straightening and bounded cohomology of hyperbolic groups* preprint (1998).

[112] I. Mineyev *Bounded cohomology characterizes hyperbolic groups* preprint (1999).

[113] N. Monod in preparation.

[114] D. Montgomery, L. Zippin *Topological transformation groups* Interscience Publishers, New York-London (1955).

[115] C. C. Moore *On the Frobenius reciprocity theorem for locally compact groups* Pacific J. Math. **12** 359–365 (1962).

[116] G. D. Mostow *Cohomology of topological groups and solvmanifolds* Ann. of Math. (2) **73** 20–48 (1961).

– N –

[117] G. A. Noskov *Bounded cohomology of discrete groups with coefficients* Leningrad Math J. **2** No. 5 (1991) 1067–1084.

[118] G. A. Noskov *The Hochschild-Serre spectral sequence for bounded cohomology* Proceedings of the International Conference on Algebra (1989) Contemp. Math. **131** Part 1 (1992) 613–629.

– O –

[119] A. Yu. Olshanskiĭ *Almost every group is hyperbolic* Internat. J. Algebra Comput. **2** No. 1 (1992) 1–17.

[120] A. L. Onishchik, E. B. Vinberg *Lie groups and algebraic groups* (translated from the Russian by D.A. Leites) Springer Series in Soviet Mathematics Springer-Verlag (1990).

– P –

[121] D. W. Paul *Theory of bounded groups and their bounded cohomology* Pac. J. Math. **134** No. 2 (1988) 313–324.

[122] J.-P. Pier *Amenable locally compact groups* Pure and Applied Mathematics Wiley-Interscience, New York (1984).

– R –

[123] H. Reiter *Classical harmonic analysis and locally compact groups* Clarendon Press, Oxford (1968).

[124] H. L. Royden *Automorphisms and isometries of Teichmller space* in : Advances in the Theory of Riemann Surfaces Proc. Conf., Stony Brook, N.Y., 1969 Ann. of Math. Studies **66** Princeton Univ. Press, Princeton (1971).

[125] W. Rudin *Invariant means on L^∞* Studia Math. **44** (1972) 219–227.

[126] W. Rudin *Functional analysis (second edition)* McGraw-Hill (1991).

– S –

[127] R. P. Savage *The space of positive definite matrices and Gromov's invariant* Trans. AMS **274** No. 1 (1982) 239–263.

[128] Z. Sela *Uniform embeddings of hyperbolic groups in Hilbert spaces* Isr. J. Math. **80** No 1–2 (1992) 171–181.

[129] J.-P. Serre *Arbres, amalgames,* SL_2 (Rédigé avec la collaboration de Hyman Bass) Astérisque **46** Société Math. de France, Paris (1977).

[130] Y. Shalom *Rigidity of commensurators and irreducible lattices* Invent. math. **141** n° 1 (2000) 1–54.

[131] M. Simonnet *Measures and probabilities* Universitext, Springer-Verlag (1996).

[132] T. Soma *The zero-norm subspace of bounded cohomology* Comment. Math. Helv. **72** N° 4 (1997) 582–592.

[133] T. Soma *Existence of non-Banach bounded cohomology* Topology **37** N° 1 (1998) 179–193.

[134] D. Sullivan *For $n > 3$ there is only one finitely additive rotationally invariant measure on the n-sphere defined on all Lebesgue measurable subsets* Bull. AMS **4** N° 1 (1981) 121–123.

[135] R. M. Switzer *Algebraic topology — homotopy and homology* Grundlehren **212** Springer-Verlag (1975).

– T –

[136] W. P. Thurston *A generalization of the Reeb stability theorem* Topology **3** No. 4 (1974) 214–231.

– W –

[137] Ch. A. Weibel *An introduction to homological algebra* Cambridge Studies in Advanced Mathematics **38** Cambridge University Press, Cambridge (1994).

[138] D. Witte *Products of similar matrices* Proc. AMS **126** No. 4 (1998) 1005–1015.

– Z –

[139] R. J. Zimmer *On the von Neumann algebra of an ergodic group action* Proc. AMS **66** No. 2 (1977) 289–293.

[140] R. J. Zimmer *Amenable ergodic group actions and an application to Poisson boundaries of random walks* J. Funct. Anal. **27** No. 3 (1978) 350–372.

[141] R. J. Zimmer *Ergodic theory and semisimple groups* Monographs in Math. **81** Birkhäuser, Basel (1984).

[142] R. J. Zimmer *Amenable actions and dense subgroups of Lie groups* J. Funct. Anal. **72** n° 1 (1987) 58–64.

[143] A. Żuk *Property (T) and Kazhdan constants for discrete groups* preprint (1999).

Index

Lecture Notes in Mathematics

For information about Vols. 1–1580
please contact your bookseller or Springer-Verlag

Vol. 1581: D. Bakry, R. D. Gill, S. A. Molchanov, Lectures on Probability Theory. Editor: P. Bernard. VIII, 420 pages. 1994.

Vol. 1582: W. Balser, From Divergent Power Series to Analytic Functions. X, 108 pages. 1994.

Vol. 1583: J. Azéma, P. A. Meyer, M. Yor (Eds.), Séminaire de Probabilités XXVIII. VI, 334 pages. 1994.

Vol. 1584: M. Brokate, N. Kenmochi, I. Müller, J. F. Rodriguez, C. Verdi, Phase Transitions and Hysteresis. Montecatini Terme, 1993. Editor: A. Visintin. VII, 291 pages. 1994.

Vol. 1585: G. Frey (Ed.), On Artin's Conjecture for Odd 2-dimensional Representations. VIII, 148 pages. 1994.

Vol. 1586: R. Nillsen, Difference Spaces and Invariant Linear Forms. XII, 186 pages. 1994.

Vol. 1587: N. Xi, Representations of Affine Hecke Algebras. VIII, 137 pages. 1994.

Vol. 1588: C. Scheiderer, Real and Étale Cohomology. XXIV, 273 pages. 1994.

Vol. 1589: J. Bellissard, M. Degli Esposti, G. Forni, S. Graffi, S. Isola, J. N. Mather, Transition to Chaos in Classical and Quantum Mechanics. Montecatini Terme, 1991. Editor: 2S. Graffi. VII, 192 pages. 1994.

Vol. 1590: P. M. Soardi, Potential Theory on Infinite Networks. VIII, 187 pages. 1994.

Vol. 1591: M. Abate, G. Patrizio, Finsler Metrics – A Global Approach. IX, 180 pages. 1994.

Vol. 1592: K. W. Breitung, Asymptotic Approximations for Probability Integrals. IX, 146 pages. 1994.

Vol. 1593: J. Jorgenson & S. Lang, D. Goldfeld, Explicit Formulas for Regularized Products and Series. VIII, 154 pages. 1994.

Vol. 1594: M. Green, J. Murre, C. Voisin, Algebraic Cycles and Hodge Theory. Torino, 1993. Editors: A. Albano, F. Bardelli. VII, 275 pages. 1994.

Vol. 1595: R.D.M. Accola, Topics in the Theory of Riemann Surfaces. IX, 105 pages. 1994.

Vol. 1596: L. Heindorf, L. B. Shapiro, Nearly Projective Boolean Algebras. X, 202 pages. 1994.

Vol. 1597: B. Herzog, Kodaira-Spencer Maps in Local Algebra. XVII, 176 pages. 1994.

Vol. 1598: J. Berndt, F. Tricerri, L. Vanhecke, Generalized Heisenberg Groups and Damek-Ricci Harmonic Spaces. VIII, 125 pages. 1995.

Vol. 1599: K. Johannson, Topology and Combinatorics of 3-Manifolds. XVIII, 446 pages. 1995.

Vol. 1600: W. Narkiewicz, Polynomial Mappings. VII, 130 pages. 1995.

Vol. 1601: A. Pott, Finite Geometry and Character Theory. VII, 181 pages. 1995.

Vol. 1602: J. Winkelmann, The Classification of Three-dimensional Homogeneous Complex Manifolds. XI, 230 pages. 1995.

Vol. 1603: V. Ene, Real Functions – Current Topics. XIII, 310 pages. 1995.

Vol. 1604: A. Huber, Mixed Motives and their Realization in Derived Categories. XV, 207 pages. 1995.

Vol. 1605: L. B. Wahlbin, Superconvergence in Galerkin Finite Element Methods. XI, 166 pages. 1995.

Vol. 1606: P.-D. Liu, M. Qian, Smooth Ergodic Theory of Random Dynamical Systems. XI, 221 pages. 1995.

Vol. 1607: G. Schwarz, Hodge Decomposition – A Method for Solving Boundary Value Problems. VII, 155 pages. 1995.

Vol. 1608: P. Biane, R. Durrett, Lectures on Probability Theory. Editor: P. Bernard. VII, 210 pages. 1995.

Vol. 1609: L. Arnold, C. Jones, K. Mischaikow, G. Raugel, Dynamical Systems. Montecatini Terme, 1994. Editor: R. Johnson. VIII, 329 pages. 1995.

Vol. 1610: A. S. Üstünel, An Introduction to Analysis on Wiener Space. X, 95 pages. 1995.

Vol. 1611: N. Knarr, Translation Planes. VI, 112 pages. 1995.

Vol. 1612: W. Kühnel, Tight Polyhedral Submanifolds and Tight Triangulations. VII, 122 pages. 1995.

Vol. 1613: J. Azéma, M. Emery, P. A. Meyer, M. Yor (Eds.), Séminaire de Probabilités XXIX. VI, 326 pages. 1995.

Vol. 1614: A. Koshelev, Regularity Problem for Quasilinear Elliptic and Parabolic Systems. XXI, 255 pages. 1995.

Vol. 1615: D. B. Massey, Le Cycles and Hypersurface Singularities. XI, 131 pages. 1995.

Vol. 1616: I. Moerdijk, Classifying Spaces and Classifying Topoi. VII, 94 pages. 1995.

Vol. 1617: V. Yurinsky, Sums and Gaussian Vectors. XI, 305 pages. 1995.

Vol. 1618: G. Pisier, Similarity Problems and Completely Bounded Maps. VII, 156 pages. 1996.

Vol. 1619: E. Landvogt, A Compactification of the Bruhat-Tits Building. VII, 152 pages. 1996.

Vol. 1620: R. Donagi, B. Dubrovin, E. Frenkel, E. Previato, Integrable Systems and Quantum Groups. Montecatini Terme, 1993. Editors:M. Francaviglia, S. Greco. VIII, 488 pages. 1996.

Vol. 1621: H. Bass, M. V. Otero-Espinar, D. N. Rockmore, C. P. L. Tresser, Cyclic Renormalization and Auto-morphism Groups of Rooted Trees. XXI, 136 pages. 1996.

Vol. 1622: E. D. Farjoun, Cellular Spaces, Null Spaces and Homotopy Localization. XIV, 199 pages. 1996.

Vol. 1623: H.P. Yap, Total Colourings of Graphs. VIII, 131 pages. 1996.

Vol. 1624: V. Brınzanescu, Holomorphic Vector Bundles over Compact Complex Surfaces. X, 170 pages. 1996.

Vol.1625: S. Lang, Topics in Cohomology of Groups. VII, 226 pages. 1996.

Vol. 1626: J. Azéma, M. Emery, M. Yor (Eds.), Séminaire de Probabilités XXX. VIII, 382 pages. 1996.

Vol. 1627: C. Graham, Th. G. Kurtz, S. Méléard, Ph. E. Protter, M. Pulvirenti, D. Talay, Probabilistic Models for Nonlinear Partial Differential Equations. Montecatini Terme, 1995. Editors: D. Talay, L. Tubaro. X, 301 pages. 1996.

Vol. 1628: P.-H. Zieschang, An Algebraic Approach to Association Schemes. XII, 189 pages. 1996.

Vol. 1629: J. D. Moore, Lectures on Seiberg-Witten Invariants. VII, 105 pages. 1996.

Vol. 1630: D. Neuenschwander, Probabilities on the Heisenberg Group: Limit Theorems and Brownian Motion. VIII, 139 pages. 1996.

Vol. 1631: K. Nishioka, Mahler Functions and Transcendence.VIII, 185 pages.1996.

Vol. 1632: A. Kushkuley, Z. Balanov, Geometric Methods in Degree Theory for Equivariant Maps. VII, 136 pages. 1996.

Vol.1633: H. Aikawa, M. Essén, Potential Theory – Selected Topics. IX, 200 pages.1996.

Vol. 1634: J. Xu, Flat Covers of Modules. IX, 161 pages. 1996.

Vol. 1635: E. Hebey, Sobolev Spaces on Riemannian Manifolds. X, 116 pages. 1996.

Vol. 1636: M. A. Marshall, Spaces of Orderings and Abstract Real Spectra. VI, 190 pages. 1996.

Vol. 1637: B. Hunt, The Geometry of some special Arithmetic Quotients. XIII, 332 pages. 1996.

Vol. 1638: P. Vanhaecke, Integrable Systems in the realm of Algebraic Geometry. VIII, 218 pages. 1996.

Vol. 1639: K. Dekimpe, Almost-Bieberbach Groups: Affine and Polynomial Structures. X, 259 pages. 1996.

Vol. 1640: G. Boillat, C. M. Dafermos, P. D. Lax, T. P. Liu, Recent Mathematical Methods in Nonlinear Wave Propagation. Montecatini Terme, 1994. Editor: T. Ruggeri. VII, 142 pages. 1996.

Vol. 1641: P. Abramenko, Twin Buildings and Applications to S-Arithmetic Groups. IX, 123 pages. 1996.

Vol. 1642: M. Puschnigg, Asymptotic Cyclic Cohomology. XXII, 138 pages. 1996.

Vol. 1643: J. Richter-Gebert, Realization Spaces of Polytopes. XI, 187 pages. 1996.

Vol. 1644: A. Adler, S. Ramanan, Moduli of Abelian Varieties. VI, 196 pages. 1996.

Vol. 1645: H. W. Broer, G. B. Huitema, M. B. Sevryuk, Quasi-Periodic Motions in Families of Dynamical Systems. XI, 195 pages. 1996.

Vol. 1646: J.-P. Demailly, T. Peternell, G. Tian, A. N. Tyurin, Transcendental Methods in Algebraic Geometry. Cetraro, 1994. Editors: F. Catanese, C. Ciliberto. VII, 257 pages. 1996.

Vol. 1647: D. Dias, P. Le Barz, Configuration Spaces over Hilbert Schemes and Applications. VII. 143 pages. 1996.

Vol. 1648: R. Dobrushin, P. Groeneboom, M. Ledoux, Lectures on Probability Theory and Statistics. Editor: P. Bernard. VIII, 300 pages. 1996.

Vol. 1649: S. Kumar, G. Laumon, U. Stuhler, Vector Bundles on Curves – New Directions. Cetraro, 1995. Editor: M. S. Narasimhan. VII, 193 pages. 1997.

Vol. 1650: J. Wildeshaus, Realizations of Polylogarithms. XI, 343 pages. 1997.

Vol. 1651: M. Drmota, R. F. Tichy, Sequences, Discrepancies and Applications. XIII, 503 pages. 1997.

Vol. 1652: S. Todorcevic, Topics in Topology. VIII, 153 pages. 1997.

Vol. 1653: R. Benedetti, C. Petronio, Branched Standard Spines of 3-manifolds. VIII, 132 pages. 1997.

Vol. 1654: R. W. Ghrist, P. J. Holmes, M. C. Sullivan, Knots and Links in Three-Dimensional Flows. X, 208 pages. 1997.

Vol. 1655: J. Azéma, M. Emery, M. Yor (Eds.), Séminaire de Probabilités XXXI. VIII, 329 pages. 1997.

Vol. 1656: B. Biais, T. Björk, J. Cvitanic, N. El Karoui, E. Jouini, J. C. Rochet, Financial Mathematics. Bressanone, 1996. Editor: W. J. Runggaldier. VII, 316 pages. 1997.

Vol. 1657: H. Reimann, The semi-simple zeta function of quaternionic Shimura varieties. IX, 143 pages. 1997.

Vol. 1658: A. Pumarino, J. A. Rodrıguez, Coexistence and Persistence of Strange Attractors. VIII, 195 pages. 1997.

Vol. 1659: V, Kozlov, V. Maz'ya, Theory of a Higher-Order Sturm-Liouville Equation. XI, 140 pages. 1997.

Vol. 1660: M. Bardi, M. G. Crandall, L. C. Evans, H. M. Soner, P. E. Souganidis, Viscosity Solutions and Applications. Montecatini Terme, 1995. Editors: I. Capuzzo Dolcetta, P. L. Lions. IX, 259 pages. 1997.

Vol. 1661: A. Tralle, J. Oprea, Symplectic Manifolds with no Kähler Structure. VIII, 207 pages. 1997.

Vol. 1662: J. W. Rutter, Spaces of Homotopy Self-Equivalences – A Survey. IX, 170 pages. 1997.

Vol. 1663: Y. E. Karpeshina; Perturbation Theory for the Schrödinger Operator with a Periodic Potential. VII, 352 pages. 1997.

Vol. 1664: M. Väth, Ideal Spaces. V, 146 pages. 1997.

Vol. 1665: E. Giné, G. R. Grimmett, L. Saloff-Coste, Lectures on Probability Theory and Statistics 1996. Editor: P. Bernard. X, 424 pages, 1997.

Vol. 1666: M. van der Put, M. F. Singer, Galois Theory of Difference Equations. VII, 179 pages. 1997.

Vol. 1667: J. M. F. Castillo, M. González, Three-space Problems in Banach Space Theory. XII, 267 pages. 1997.

Vol. 1668: D. B. Dix, Large-Time Behavior of Solutions of Linear Dispersive Equations. XIV, 203 pages. 1997.

Vol. 1669: U. Kaiser, Link Theory in Manifolds. XIV, 167 pages. 1997.

Vol. 1670: J. W. Neuberger, Sobolev Gradients and Differential Equations. VIII, 150 pages. 1997.

Vol. 1671: S. Bouc, Green Functors and *G*-sets. VII, 342 pages. 1997.

Vol. 1672: S. Mandal, Projective Modules and Complete Intersections. VIII, 114 pages. 1997.

Vol. 1673: F. D. Grosshans, Algebraic Homogeneous Spaces and Invariant Theory. VI, 148 pages. 1997.

Vol. 1674: G. Klaas, C. R. Leedham-Green, W. Plesken, Linear Pro-*p*-Groups of Finite Width. VIII, 115 pages. 1997.

Vol. 1675: J. E. Yukich, Probability Theory of Classical Euclidean Optimization Problems. X, 152 pages. 1998.

Vol. 1676: P. Cembranos, J. Mendoza, Banach Spaces of Vector-Valued Functions. VIII, 118 pages. 1997.

Vol. 1677: N. Proskurin, Cubic Metaplectic Forms and Theta Functions. VIII, 196 pages. 1998.

Vol. 1678: O. Krupková, The Geometry of Ordinary Variational Equations. X, 251 pages. 1997.

Vol. 1679: K.-G. Grosse-Erdmann, The Blocking Technique. Weighted Mean Operators and Hardy's Inequality. IX, 114 pages. 1998.

Vol. 1680: K.-Z. Li, F. Oort, Moduli of Supersingular Abelian Varieties. V, 116 pages. 1998.

Vol. 1681: G. J. Wirsching, The Dynamical System Generated by the 3n+1 Function. VII, 158 pages. 1998.

Vol. 1682: H.-D. Alber, Materials with Memory. X, 166 pages. 1998.

Vol. 1683: A. Pomp, The Boundary-Domain Integral Method for Elliptic Systems. XVI, 163 pages. 1998.

Vol. 1684: C. A. Berenstein, P. F. Ebenfelt, S. G. Gindikin, S. Helgason, A. E. Tumanov, Integral Geometry, Radon Transforms and Complex Analysis. Firenze, 1996. Editors: E. Casadio Tarabusi, M. A. Picardello, G. Zampieri. VII, 160 pages. 1998

Vol. 1685: S. König, A. Zimmermann, Derived Equivalences for Group Rings. X, 146 pages. 1998.

Vol. 1686: J. Azéma, M. Émery, M. Ledoux, M. Yor (Eds.), Séminaire de Probabilités XXXII. VI, 440 pages. 1998.

Vol. 1687: F. Bornemann, Homogenization in Time of Singularly Perturbed Mechanical Systems. XII, 156 pages. 1998.

Vol. 1688: S. Assing, W. Schmidt, Continuous Strong Markov Processes in Dimension One. XII, 137 page. 1998.

Vol. 1689: W. Fulton, P. Pragacz, Schubert Varieties and Degeneracy Loci. XI, 148 pages. 1998.

Vol. 1690: M. T. Barlow, D. Nualart, Lectures on Probability Theory and Statistics. Editor: P. Bernard. VIII, 237 pages. 1998.

Vol. 1691: R. Bezrukavnikov, M. Finkelberg, V. Schechtman, Factorizable Sheaves and Quantum Groups. X, 282 pages. 1998.

Vol. 1692: T. M. W. Eyre, Quantum Stochastic Calculus and Representations of Lie Superalgebras. IX, 138 pages. 1998.

Vol. 1694: A. Braides, Approximation of Free-Discontinuity Problems. XI, 149 pages. 1998.

Vol. 1695: D. J. Hartfiel, Markov Set-Chains. VIII, 131 pages. 1998.

Vol. 1696: E. Bouscaren (Ed.): Model Theory and Algebraic Geometry. XV, 211 pages. 1998.

Vol. 1697: B. Cockburn, C. Johnson, C.-W. Shu, E. Tadmor, Advanced Numerical Approximation of Nonlinear Hyperbolic Equations. Cetraro, Italy, 1997. Editor: A. Quarteroni. VII, 390 pages. 1998.

Vol. 1698: M. Bhattacharjee, D. Macpherson, R. G. Möller, P. Neumann, Notes on Infinite Permutation Groups. XI, 202 pages. 1998.

Vol. 1699: A. Inoue, Tomita-Takesaki Theory in Algebras of Unbounded Operators. VIII, 241 pages. 1998.

Vol. 1700: W. A. Woyczyński, Burgers-KPZ Turbulence, XI, 318 pages. 1998.

Vol. 1701: Ti-Jun Xiao, J. Liang, The Cauchy Problem of Higher Order Abstract Differential Equations. XII, 302 pages. 1998.

Vol. 1702: J. Ma, J. Yong, Forward-Backward Stochastic Differential Equations and Their Applications. XIII, 270 pages. 1999.

Vol. 1703: R. M. Dudley, R. Norvaiša, Differentiability of Six Operators on Nonsmooth Functions and p-Variation. VIII, 272 pages. 1999.

Vol. 1704: H. Tamanoi, Elliptic Genera and Vertex Operator Super-Algebras. VI, 390 pages. 1999.

Vol. 1705: I. Nikolaev, E. Zhuzhoma, Flows in 2-dimensional Manifolds. XIX, 294 pages. 1999.

Vol. 1706: S. Yu. Pilyugin, Shadowing in Dynamical Systems. XVII, 271 pages. 1999.

Vol. 1707: R. Pytlak, Numerical Methods for Optimal Control Problems with State Constraints. XV, 215 pages. 1999.

Vol. 1708: K. Zuo, Representations of Fundamental Groups of Algebraic Varieties. VII, 139 pages. 1999.

Vol. 1709: J. Azéma, M. Émery, M. Ledoux, M. Yor (Eds), Séminaire de Probabilités XXXIII. VIII, 418 pages. 1999.

Vol. 1710: M. Koecher, The Minnesota Notes on Jordan Algebras and Their Applications. IX, 173 pages. 1999.

Vol. 1711: W. Ricker, Operator Algebras Generated by Commuting Projections: A Vector Measure Approach. XVII, 159 pages. 1999.

Vol. 1712: N. Schwartz, J. J. Madden, Semi-algebraic Function Rings and Reflectors of Partially Ordered Rings. XI, 279 pages. 1999.

Vol. 1713: F. Bethuel, G. Huisken, S. Müller, K. Steffen, Calculus of Variations and Geometric Evolution Problems. Cetraro, 1996. Editors: S. Hildebrandt, M. Struwe. VII, 293 pages. 1999.

Vol. 1714: O. Diekmann, R. Durrett, K. P. Hadeler, P. K. Maini, H. L. Smith, Mathematics Inspired by Biology. Martina Franca, 1997. Editors: V. Capasso, O. Diekmann. VII, 268 pages. 1999.

Vol. 1715: N. V. Krylov, M. Röckner, J. Zabczyk, Stochastic PDE's and Kolmogorov Equations in Infinite Dimensions. Cetraro, 1998. Editor: G. Da Prato. VIII, 239 pages. 1999.

Vol. 1716: J. Coates, R. Greenberg, K. A. Ribet, K. Rubin, Arithmetic Theory of Elliptic Curves. Cetraro, 1997. Editor: C. Viola. VIII, 260 pages. 1999.

Vol. 1717: J. Bertoin, F. Martinelli, Y. Peres, Lectures on Probability Theory and Statistics. Saint-Flour, 1997. Editor: P. Bernard. IX, 291 pages. 1999.

Vol. 1718: A. Eberle, Uniqueness and Non-Uniqueness of Semigroups Generated by Singular Diffusion Operators. VIII, 262 pages. 1999.

Vol. 1719: K. R. Meyer, Periodic Solutions of the N-Body Problem. IX, 144 pages. 1999.

Vol. 1720: D. Elworthy, Y. Le Jan, X-M. Li, On the Geometry of Diffusion Operators and Stochastic Flows. IV, 118 pages. 1999.

Vol. 1721: A. Iarrobino, V. Kanev, Power Sums, Gorenstein Algebras, and Determinantal Loci. XXVII, 345 pages. 1999.

Vol. 1722: R. McCutcheon, Elemental Methods in Ergodic Ramsey Theory. VI, 160 pages. 1999.

Vol. 1723: J. P. Croisille, C. Lebeau, Diffraction by an Immersed Elastic Wedge. VI, 134 pages. 1999.

Vol. 1724: V. N. Kolokoltsov, Semiclassical Analysis for Diffusions and Stochastic Processes. VIII, 347 pages. 2000.

Vol. 1725: D. A. Wolf-Gladrow, Lattice-Gas Cellular Automata and Lattice Boltzmann Models. IX, 308 pages. 2000.

Vol. 1726: V. Marić, Regular Variation and Differential Equations. X, 127 pages. 2000.

Vol. 1727: P. Kravanja, M. Van Barel, Computing the Zeros of Analytic Functions. VII, 111 pages. 2000.

Vol. 1728: K. Gatermann, Computer Algebra Methods for Equivariant Dynamical Systems. XV, 153 pages. 2000.

Vol. 1729: J. Azéma, M. Émery, M. Ledoux, M. Yor, Séminaire de Probabilités XXXIV. VI, 431 pages. 2000.

Vol. 1730: S. Graf, H. Luschgy, Foundations of Quantization for Probability Distributions. X, 230 pages. 2000.

Vol. 1731: T. Hsu, Quilts: Central Extensions, Braid Actions, and Finite Groups,. XII, 185 pages. 2000.

Vol. 1732: K. Keller, Invariant Factors, Julia Equivalences and the (Abstract) Mandelbrot Set. X, 206 pages. 2000.

Vol. 1733: K. Ritter, Average-Case Analysis of Numerical Problems. IX, 254 pages. 2000.

Vol. 1734: M. Espedal, A. Fasano, A. Mikelić, Filtration in Porous Media and Industrial Applications. Cetraro 1998. Editor: A. Fasano. 2000.

Vol. 1735: D. Yafaev, Scattering Theory: Some Old and New Problems. XVI, 169 pages. 2000.

Vol. 1736: B. O. Turesson, Nonlinear Potential Theory and Weighted Sobolev Spaces. XIV, 173 pages. 2000.

Vol. 1737: S. Wakabayashi, Classical Microlocal Analysis in the Space of Hyperfunctions. VIII, 367 pages. 2000.

Vol. 1738: M. Émery, A. Nemirovski, D. Voiculescu, Lectures on Probability Theory and Statistics. XI, 356 pages. 2000.

Vol. 1739: R. Burkard, P. Deuflhard, A. Jameson, J.-L. Lions, G. Strang, Computational Mathematics Driven by Industrial Problems. Martina Franca, 1999. Editors: V. Capasso, H. Engl, J. Periaux. VII, 418 pages. 2000.

Vol. 1740: B. Kawohl, O. Pironneau, L. Tartar, J.-P. Zolesio, Optimal Shape Design. Tróia, Portugal 1999. Editors: A. Cellina, A. Ornelas. IX, 388 pages. 2000.

Vol. 1741: E. Lombardi, Oscillatory Integrals and Phenomena Beyond all Algebraic Orders. XV, 413 pages. 2000.

Vol. 1742: A. Unterberger, Quantization and Non-holomorphic Modular Forms. VIII, 253 pages. 2000.

Vol. 1743: L. Habermann, Riemannian Metrics of Constant Mass and Moduli Spaces of Conformal Structures. XII, 116 pages. 2000.

Vol. 1744: M. Kunze, Non-Smooth Dynamical Systems. X, 228 pages. 2000.

Vol. 1745: V. D. Milman, G. Schechtman, Geometric Aspects of Functional Analysis. VIII, 289 pages. 2000.

Vol. 1746: A. Degtyarev, I. Itenberg, V. Kharlamov, Real Enriques Surfaces. XVI, 259 pages. 2000.

Vol. 1747: L. W. Christensen, Gorenstein Dimensions. VIII, 204 pages. 2000.

Vol. 1748: M. Růžička, Electrorheological Fluids: Modeling and Mathematical Theory. XV, 176 pages. 2001.

Vol. 1749: M. Fuchs, G. Seregin, Variational Methods for Problems from Plasticity Theory and for Generalized Newtonian Fluids. VI, 269 pages. 2001.

Vol. 1750: B. Conrad, Grothendieck Duality and Base Change. X, 296 pages. 2001.

Vol. 1751: N. J. Cutland, Loeb Measures in Practice: Recent Advances. XI, 111 pages. 2001.

Vol. 1752: Y. V. Nesterenko, P. Philippon, Introduction to Algebraic Independence Theory. XIII, 256 pages. 2001.

Vol. 1753: A. I. Bobenko, U. Eitner, Painlevé Equations in the Differential Geometry of Surfaces. VI, 120 pages. 2001.

Vol. 1754: W. Bertram, The Geometry of Jordan and Lie Structures. XVI, 269 pages. 2001.

Vol. 1755: J. Azéma, M. Émery, M. Ledoux, M. Yor, Séminaire de Probabilités XXXV. VI, 427 pages. 2001.

Vol. 1756: P. E. Zhidkov, Korteweg de Vries and Nonlinear Schrödinger Equations: Qualitative Theory. VII, 147 pages. 2001.

Vol. 1757: R. R. Phelps, Lectures on Choquet's Theorem. VII, 124 pages. 2001.

Vol. 1758: N. Monod, Continuous Bounded Cohomology of Locally Compact Groups. X, 214 pages. 2001.

Vol. 1759: Y. Abe, K. Kopfermann, Toroidal Groups. VIII, 133 pages. 2001.

Vol. 1760: D. Filipović, Consistency Problems for Heath-Jarrow-Morton Interest Rate Models. VIII, 134 pages. 2001.

Vol. 1761: C. Adelmann, The Decomposition of Primes in Torsion Point Fields. VI, 142 pages. 2001.

Recent Reprints and New Editions

Vol. 1200: V. D. Milman, G. Schechtman, Asymptotic Theory of Finite Dimensional Normed Spaces - Corrected Second Printing 2001. X, 156 pages. 1986.

Vol. 1618: G. Pisier, Similarity Problems and Completely Bounded Maps - Second, Expanded Edition VII, 198 pages. 2001.

Vol. 1629: J. D. Moore. Lectures on Seiberg-Witten Invariants - Second Edition. VIII, 121 pages. 2001.

Vol. 1702: J. Ma, J. Yong. Forward-Backward Stochastic Differential Equations and Their Applications - Corrected Second Printing 2000. XIII, 270 pages. 1999.

4. Lecture Notes are printed by photo-offset from the master-copy delivered in camera-ready form by the authors. Springer-Verlag provides technical instructions for the preparation of manuscripts. Macro packages in TEX, LATEX2e, LATEX2.09 are available from Springer's web-pages at

http://www.springer.de/math/authors/b-tex.html.

Careful preparation of the manuscripts will help keep production time short and ensure satisfactory appearance of the finished book.

The actual production of a Lecture Notes volume takes approximately 12 weeks.

5. Authors receive a total of 50 free copies of their volume, but no royalties. They are entitled to a discount of 33.3 % on the price of Springer books purchase for their personal use, if ordering directly from Springer-Verlag.

Commitment to publish is made by letter of intent rather than by signing a formal contract. Springer-Verlag secures the copyright for each volume. Authors are free to reuse material contained in their LNM volumes in later publications: A brief written (or e-mail) request for formal permission is sufficient.

Addresses:

Professor J.-M. Morel
CMLA, Ecole Normale Supérieure de Cachan
61 Avenue du Président Wilson
94235 Cachan Cedex France
E-mail: Jean-Michel.Morel@cmla.ens-cachan.fr

Professor B. Teissier
Université Paris 7
UFR de Mathématiques
Equipe Géométrie et Dynamique
Case 7012
2 place Jussieu
75251 Paris Cedex 05
E-mail: Teissier@ens.fr

Professor F. Takens, Mathematisch Instituut,
Rijksuniversiteit Groningen, Postbus 800,
9700 AV Groningen, The Netherlands
E-mail: F.Takens@math.rug.nl

Springer-Verlag, Mathematics Editorial, Tiergartenstr. 17
D-69121 Heidelberg, Germany
Tel.: *49 (6221) 487-701
Fax: *49 (6221) 487-355
E-mail: lnm@Springer.de

9 783540 420545